T0229435

SUSTAINABLE BIOLOGICAL SYSTEMS FOR AGRICULTURE

Emerging Issues in Nanotechnology, Biofertilizers, Wastewater, and Farm Machines

Innovations in Agricultural and Biological Engineering

SUSTAINABLE BIOLOGICAL SYSTEMS FOR AGRICULTURE

Emerging Issues in Nanotechnology, Biofertilizers,
Wastewater, and Farm Machines

Edited by
Megh R Goyal, PhD, PE

CRC Press
Taylor & Francis Group
Boca Raton London New York

CRC Press is an imprint of the
Taylor & Francis Group, an **informa** business

First published 2018 by CRC Press

Published 2019 by CRC Press
Taylor & Francis Group
6000 Broken Sound Parkway NW, Suite 300
Boca Raton, FL 33487-2742

© 2018 by Taylor & Francis Group, LLC

First issued in paperback 2021

CRC Press is an imprint of the Taylor & Francis Group, an informa business

No claim to original U.S. Government works

ISBN 13: 978-1-77-463053-2 (pbk)
ISBN 13: 978-1-77-188614-7 (hbk)

This book contains information obtained from authentic and highly regarded sources. Reasonable efforts have been made to publish reliable data and information, but the author and publisher cannot assume responsibility for the validity of all materials or the consequences of their use. The authors and publishers have attempted to trace the copyright holders of all material reproduced in this publication and apologize to copyright holders if permission to publish in this form has not been obtained. If any copyright material has not been acknowledged please write and let us know so we may rectify in any future reprint.

Except as permitted under U.S. Copyright Law, no part of this book may be reprinted, reproduced, transmitted, or utilized in any form by any electronic, mechanical, or other means, now known or hereafter invented, including photocopying, microfilming, and recording, or in any information storage or retrieval system, without written permission from the publishers.

For permission to photocopy or use material electronically from this work, please access www.copyright.com (http://www.copyright.com/) or contact the Copyright Clearance Center, Inc. (CCC), 222 Rosewood Drive, Danvers, MA 01923, 978-750-8400. CCC is a not-for-profit organization that provides licenses and registration for a variety of users. For organizations that have been granted a photocopy license by the CCC, a separate system of payment has been arranged.

Trademark Notice: Product or corporate names may be trademarks or registered trademarks, and are used only for identification and explanation without intent to infringe.

Library and Archives Canada Cataloguing in Publication

Sustainable biological systems for agriculture : emerging issues in nanotechnology, biofertilizers, wastewater, and farm machines / edited by Megh R Goyal, PhD, PE.

(Innovations in agricultural and biological engineering)
Includes bibliographical references and index.
Issued in print and electronic formats.
ISBN 978-1-77188-614-7 (hardcover).--ISBN 978-1-315-16526-4 (PDF)

1. Agricultural engineering. 2. Sustainable agriculture. 3. Nanotechnology. 4. Biofertilizers. 5. Agricultural machinery. I. Goyal, Megh Raj, editor II. Series: Innovations in agricultural and biological engineering

| S675.S87 2018 | 631.5'233 | C2018-900009-0 | C2018-900010-4 |

Library of Congress Cataloging-in-Publication Data

Names: Goyal, Megh Raj, editor.
Title: Sustainable biological systems for agriculture : emerging issues in nanotechnology, biofertilizers, wastewater, and farm machines / editor: Megh R. Goyal.
Description: Waretown, NJ : Apple Academic Press, 2018. | Includes bibliographical references and index.
Identifiers: LCCN 2017061142 (print) | LCCN 2017061824 (ebook) | ISBN 9781315165264 (ebook) | ISBN 9781771886147 (hardcover : alk. paper)
Subjects: LCSH: Agricultural engineering. | Nanotechnology. | Biofertilizers. | Agricultural machinery.
Classification: LCC S675 (ebook) | LCC S675 .S93 2018 (print) | DDC 631.5/233--dc23
LC record available at https://lccn.loc.gov/2017061142

**Visit the Taylor & Francis Web site at
http://www.taylorandfrancis.com**

**and the Psychology Press Web site at
http://www.psypress.com**

CONTENTS

LIST OF CONTRIBUTORS

Anurag Jyoti, PhD
Assistant Professor, Amity Institute of Biotechnology, Amity University Madhya Pradesh, Gwalior 474005, India, Mobile: +91-7898805402, E-mail: ajyoti@gwa.amity.edu; anurag.bt@gmail.com

Arti Goel, PhD
Assistant Professor, Amity Institute of Microbial Biotechnology, Amity University, J-3 Block, 3rd Floor, Room 307, Sector 125, Noida (U.P.), 201313, India. Mobile: +91-8800422339, E-mail: agoel2@amity.edu

Ashwani Kumar, PhD
Assistant Professor, Department of Nutrition Biology, School of Interdisciplinary and Applied Life Sciences, Central University of Haryana, Mahendergarh 123029 Haryana, India, Tel.: +91-9813968380, E-mail: ashwanindri@gmail.com

B. C. Mal, PhD
Former Vice-Chancellor, Chhattisgarh Swami Vivekananda Technical University, Bhilai, Chhattisgarh and Ex-Head of the Department of Agricultural and Food Engineering, Indian Institute of Technology Kharagpur 721302, West Bengal, India, E-mail: bmal@agfe.iitkgp.ernet.in

B. Krishna Prasad, M. Pharm.
Drug Safety Associate, Bioclinica Safety and Regulatory Solutions, Metagalli, Mysore 570016, India, Mobile: +91-9986263912, E-mail: krishnaprasad8710@gmail.com

Bhavesh D. Kevadiya, PhD
Postdoctoral Fellow, University of Nebraska Medical Center, Omaha, NE, USA, Mobile: +001-8454809317, E-mail: bbhaveshpatel@gmail.com

Bhumika P. Chadamiya, M.Sc.
Student, Department of Biochemistry, Saurashtra University, Rajkot, Gujarat, India, Mobile: +918511209023, E-mail: bhumikachadamiya32@gmail.com

Chhaya R. Kasundra, M.Sc.
Student, Department of Biochemistry, Saurashtra University, Rajkot, Gujarat, India, Mobile: +91-9998958534, E-mail:chhayachemist@gmail.com

Deepika Choudhary, PhD
Assistant Professor, Department of Chemistry, Banasthali University, 514 Gautam Budh, Banasthali, Rajasthan 304022, India, Tel.: (office: Head) +91-1438-228341 Ext. 340

Devendra Vaishnav, M. Pharm.
Assistant Professor, Department of Pharmaceutical Sciences, Saurashtra University, Rajkot 360005, Mobile: +919725251705, E-mail: devvaishnav@gmail.com

Divya Lakshminarayanan, PhD
Senior Researcher, Department of Water and Health, Faculty of Life Sciences, Jagadguru Sri Shivaathreeswara University, SS Nagar, Mysore 570015, Karnataka, India, Mobile: +91-9037893395, Email: divyalakshminarayanan@gmail.com

Elias Munapo, PhD
Associate Professor, School of Economics and Decision Sciences, North West University, Mafeking-2735, South Africa, Mobile: +27-732325211, E-mail: emunapo@gmail.com

G. V. S. Rama Krishna, PhD
Assistant Professor, Department of Biotechnology, School of Life Sciences, K. L. University, Green Fields, Vaddeswaram 522502, District Guntur, Andhra Pradesh, India, Mobile: +91-9491186246, E-mail: krishna.ganduri@kluniversity.in

Gaurav S. Dave, PhD
Assistant Professor, Department of Biochemistry, Saurashtra University, Rajkot 360005, Gujarat, India, Mobile: +91-9428275894, E-mail: gsdspu@gmail.com

Gaurav V. Sanghvi, PhD
Assistant Professor, Department of Pharmaceutical Sciences, Saurashtra University, Rajkot 360005, Current Address: Postdoctoral Fellow, Max Planck Institute of Developmental Biology, Tubingen, Germany, Mobile: +491-7672599867, E-mail: sanghavi83@gmail.com

Hetal J. Chadamiya, M.Sc.
Student, Department of Biochemistry, Saurashtra University, Rajkot, Gujarat, India, Mobile: +91-8511209023, E-mail: hetalchadmiya@gmail.com

Jessen George, PhD
Senior Researcher, Department of Water and Health, Faculty of Life Sciences, Jagadguru Sri Shivaathreeswara University, SS Nagar, Mysore 570015, Karnataka, India, Mobile: +91-8123552370; Email: georgejessen@gmail.com

Kirti Kumari, PhD
Subject Matter Specialist, University of Horticulture and Forestry, Krishi Vigyan Kendra, Bharsar Ranichauri 249199, Uttarakhand, India; Mobile: +91-8476004175, E-mail: kumarikirti95@gmail.com

Kirubanandan Shanmugam, M. Tech.
Former Graduate Student (Biopharmaceutical Technology), Centre for Biotechnology, Anna University, Chennai 600020, India, Mobile:+61420804215, E-mail: skirubanandan80@gmail.com

Lala I. P. Ray, PhD
Assistant Professor, School of Natural Resource Management, College of Postgraduate Studies (Central Agricultural University, Imphal), Umiam, Barapani 793103, Meghalaya, India, Tel.: +91-364-2570031/2570614, Mobile: + 91-9436336021, E-mail: lalaipray@rediffmail.com

Lohith Kumar Dasarahalli Huligowda, M.Tech.
PhD Research Scholar, Department of Food Process Engineering, National Institute of Technology at Rourkela, Rourkela 769008, India, Mobile: +91-7064655392, E-mail: lohithhanum8@gmail.com

Mrudulata Deshmukh, PhD
Assistant Professor, Department of Farm Power and Machinery, Dr. Panjabrao Deshmukh Krishi Vidyapeeth (PRDKV), Akola (MS) 444104, Mobile: +91-9850192915, E-mail: mrudulatad@rediffmail.com

Nikhil Kumar
Undergraduate Student, Department of Agricultural Engineering, North Eastern Regional Institute of Science and Technology (NERIST), Nirjuli (Itanagar), Arunachal Pradesh 791109, India, E-mail: nikhilnerist@gmail.com

P. H. Rajasree, M. Pharm., PhD
Research Scholar, Department of Pharmaceutics, JSS College of Pharmacy, Mysore, Karnataka, India 570015, Mobile: +91-7829070020, E-mail: rajasreeph@gmail.com

P. K. Panigrahi, M.Sc. (Fisheries)
Ex-Research Scholar, Department of Agricultural and Food Engineering, Indian Institute of Technology Kharagpur 721302; West Bengal and Fisheries Development Officer, Government of Odisha, India, E-mail: panigrahi.prasanta@gmail.com.

P. K. Pranav, PhD
Assistant Professor, Department of Agricultural Engineering, North Eastern Regional Institute of Science and Technology (NERIST), Nirjuli (Itanagar), Arunachal Pradesh– 791109, India. E-mail: pkjha78@gmail.com

Prasanna Subramanian, M. Tech.
Former Graduate Student (Biopharmaceutical technology), Centre for Biotechnology, Anna University, Chennai 600020, India, E-mail: prasanarengarajan@gmail.com

Prashant D. Kunjadia, PhD
Assistant Professor, B. N. Patel Institute of Paramedical and Science, Bhalej Road, Anand, Gujarat, India, Mobile: +91-9824252544,
E-mail: pdkunjadia@yahoo.com

Preetam Sarkar, PhD
Assistant Professor, Department of Food Process Engineering, National Institute of Technology at Rourkela, Rourkela 769008, India, Mobile: +91-7064031514, +91-6612462906,
E-mail: sarkarpreetam@nitrkl.ac.in; preetamdt@gmail.com;

Rajesh Singh Tomar, PhD
Professor, Amity Institute of Biotechnology, Amity University Madhya Pradesh, Gwalior 474005, India, Mobile: +91-9301117515, E-mail: rstomar@amity.edu

Rita S. Majumdar, PhD
Professor and Head, Department of Biotechnology, School of Basic Science & Research, Sharda University, Plot No. 32-34, Knowledge Park III, Greater Noida, UP 201306, Mobile: +91-9971168120, E-mail: rita.singh@sharda.ac.in

S. K. Thakar, PhD
Assistant Professor, Department of Farm Power and Machinery, Dr. Panjabrao Deshmukh Krishi Vidyapeeth, Akola (MS) 444104; Mobile: +91-9822681722, E-mail: skthakare@gmail.com

S. Moulick, PhD
Associate Professor, Department of Civil Engineering, Kalinga Institute of Industrial Technology (KIIT) University, Bhubaneswar, Odisha, India, E-mail: sanjib_moulick72@yahoo.co.uk

Saurabhh Jain, M.Tech.
Assistant Professor and Head, Department of Applied Sciences, Seth Jai Parkash Mukandlal Innovative Engineering and Technology Institute (JMIETI), Chota Bans, Radaur 135133, Yamuna Nagar District, Haryana, Tel.: +91-9991106838, E-mail: Saurabh.Jain83@gmail.com

Savan D. Fasara, M.Sc.
Research Scholar, Department of Biochemistry, Saurashtra University, Rajkot, Gujarat, India, Mobile: +91-9137668720, E-mail: sdfasara@gmail.com

Severeni Ashili, M.Sc.
Laboratory Technologist, Hifikepunye Pohamba Campus, University of Namibia, Ongwediva, Namibia, Mobile +264812444588, E-mail: severeni.ashili@gmail.com

Shweta Jha, PhD
Assistant Professor and Young Scientist (DST - SERB), Biotechnology Lab, Department of Botany (UGC-Centre of Advance Study), Jai Narain Vyas University, Jodhpur 342001, India, Mobile: 91-9602480618; 91-9530376943, E-mail: jha.shweta80@gmail.com; sj.bo@jnvu.edu.in

Shyamtanu Chaudhuri
Undergraduate Student, Department of Agricultural Engineering, North Eastern Regional Institute of Science and Technology (NERIST), Nirjuli (Itanagar), Arunachal Pradesh 791109, India, E-mail: nikhilnerist@gmail.com

Sonal V. Panara, M.Sc.
Student, Department of Biochemistry, Saurashtra University, Rajkot, Gujarat, India, Mobile: +91-9712584684, E-mail: sonalipanara3@gmail.com

Sudesh Kumar, PhD
Professor, Department of Chemistry, Banasthali University, 514 Gautam Budh, Banasthali, Rajasthan, 304022, India, Mobile: +91-9461594889, +91-9509111149, E-mail: sudeshneyol@gmail.com

Surbhi Panwar, M.Sc.
Assistant Professor, Department of Genetics & Plant Breeding, Chaudhary Charan Singh University, Meerut, 200005, Uttar Pradesh; Tel.: +91-9050873488, E-mail: surbhipanwar11086@gmail.com

Suriyanarayanan Sarvajayakesavalu, PhD
Science Officer, SCOPE Beijing Office, #18 Shuangqing Road, Haidian District, Beijing 100085, China, Mobile: +86-15712886063, E-mail: sunsjk@gmail.com

Tejpal Dhewa, PhD
Assistant Professor, Department of Nutrition Biology, School of Interdisciplinary and Applied Life Sciences, Central University of Haryana, Jant Pali, Mahendergarh 123029 Haryana, India, Tel: +91-8826325454, E-mail: tejpaldhewa@gmail.com, tejpal_dhewa07@rediffmail.com

Thaneswer Patel, PhD
Assistant Professor, Department of Agricultural Engineering, North Eastern Regional Institute of Science and Technology (NERIST), Nirjuli (Itanagar) 791109, Arunachal Pradesh, India, Mobile: +91-9436228996, E-mail: thaneswer@gmail.com

Vaishali D. Patel, M.Sc.
Student, Department of Biochemistry, Saurashtra University, Rajkot, Gujarat, India, Mobile: +91-8511209023, E-mail: vaishalichadmiya@gmail.com

LIST OF ABBREVIATIONS

2D	two dimensional
2-DE	two-dimensional gel electrophoresis
3D	three dimensional
A. P.	Andra Pradesh
Ag	silver
Agri	agriculture
Am Cl	amoxicillin Clavulinic acid
Am	amoxicillin
AMA	Agricultural Mechanization in Asia, Africa and Latin America
Ami	amikacin
Amp	ampicillin
API	active pharmaceutical ingredient
APX	ascorbate peroxidase
ASABE	American Society of Agricultural and Biological Engineers
ATCC	American Type Culture Collection
ATP	adenosine tri phosphate
BIS	Bureau of Indian Standards
BOD	biochemical oxygen demand
BPDS	body part discomfort score
BPW	buffered peptone water
C. I.	Cast iron
CAD	computer aided design
CAET	College of Agricultural Engineering and Technology
CAT	Catalase
CATIA	Computer-Aided Three dimensional Interactive Application
CCD	central composite design
CCDS	corn condensed distillers solubles
CdS	cadmium sulfide
Ce	cephalexin

Cef	cefazolin
Cep	cephalothin
CFIA	Canadian Food Inspection Agency
Cfm	cefotaxime
Cft	ceftazidime
CFU	colony-forming unit
Cfx	cefoxitin
CHIP	Chicken and Hen Infection Program
Chl	chloramphenicol
CIFA	Central Institute of Freshwater Aquaculture
Cip	ciprofloxacin
Cl	cloxacillin
CLA	conjugated linolenic acid
Clin	clindamycin
COD	chemical oxygen demand
Co-Tri	Co-Trimoxazole
CPPs	caseinophosphopeptides
CRS	Center Research Station
CSL	corn steep liquor
Cu	Copper
Cu_2O	cuprous oxide
CuO	copper oxide
$CuSO_4$	copper sulfate
DAEC	diffusely adherent E. coli
DEC	diarrheagenic E. coli
DNA	deoxyribonucleic acid
DNSA	dinitrosalicylic acid
DOJSC	de-oiled Jatropha seed cake
Dox	doxycycline
EAggEC	enteroaggregative E. coli
EDTA	ethylenediaminetetraacetic acid
EIEC	enteroinvasive E. coli
ELISA	enzyme-linked immunosorbent assay
EMB	eosin methylene blue
ENP	engineered nanoparticles
EPEC	enteropathogenic E. coli
EPS	exopolysaccharide
ER	endoplasmic reticulum

Ery	erythromycin
ETEC	enterotoxigenic E. coli
FPM	farm power and machinery
F1	suboptimal doses of nitrogen at 90% of recommended dose
F2	suboptimal doses of nitrogen at 80% of recommended dose
F3	suboptimal doses of nitrogen at 70% of recommended dose
FAO	Food and Agriculture Organization
FC	fecal coliform
FCC	fecal coliform count
Fe-NPs	Iron Nanoparticles
FPM	farm power and machinery
FS	fullerene soot
FTIR	Fourier transform infra-red
GDP	gross domestic product
Gen	gentamicin
GIS	geographical information system
GMP	glycomacropeptide
GNP	gold nanoparticle
GO	graphene oxide
GPS	geosynchronous positioning system
GPX	glutathione peroxidase
GR	glutathione reductase
GRAS	generally regarded as safe
GWI	Global Water Intelligence
HEK 293	embryonic kidney cell
HeLA	cervical cancer cell line
HEPES	4-(2-hydroxyethyl)-1-piperazineethanesulfonic acid
HGT	horizontal gene transfer
hp	horse power
$\mathbf{I_0}$	Irrigation with direct tube-well water
$\mathbf{I_1}$	Irrigation from fishpond with stocking density of $5.0/m^2$
$\mathbf{I_2}$	Irrigation from fishpond with stocking density of $3.5/m^2$
$\mathbf{I_3}$	Irrigation from fishpond with stocking density of $2.0/m^2$
ICAR	Indian Council of Agricultural Research
IIT	Indian Institute of Technology

IMC	Indian major carps
IR	Infrared
ISAE	Indian Society of Agricultural Engineers
IWMI	International Water Management Institute
JBM	jaggery based medium
JMIETI	Seth Jai Parkash Mukandlal Innovative Engineering and Technology Institute
KSTP	Kesare Sewage Treatment Plant
LBA	Luria Bertani agar
LC–ESI–MS/MS	Liquid chromatography–electrospray ionization–tandem mass spectrometry
Lev	Levofloxacin
LILP	logistic incorporated Luedeking–Piret
LIMLP	logistic incorporated modified Luedeking–Piret
LMP	low methoxyl pectin
LT	heat labile toxin
MBC	minimum bactericidal concentration
mc	moisture content
MDHAR	monodehydroascorbate reductase
MDR	multi drug resistance
MFGM	milk fat globule membrane
mg	milligram
Mha	million hectare
MIC	minimum inhibitory concentration
miRNA	micro RNA
MLD	millions of liters per day
mm	millimeter
MMR	mismatch repair gene
Mn	manganese
MnSO$_4$	manganese Sulfate
Mo	molybdenum
MPN	most probable number
MRSA	methicillin-resistant Staphylococcus aureus
MS	mass spectrometry
MS	mild steel
MT	million ton
mt/ha	metric ton/Hectare
MTCC	Microbial Type Culture Collection and Gene Bank

Mup	Mupirocin
MWCNT	Multi-walled carbon nanotubes
N	nitrogen
Nax	Nalidixic Acid
Net	Netilmicin
NGS	next generation sequencing
NH$_3$	ammonia
Nm	nanometer
NMs	nanomaterials
NO$_2$−	nitrite
NO$_3$−	nitrate
Nor	Norfloxacin
NPs	nanoparticles
NSS	nanosized Ag-silica hybrid
OC	organic carbon
ODR	Overall discomfort rating
OFAT	one factor at time
Ofl	Ofloxacin
Ox	Oxacillin
PTO	power take off
PB	Plackett–Burman
PCNA	proliferating cell nuclear antigen
PCR	polymerase chain reaction
PCS1	phytochelatin synthase1
PDA	potato dextrose agar
Pg	pencillin G
Pip	piperacillin
PIP	plasma membrane intrinsic protein
POX	peroxidase
ppm	parts per million
PPP	processing, packaging, post processing
PR	pathogenesis-related
PTFE	polytetrafluorethylene
PTMs	post-translational modifications
PVC	poly vinyl chloride
QDs	quantum dots
qRT-PCR	quantitative real-time polymerase chain reaction
RER	rare earth element

rev	revolutions
RH	relative humidity
Rif	rifampin
RNA	ribose nucleic acid
ROS	reactive oxygen species
rpm	revolutions per minute
RS	remote sensing
RSM	response surface methodology
RSTP	Rayankere sewage treatment plant
RTi-PCR	real time polymerase chain reaction
SAR	systemic acquired resistance
SCM	standard cultivation medium
SD	stocking density
SiHa	Cervix human cell line
SIP	small and basic intrinsic protein
SMAC	sorbitol MacConkey agar
SOD	superoxide dismutase
SOM	soil organic matter
SPI	Salmonella pathogenicity Island
SR	survival rate
ST	heat stable toxin
STEC	shiga toxin-producing E. coli
STPs	sewage treatment plants
Str	streptomycin
Sul	sulfamethoxazole
SWCNT	single-walled carbon nanotubes
TAN	total available nitrogen
TC	total coliform
TCC	total coliform count
Tei	teicoplanin
TEM	transmission electron microscopy
Tet	tetracycline
TiO$_2$	titanium dioxide
TIP	tonoplast intrinsic protein
Tri	trimethoprim
TSS	total suspended solid
UDPG	uridine diphospho glucose
US EPA	United States Environmental Protection Agency

USFDA	United States Food and Drug Administration
UV-Visible	ultraviolet-visible
UW	urban wastewater
VAD	Visual analog discomfort
Van	Vancomycin
VBNC	viable but non-culturable
VSTP	Vidyaranyapuram Sewage Treatment Plant
VTEC	verocytotoxin producing E. coli
W	Watts
Zn	zinc
Zn-NPs	zinc nanoparticles
ZnO	zinc oxide
ZnS	zinc sulfide

LIST OF SYMBOLS

%	percent
°C	degree Celsius
α	growth-associated product formation constant, g/g
β	nongrowth-associated product formation constant, g/(g.hr)
β_0	intercept coefficient
β_i	linear coefficient
β_{ii}	quadratic coefficients
β_{ij}	interaction coefficients
cm	centimeter
γ	growth-associated substrate utilization constant, g/g
η	nongrowth-associated substrate utilization constant, g/(g.hr)
μ	specific growth rate, hr^{-1}
μ_{max}	maximum specific growth rate, hr^{-1}
F	flow rate, m^3/sec
m_S	maintenance energy coefficient
P	product
P_0	initial product concentration, g/l
P_t	final product concentration, g/l
r_P	rate product formation, g/(l hr)
r_S	rate of substrate utilization, g/(l hr)
S	substrate
S_0	initial substrate concentration, g/l
S_t	final substrate concentration, g/l
t	time, hr.
V	volume, m^3
X	biomass
X_0	initial biomass concentration, g/l
X_i, X_j	coded values
X_m	maximum attainable biomass concentration, g/l
X_t	final biomass concentration, g/l
Y	predicted response, g/l
Y_P/S	product Yield, g/g

Y_X/S biomass Yield, g/g
Fe   iron
Fe_3O_4 ferric Oxide
Fe_3O_4 magnetite

PREFACE BY MEGH R. GOYAL

Resource, ASABE, November/December 2015, 22(6), 20–21 introduces *"cubic farming™ that is a form of controlled environment agriculture in a commercial reality for the urban sector. It is referred to as vertical farming. It allows growth of fresh produce year round in locations that are challenging for conventional forms of agriculture, including arid urban areas and harsh northern climates—This technology might seem far out for some consumers, but is an insight into the future fresh food production"*. <https://en.wikipedia.org/wiki/Vertical_farming> indicates following potential advantages of vertical farming. Vertical farming relies on the use of various physical methods to become effective. The most common technologies suggested are: Greenhouse, trunk injection, the folkewall and other vertical growing architectures, aeroponics/hydroponics/aquaponics, composting, phytoremediation, grow light, and skyscraper.

This new technology poses challenges to agricultural engineers to come up with appropriate solutions to vertical farming. Apple Academic Press Inc. (AAP) published my first book on *"Management of Drip/Trickle or Micro Irrigation"*, 10-volume set under book series *"Research Advances in Sustainable Micro Irrigation,"* in addition to other books in the focus areas of *"Agricultural and Biological Engineering."* The mission of this book volume is to introduce the profession of agricultural and biological engineering. I cannot guarantee the information in this book series will be enough for all situations.

At 49th annual meeting of Indian Society of Agricultural Engineers at Punjab Agricultural University during February 22–25 of 2015, a group of ABEs convinced me that there is a dire need to publish book volumes on focus areas of agricultural and biological engineering (ABE). This is how the idea was born on new book series titled, *"Innovations in Agricultural & Biological Engineering."* AAP has published 19 book volumes under this book series. This book volume is titled, *"Sustainable Biological Systems for Agriculture: Emerging Issues in Nanotechnology, Biofertilizers, Wastewater, and Farm Machines."*

The contribution by all cooperating authors to this book volume has been most valuable in the compilation. Their names are mentioned in each chapter and in the list of contributors. This book would not have been written without the valuable cooperation of these investigators, many of them are renowned scientists who have worked in the field of ABE throughout their professional careers. I have inherited many ethical qualities of a successful educator from clean testimony of Dr R. K. Sivanappan, former Dean and professor at Tamil Nadu Agricultural University (TNAU)/ Founding Director of Water Technology Institute at TNAU/author of more than 1000 professional articles and more than 30 books. He is a frequent contributor to my book series and staunch supporter of my profession. His suggestions to the contents and quality of this book have been invaluable.

I thank editorial staff, Sandy Jones Sickles, Vice President, and Ashish Kumar, Publisher and President at Apple Academic Press, Inc., for making every effort to publish the book when the diminishing water resources are a major issue worldwide. Special thanks are due to the AAP production staff for typesetting the entire manuscript and for the quality production of this book. I request the reader to offer your constructive suggestions that may help to improve the next edition. The reader can order a copy of this book for the library, the institute or for a gift from "http://appleacademic-press.com."

I express my deep admiration to my family for understanding and collaboration during the preparation of this book. As an educator, there is a piece of advice to one and all in the world: *"Permit that our Almighty God, our Creator and excellent Teacher, irrigate the life with His Grace of rain trickle by trickle, because our life must continue trickling on... and Get married to my profession of Agricultural Engineering —"*

—**Megh R. Goyal, Ph.D., P.E.**
Senior Editor-in-Chief

FOREWORD BY N. C. PATEL

—Healthy Soil for Healthy Life

"World Class Agricultural Engineering" denotes the engineer of tomorrow with a spectrum of competencies and skills to formulate new ideas; and develop new processes and technologies in order to address the challenges anywhere across the globe. We know that our engineers must have the wealth of knowledge and proficiency in the diverse agricultural situations/sectors and agribusiness industries. Agricultural engineers are committed to new developments, techniques, products, processes, and other related issues of interest and concern. We are proud that as agricultural engineers, we play an essential role in sustainable agriculture through agricultural engineering interventions.

Our ultimate goal as agricultural & biological engineers is to contribute our genuine inputs in the sustainable food production to meet the food requirement of the country along with meeting the energy demand and improving the environment at large.

I wish all budding engineers a very good time and challenges ahead. I encourage Dr. Megh R Goyal and Apple Academic Press Inc. to continue publishing quality books on focus areas of agricultural & biological engineering.

N. C. Patel, PhD
President of Indian Society of
Agricultural Engineers and Vice Chancellor
at Anand Agricultural University
Anand 388110, Gujarat State (INDIA)
E-mail: vc@aau.in; ncpatel@aau.in
Tel.: +91 2692 261273 (O)
Fax: +91 2692 261520 (O)
Mobile: +91 9998009960

WARNING/DISCLAIMER

USER MUST READ IT CAREFULLY

The goal of this compendium on, *"Sustainable Biological Systems for Agriculture*: *Emerging Issues in Nanotechnology, Biofertilizers, Wastewater, and Farm Machines"* is to: guide the world engineering community on how to efficiently manage crop production; and introduce applications of nanotechnology. The reader must be aware that the dedication, commitment, honesty, and sincerity are most important factors in a dynamic manner for a complete success. It is not a one-time reading of this compendium. Read and follow every time, it is needed. To err is human. However, we must do our best. Always, there is a space for learning new experiences.

The editor, the contributing authors, the publisher and the printer have made every effort to make this book as complete and as accurate as possible. However, there still may be grammatical errors or mistakes in the content or typography. Therefore, the contents in this book should be considered as a general guide and not a complete solution to address any specific situation in irrigation. For example, one size of irrigation pump does not fit all sizes of agricultural land and to all crops; and the selection of a production technology will depend on the crop type and farm size.

The editor, the contributing authors, the publisher and the printer shall have neither liability nor responsibility to any person, any organization, or entity with respect to any loss or damage caused, or alleged to have caused, directly or indirectly, by information or advice contained in this book. Therefore, the purchaser/reader must assume full responsibility for the use of the book or the information therein.

The mentioning of commercial brands and trade names are only for technical purposes. It does not mean that a particular product is endorsed over to another product or equipment not mentioned. Author, cooperating authors, educational institutions, and the publisher Apple Academic Press Inc. do not have any preference for a particular product.

All weblinks that are mentioned in this book were active on December 31, 2016. The editors, the contributing authors, the publisher and the printing company shall have neither liability nor responsibility, if any of the weblinks is inactive at the time of reading of this book.

ABOUT SENIOR EDITOR-IN-CHIEF

Megh R Goyal, PhD, PE
*Retired Professor in Agricultural and Biomedical Engineering, University of Puerto Rico,
Mayaguez Campus Senior Acquisitions Editor,
Biomedical Engineering and Agricultural Science,
Apple Academic Press, Inc.*

Megh R. Goyal, Ph.D., P.E., is a retired professor in agricultural and biomedical engineering from the General Engineering Department in the College of Engineering at University of Puerto Rico – Mayaguez Campus; and senior acquisitions editor and senior technical editor-in-chief in agricultural and biomedical engineering for Apple Academic Press Inc.

He received his B.Sc. degree in engineering in 1971 from Punjab Agricultural University, Ludhiana, India; his M.Sc. degree in 1977 and Ph.D. in 1979 from the Ohio State University, Columbus; and his Master of Divinity degree in 2001 from Puerto Rico Evangelical Seminary, Hato Rey, Puerto Rico, USA.

Since 1971, he has worked as soil conservation inspector (1971); research assistant at Haryana Agricultural University (1972–1975) and the Ohio State University (1975–1979); research agricultural engineer/professor at Department of Agricultural Engineering of UPRM (1979–1997); and professor in agricultural and biomedical engineering at General Engineering Department of UPRM (1997–2012). He spent one-year sabbatical leave in 2002–2003 at Biomedical Engineering Department, Florida International University, Miami, USA.

He was first agricultural engineer to receive the professional license in agricultural engineering in 1986 from College of Engineers and Surveyors of Puerto Rico. On September 16, 2005, he was proclaimed as "Father of Irrigation Engineering in Puerto Rico for the twentieth century" by the ASABE, Puerto Rico Section, for his pioneer work on micro irrigation, evapotranspiration, agroclimatology, and soil & water engineering. During his professional career of 45 years, he has received awards such as: Scientist of the Year, Blue Ribbon Extension Award, Research Paper Award, Nolan Mitchell Young Extension Worker Award, Agricultural Engineer of the Year, Citations by Mayors of Juana Diaz and Ponce, Membership Grand Prize for ASAE Campaign, Felix Castro Rodriguez Academic Excellence, Rashtrya Ratan Award and Bharat Excellence Award and Gold Medal, Domingo Marrero Navarro Prize, Adopted son of Moca, Irrigation Protagonist of UPRM, Man of Drip Irrigation by Mayor of Municipalities of Mayaguez/Caguas/Ponce, and Senate/Secretary of Agriculture of ELA, Puerto Rico.

He has authored more than 200 journal articles and 45 books, namely: "Elements of Agroclimatology (Spanish) by UNISARC, Colombia"; two "Bibliographies on Drip Irrigation". Apple Academic Press Inc. (AAP) has published "Management of Drip/Trickle or Micro Irrigation," "Evapotranspiration: Principles and Applications for Water Management," ten-volume set in "Research Advances in Sustainable Micro Irrigation." During 2016–2020, AAP will be publishing book volumes on emerging technologies/issues/challenges under book series, "Innovations and Challenges in Micro Irrigation," and "Innovations in Agricultural & Biological Engineering." Readers may contact him at: goyalmegh@gmail.com.

OTHER BOOKS ON AGRICULTURAL AND BIOLOGICAL ENGINEERING BY APPLE ACADEMIC PRESS, INC.

Management of Drip/Trickle or Micro Irrigation
Megh R. Goyal, PhD, P.E., Senior Editor-in-Chief

Evapotranspiration: Principles and Applications for Water Management
Megh R. Goyal, PhD, P.E., and Eric W. Harmsen, Editors

Book Series: Research Advances in Sustainable Micro Irrigation
Senior Editor-in-Chief: Megh R. Goyal, PhD, P.E.
  Volume 1: Sustainable Micro Irrigation: Principles and Practices
  Volume 2: Sustainable Practices in Surface and Subsurface Micro Irrigation
  Volume 3: Sustainable Micro Irrigation Management for Trees and Vines
  Volume 4: Management, Performance, and Applications of Micro Irrigation Systems
  Volume 5: Applications of Furrow and Micro Irrigation in Arid and Semi-Arid Regions
  Volume 6: Best Management Practices for Drip Irrigated Crops
  Volume 7: Closed Circuit Micro Irrigation Design: Theory and Applications
  Volume 8: Wastewater Management for Irrigation: Principles and Practices
  Volume 9: Water and Fertigation Management in Micro Irrigation
  Volume 10: Innovation in Micro Irrigation Technology

Book Series: Innovations and Challenges in Micro Irrigation

Senior Editor-in-Chief: Megh R. Goyal, PhD, P.E.

- Micro Irrigation Engineering for Horticultural Crops: Policy Options, Scheduling and Design
- Micro Irrigation Management Technological Advances and Their Applications
- Micro Irrigation Scheduling and Practices
- Performance Evaluation of Micro Irrigation Management: Principles and Practices
- Potential of Solar Energy and Emerging Technologies in Sustainable Micro Irrigation
- Principles and Management of Clogging in Micro Irrigation
- Sustainable Micro Irrigation Design Systems for Agricultural Crops: Methods and Practices

Book Series: Innovations in Agricultural & Biological Engineering

Senior Editor-in-Chief: Megh R. Goyal, PhD, P.E.

- Dairy Engineering: Advanced Technologies and Their Applications
- Developing Technologies in Food Science: Status, Applications, and Challenges
- Emerging Technologies in Agricultural Engineering
- Engineering Practices for Agricultural Production and Water Conservation: An Inter-disciplinary Approach
- Flood Assessment: Modeling and Parameterization
- Food Engineering: Emerging Issues, Modeling, and Applications
- Food Process Engineering: Emerging Trends in Research and Their Applications
- Food Technology: Applied Research and Production Techniques
- Engineering Interventions in Agricultural Processing
- Modeling Methods and Practices in Soil and Water Engineering
- Processing Technologies for Milk and Dairy Products: Methods Application and Energy Usage
- Soil and Water Engineering: Principles and Applications of Modeling
- Soil Salinity Management in Agriculture: Technological Advances and Applications

EDITORIAL

Apple Academic Press Inc., (AAP) will be publishing various book volumes on the focus areas under book series titled *Innovations in Agricultural and Biological Engineering*. Over a span of 8–10 years, Apple Academic Press Inc., will publish subsequent volumes in the specialty areas defined by *American Society of Agricultural and Biological Engineers* (asabe.org). **We need book proposals from the readers in area of their expertise**.

The mission of this series is to provide knowledge and techniques for agricultural and biological engineers (ABEs). The series aims to offer high-quality reference and academic content in ABE that is accessible to academicians, researchers, scientists, university faculty, and university-level students and professionals around the world.

The following material has been edited/modified and reproduced below ["*Goyal, Megh R., 2006. Agricultural and biomedical engineering: Scope and opportunities. Paper Edu_47 at the Fourth LACCEI International Latin American and Caribbean Conference for Engineering and Technology (LACCEI' 2006): Breaking Frontiers and Barriers in Engineering: Education and Research by LACCEI University of Puerto Rico – Mayaguez Campus, Mayaguez, Puerto Rico, June 21 – 23*"]:

WHAT IS AGRICULTURAL AND BIOLOGICAL ENGINEERING (ABE)?

According to (isae.in), "*Agricultural engineering (AE) involves application of engineering to production, processing, preservation and handling of food, fiber, and shelter. It also includes transfer of technology for the development and welfare of rural communities.*" According to (asabe.org), "*ABE is the discipline of engineering that applies engineering principles and the fundamental concepts of biology to agricultural and biological systems and tools, for the safe, efficient and environmentally sensitive production, processing, and management of agricultural, biological, food, and natural resources systems.*" "*AE is the branch of engineering involved with the design of farm machinery, with soil management, land development, and mechanization and automation of livestock farming, and with the efficient planting, harvesting, storage, and processing of farm commodities,"* definition by: http://dictionary.reference.com/browse/agricultural+engineering.

"AE incorporates many science disciplines and technology practices to the efficient production and processing of food, feed, fiber and fuels. It involves disciplines like mechanical engineering (agricultural machinery and automated machine systems), soil science (crop nutrient and fertilization, etc.), environmental sciences (drainage and irrigation), plant biology (seeding and plant growth management), and animal science (farm animals and housing) etc." by http://www.ABE.ncsu.edu/academic/agricultural-engineering.php.

According to https://en.wikipedia.org/wiki/Biological_engineering: *"BE (Biological engineering) is a science-based discipline that applies concepts and methods of biology to solve real-world problems related to the life sciences or the application thereof. In this context, while traditional engineering applies physical and mathematical sciences to analyze, design and manufacture inanimate tools, structures and processes, biological engineering uses biology to study and advance applications of living systems".*

SPECIALTY AREAS OF ABE

Agricultural and biological engineers (ABEs) ensure that the world has the necessities of life including safe and plentiful food, clean air and water, renewable fuel and energy, safe working conditions, and a healthy environment by employing knowledge and expertise of sciences, both pure and applied, and engineering principles. Biological engineering applies engineering practices to problems and opportunities presented by living things and the natural environment in agriculture. BA engineers understand the interrelationships between technology and living systems, have available a wide variety of employment options. *"ABE embraces a variety of following specialty areas"*, (asabe.org). As new technology and information emerge, specialty areas are created, and many overlap with one or more other areas.

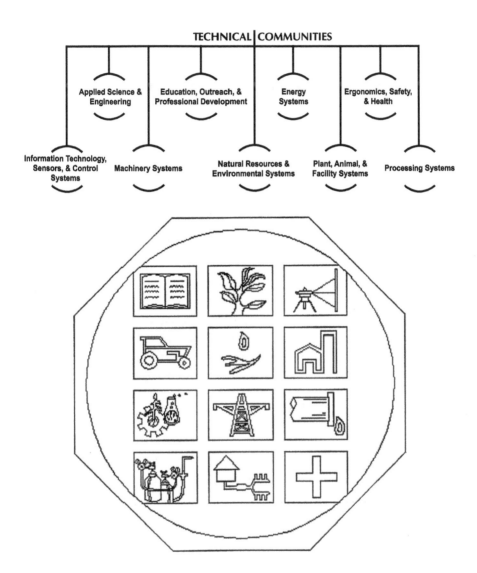

1. **Aquacultural Engineering:** ABEs help design farm systems for raising fish and shellfish, as well as ornamental and bait fish. They specialize in water quality, biotechnology, machinery, natural resources, feeding and ventilation systems, and sanitation. They seek ways to reduce pollution from aquacultural discharges, to reduce excess water use, and to improve farm systems. They also work with aquatic animal harvesting, sorting, and processing.

2. **Biological Engineering** applies engineering practices to problems and opportunities presented by living things and the natural environment.

3. **Energy:** ABEs identify and develop viable energy sources—biomass, methane, and vegetable oil, to name a few—and to make these and other systems cleaner and more efficient. These specialists also develop energy conservation strategies to reduce costs and protect the environment, and they design traditional and alternative energy systems to meet the needs of agricultural operations.

4. **Farm Machinery and Power Engineering:** ABEs in this specialty focus on designing advanced equipment, making it more efficient and less demanding of our natural resources. They develop equipment for food processing, highly precise crop spraying, agricultural commodity and waste transport, and turf and landscape maintenance, as well as equipment for such specialized tasks as removing seaweed from beaches. This is in addition to the tractors, tillage equipment, irrigation equipment, and harvest equipment that have done so much to reduce the drudgery of farming.

5. **Food and Process Engineering:** Food and process engineers combine design expertise with manufacturing methods to develop economical and responsible processing solutions for industry. Furthermore, food and process engineers look for ways to reduce waste by devising alternatives for treatment, disposal, and utilization.

6. **Forest Engineering:** ABEs apply engineering to solve natural resource and environment problems in forest production systems and related manufacturing industries. Engineering skills and expertise are needed to address problems related to equipment design and manufacturing, forest access systems design and construction; machine–soil interaction and erosion control; forest operations analysis and improvement; decision modeling; and wood product design and manufacturing.

7. **Information & Electrical Technologies Engineering** is one of the most versatile areas of the ABE specialty areas, because it is applied to virtually all the others, from machinery design to soil testing to food quality and safety control. Geographic information systems, global positioning systems, machine instrumentation and controls, electromagnetics, bioinformatics, biorobotics, machine

vision, sensors, spectroscopy are some of the exciting information and electrical technologies being used today and being developed for the future.

8. **Natural Resources:** ABEs with environmental expertise work to better understand the complex mechanics of these resources so that they can be used efficiently and without degradation. ABEs determine crop water requirements and design irrigation systems. They are experts in agricultural hydrology principles, such as controlling drainage, and they implement ways to control soil erosion and study the environmental effects of sediment on stream quality. Natural resources engineers design, build, operate and maintain water control structures for reservoirs, floodways and channels. They also work on water treatment systems, wetlands protection, and other water issues.

9. **Nursery and Greenhouse Engineering:** In many ways, nursery and greenhouse operations are microcosms of large-scale production agriculture, with many similar needs—irrigation, mechanization, disease and pest control, and nutrient application. However, other engineering needs also present themselves in nursery and greenhouse operations: equipment for transplantation; control systems for temperature, humidity, and ventilation; and plant biology issues, such as hydroponics, tissue culture, and seedling propagation methods. Sometimes the challenges are extraterrestrial: ABEs at NASA are designing greenhouse systems to support a manned expedition to Mars!

10. **Safety and Health:** ABEs analyze health and injury data, the use and possible misuse of machines, and equipment compliance with standards and regulation. They constantly look for ways in which the safety of equipment, materials and agricultural practices can be improved and for ways in which safety and health issues can be communicated to the public.

11. **Structures and Environment:** ABEs with expertise in structures and environment design animal housing, storage structures, and greenhouses, with ventilation systems, temperature and humidity controls, and structural strength appropriate for their climate and purpose. They also devise better practices and systems for storing, recovering, reusing, and transporting waste products.

CAREER IN AGRICULTURAL AND BIOLOGICAL ENGINEERING

One will find that university ABE programs have many names, such as biological systems engineering, bioresource engineering, environmental engineering, forest engineering, or food and process engineering. Whatever the title, the typical curriculum begins with courses in writing, social sciences, and economics, along with mathematics (calculus and statistics), chemistry, physics, and biology. Student gains a fundamental knowledge of the life sciences and how biological systems interact with their environment. One also takes engineering courses, such as thermodynamics, mechanics, instrumentation and controls, electronics and electrical circuits, and engineering design. Then student adds courses related to particular interests, perhaps including mechanization, soil and water resource management, food and process engineering, industrial microbiology, biological engineering or pest management. As seniors, engineering students team up to design, build, and test new processes or products.

For more information on this series, readers may contact:

Ashish Kumar, Publisher and President
Sandy Sickles, Vice President
Apple Academic Press, Inc.,
Fax: 866–222–9549
E-mail: ashish@appleacademicpress.com
http://www.appleacademicpress.com/
publishwithus.php

Megh R Goyal, PhD, P.E.
Book Series Senior
Editor-in-Chief
Innovations in Agricultural and Biological Engineering
E-mail: goyalmegh@gmail.com

PART I

Potential Applications of
Nanotechnology in Biological Systems

APPLICATIONS OF NANOTECHNOLOGY IN AGRICULTURE—A REVIEW: CONCEPTS, ASPECTS, PROSPECTS, AND CONSTRAINTS

SAURABHH JAIN[1], ASHWANI KUMAR[2], TEJPAL DHEWA[2,*], SURBHI PANWAR[3] AND RITA S. MAJUMDAR[4]

[1]*Department of Applied Sciences, Seth Jai Parkash Mukandlal Innovative Engineering and Technology Institute (JMIETI), Chota Bans, Radaur Yamuna Nagar, Haryana 135133, India*

[2]*Department of Nutrition Biology, School of Interdisciplinary and Applied Life Sciences, Central University of Haryana, Mahendergarh, Haryana 123029, India*
[3]*Department of Genetics & Plant Breeding, Chaudhary Charan Singh University, Meerut, Uttar Pradesh 200005, India*

[4]*Department of Biotechnology, School of Basic Science & Research, Sharda University, Plot No. 32-34, Knowledge Park III, Greater Noida, Uttar Pradesh 201306, India*

* *Corresponding author. E-mail: tejpaldhewa@gmail.com;*

tejpal_dhewa07@rediffmail.com

CONTENTS

1.1   INTRODUCTION

The term "nanotechnology" is based on the prefix "nano" which is from the Greek word meaning "dwarf." It is usually employed for materials having a size ranging from 1 to 100 nm. Nanotechnology is the manipulation or self-assembly of individual atoms, molecules, or molecular clusters into structures to create new materials and devices with totally different properties. Generally, it works by following the top down (includes reducing the size of the smallest structures to the nanoscale) or the bottom up (comprises manipulating individual atoms and molecules into nanostructures with nearly similar chemistry or biology) approach.

It is a unique technological intervention that comprises things and tools accomplished for manipulating physical as well as chemical properties of a substance at molecular levels. Agriculture plays a crucial role in developing nations, where majority of people depend on it for generating revenue and support.[5] Nanotechnology plays a great role in the fast/accurate detection of management and diseases, improving the capability of plants for the utilization of nutrients. Therefore, it has immense possibility to revolutionize the agricultural and food sectors.

The supervision of safety aspects, ailment, transport systems, disease management, molecular methods for pathogen identification, and environment protection are few of the applications of the nanotechnology in agriculture and food sectors.[29,34] This sector is going to face a lot of challenges, that is, an increasing demand for healthy food and its safety, growing threat of diseases, and agricultural production particularly in varying climate conditions.[3] Further, nanobiotechnology also holds the potential of sensing and delivering systems to fight against viruses and other high-risk crop pathogens even before the onset of indications. The combined sensing, monitoring, and controlling system could distinguish the incidence of disease and inform the farmer, and subsequently trigger the bioactive systems such as drugs, pesticides, nutrients, probiotics, and nutraceuticals.

In the future, nanosized catalysts will allow administration of lower doses of pesticides and herbicides and will augment the efficiency. Environment protection by employing alternative energy provisions and filters/catalysts to diminish pollution, and clear-out prevailing noxious waste in soil and water are the other fascinating focus areas of this technology. The likely addition to these would be framing the subsequent steps of the development of genetically modified crops, animal production efforts, chemical pesticides, and precision farming techniques.[20] All these tools, techniques, and subsequent advancements would certainly transform the existing structure of food and agricultural production systems. These applications include fine-tuning and more specific micro administration of soils, novel toxin preparations for pest control, innovative crop and animal traits, and the broadening and distinction of farming practices and products keeping in view the large scale and extremely undeviating systems of production.

Therefore, this chapter discusses and reviews the concepts and applications of nanotechnology, particularly in agriculture along with its constraints/limitations.

1.2   APPLICATIONS OF NANOTECHNOLOGY IN AGRICULTURE

Potential applications of nanotechnology in the field of agriculture are given in Tables 1.1 and 1.2.

1.2.1   PRECISION AGRICULTURE

Precision agriculture is defined as an inclusive structure to enhance agricultural production through cautiously modifying soil and crop management to resemble to the inimitable situation present in each soil despite retaining

TABLE 1.1   Potential Applications of Nanotechnology in Agriculture.

Agro-Nanotechnology				
Food packaging	Nano-based products	Precision farming	Gene transfer	Nono-foods
	Nano-fungicides	Remote sensing devices	Crop improvement	
	Nano-pesticides			
	Nano-fertilizers			

TABLE 1.2   Selected Examples of Nanotechnology Applications in Agriculture.

Areas	Definition	Example
Crop production		
Plant protection products	Nanocapsules, nanoparticles, nanoemulsions, and viral capsids as smart delivery systems of active ingredients for disease and pest control in plants	Neem oil (*Azadirachta indica*) nanoemulsion as larvicidal agent (VIT University, IN)
Fertilizers	Nanocapsules, nanoparticles, and viral capsids for the enhancement of nutrients absorption by plants and the delivery of nutrients to specific sites	Macronutrient fertilizers coated with zinc oxide nanoparticles (University of Adelaide, AUCSIRO Land and Water–AU, Kansas State University, US)
Soil improvement (http://www.geohumus.com/us/products.html)		
Water/liquid retention	Nanomaterials, e.g., zeolites and nanoclays, for water or liquid agro-chemicals retention in the soil for their slow release to the plants	Soil-enhancer product, based on a nanoclay component, for water retention and release (Geohumus-Frankfurt, DE)
Water purification		
Water purifica-tion and pollutant remediation	Nanomaterials, e.g., nanoclays, filtering and binding to a variety of toxic substances, including pesticides, to be removed from the environment	Filters coated with TiO_2 nanoparticles for the photo-catalytic degradation of agrochemicals in contami-nated waters (University of Ulster, UK)
Diagnostic		
Nanosensors and diagnostic devices	Nanomaterials and nanostructures (e.g., electrochemically active carbon nanotubes, nanofibers, and fullerenes) that are highly sensitive bio-chemical sensors to closely monitor environmental conditions, plant health and growth	Pesticide detection with a liposome-based nano-biosensor (University of Crete, GR)
Plant breeding		
Plant genetic modifications	Nanoparticles carrying DNA or RNA to be delivered to plant cells for their genetic transformation or to trigger defense response, acti-vated by pathogens.	Mesoporous silica nanopar-ticles transporting DNA to transform plant cells (Iowa State university, USA)

TABLE 1.2 *(Continued)*

Areas	Definition	Example
Nanomaterials from plant		
Nanoparticles from plants	Production of nanomaterials through the use of engineered plants or microbes and through the processing of waste agricultural products	Nanofibers from wheat straw and soy hulls for bio-nanocomposite production (Ontario Ministry of Agriculture, Food and Rural Affairs, CA)

environmental worth.[4] It is employed to maximize the output in terms of crop yields and at the same time minimizing the inputs (i.e., fertilizers, pesticides, herbicides, etc.) by analyzing various environmental variables and subsequently employing the targeted action. The technologies can be adjusted in order to lower the production costs and subsequently increasing the production by using remote sensing (RS), geosynchronous positioning system (GPS), and geographical information system (GIS). Identification of soil conditions and plant development can be done through centralized data; the seeding, fertilizer, chemical, and water use.[21] Generation of soil map and profile are the prerequisite steps in precision agriculture. All these include grid soil sampling, yield monitoring, and crop scouting. Remote sensing in combination with GPS can provide precise maps and models of the agricultural fields. Precision farming promises to decrease agricultural waste; therefore, helps to keep the environmental pollution level to a minimum. Other important functions for nanotechnology-enabled devices would be GPS coupled autonomous sensing for real-time monitoring of soil conditions and crop growth throughout the field.

1.2.2 DETECTION OF NUTRIENTS AND PATHOGENS

Nanotechnology-based biosensors have been designed for fast and accurate determination of contaminants or pollutants in agricultural wastes. Furthermore, they are cost effective, operator-friendly, and small in size. The enzyme-coupled biosensor technology has also been designed and developed for the specific sensing of many elements. The electronic nose (E-nose) device incorporates ZnO nanowires made of gas sensor for the determination of concentration (quantity) and characteristics of odors.

The detection depends on the action of human nose. In the presence of a particular gas, the resistance of the device changes and subsequently produces an altered electrical signal and that in turn generates a fingerprint pattern corresponding to type, quality, and quantity of gas.[12] The device can also be employed for rapid and sensitive detection of various products and organisms, for example, alcohol production in the course of fermentation, growth pattern of microorganisms, bakery and dairy products, foul odor during bacterial rotting, and so forth. The widely used bio-labels (i.e., organic dyes) are helpful in identification of bacteria. However, these labels are costly and their fluorescence vanishes with time. Latest progress in this area is the development of luminescent nanocrystals, that is, quantum dots (QDs) are being used in fluorescent labeling for biological recognition of molecules. Quantum dots have more effective luminescence, contracted emission spectra, and brilliant photo stability properties; therefore are better in comparison to conventional organic dyes.

1.2.3 NANOSCALE CARRIERS FOR TARGETED DELIVERY

Nanotechnology has developed ways of delivering important compounds to the plants for improving their yields. The nanotubes can be employed for administration of pesticides, herbicides, fertilizers, plant growth regulators, and chemicals effectively to the targeted site in plants.[18] The agents are attached at the outward of these polymers and dendrimers by ionic and weak bonds. These carriers generally bind to the roots of plants effectively. The method of the delivery of these chemicals helps in improving the steadiness of compounds by reducing the degradation in the environment. Hence, the increased steadiness of these nanocomposites in the natural environment reduces the dose, thus, minimizing environmental pollution as well as production; ultimately, this will also reduce the waste production. The developments and innovations in nanofabrication have developed better realization of communication patterns between plant cells and the pathogens. Therefore, it has been able to develop better means for treatment of diseases.

1.2.4   WASTEWATER TREATMENT AND DISINFECTION

The pollution or contamination of natural water bodies is presently a major problem throughout world. The wastewater has negative influence not only on the environment but also on the health of animals and humans. Therefore, the management of wastewater is now a serious issue that must be solved with an immediate effect using eco-friendly technologies, that is, nanotechnology. For example, the practice of photocatalysis can be used to treat the contaminated water. It has been scientifically proven that the organic compounds, destruction of malignant cells, removal of bacterial cells and viruses can be achieved by using semiconductor sensitized photosynthetic and photocatalytic processes. Nanotechnology is also useful in cleaning wastewater. During disinfection of water, the nanoparticles are excited with the light source and the negative electrons are released. These electrons can be applied for removing the bacterial cells from contaminated water. Further, these nanoparticles can also be employed as disinfectants in food packaging industries.[27]

1.2.5   BIOREMEDIATION

Nanotechnology has played a significant role in microbial remediation. In agricultural system, some chemicals such as pesticides are slow or resistant to degrade in nature; hence, they persist longer in the environment and cause contamination problems. If they are not degraded, they may invade into the food chain and cause serious health problems. Recent developments in agricultural nanotechnology have shown a promising step in degradation. For example, nanoparticle-water slurry can be mixed with polluted soil and in due course of time, these particles will reduce the toxicity of slowly degradable or resistant pesticides.

1.2.6   RECYCLING OF AGRICULTURAL WASTE

Currently, the continuous deposition of agricultural wastes or by-products in nature is a challenging problem. Nanotechnology can be employed in the reduction of wastes during agricultural manufacturing such as cotton, beverage, rice milling industries, and so forth. In the cotton industry, during the processing of cotton into fabric, some by-products like cellulose or fibers are discarded or used as low value products. By using electrospinning and newly developed solvents, researchers are producing nanofibers (100 nm diameter) that can be applied as a fertilizer or pesticide absorbent. Beverage industries, mainly engaged in ethanol production, are using maize feedstock continuations. Therefore, the global price of maize has increased rapidly recently. Moreover, cellulosic feedstocks are now considered as a feasible alternative for biofuel production. Nanotechnology can be applied in augmenting the efficiency of enzymes involved in the conversion of cellulose to ethanol. Recently, scientists have developed nano-engineered enzymes, which can allow easy and low-cost conversion of cellulose from waste plant materials. Rice milling industries produce rice husk as a by-product which has a prospective alternative for renewable energy. A massive quantity of superior quality nanosilica is generated when rice husk is burned into thermal energy or biofuel. This nanosilica can further be utilized in making several useful materials like concrete, glass, and so forth. Hence, nanosilica can give an effective and useful solution to rice husk disposal issues.

1.2.7 QUALITY ENHANCEMENT OF AGRICULTURAL PRODUCTS

The nutritive properties and health-related benefits of the agricultural products by applying the nanotechnology have attracted interest of the stakeholders and the agro-food industry. The zinc spray of the nanoparticles has been reported to be effective in increasing protein, fat, and fibers in the Indian diets. Numerous studies are underway for testing the genotoxicity of nanomaterials and for developing and testing various nanoparticles to protect crops from powdery mildew disease.[15] From early years, the gold was also an attractive and useful metal because of its unique and valuable nature. The development of gold-nanoparticles has many commercial applications, such as identification of biomolecules. The detection is based on shape, size, refractive index of the nearby medium, and the distance between the gold nanoparticles; and these are the major factors on which the color of these colloids is dependent. Even a small variation in the abovementioned factors may cause the reckonable alteration in the surface plasmon response (SPR) absorption peak. The specific molecules are attached to the gold nanoparticles by their adsorption on the external site of the particle that subsequently changes the RI (refractive index) of gold nanoparticles. If the biomolecules to be attached are bigger than the gold nanoparticles, only few molecules will be adsorbed at the outward of the nanoparticles and will lead to the development of lumps and this ultimately changes the color of gold nanoparticles. The color alteration of nanoparticles is resulted from the shift in SPR that cause reduction in spacing of particles eventually. A very interesting advancement in the area of nanotechnology is the "smart dust," which can be applied to monitor various factors in food or environment like temperature, humidity and so forth.[12]

1.2.8   IDENTIFICATION AND TRACKING OF AGRO-FOODS

Identification (ID) tags have been useful in wholesale trading of food and agricultural products. Because of the tiny size nanomaterials have been applied in different sectors for encoding of agricultural products. Nanobarcodes have been employed in multiplexed bioassays and usual encoding as they have great potential of the development of numerous blends. Particles that are used in the nanobarcodes should be encodable, readable by the machine, robust enough to be employed for longer time and sub-micro nanometer in size. These nanobarcode particles are manufactured by the process that is extremely scalable and not fully automated. This involves electroplating of inert metals such as gold, silver, and so forth, into the prototypes that define the diameter of the particles. Finally, the resulting stripped nanorods are liberated from the template.

The nanobarcodes have both biological and nonbiological applications in the agricultural fields.[12] The significant biological applications of the nanobarcodes are in the use of ID tags for multiplexed analysis of the gene expression and the intracellular histopathology. The advancements and developments in the nanotechnology have directed to the improvement in the plant resistance properties against many environmental stresses, that is, resistance to drought, salinity and various infections, and so forth.[2] The nanobarcodes ID tags have also been applied to nonbiological systems. It has been utilized for the authentication (or tracking) in the agricultural foods and other goods such as husbandry products. Thus, nanobarcode technology is improving traceability in the food trade and will be a promising tool for promoting bio safe global agro-food business. It also enables us to develop methods for the preservation of not only the freshness of the agricultural products (vegetables, fruits, etc.) but also their quality and safety. In nanofood systems, packaging plays significant role in preventing post-production or post-harvest damages, in addition to prolonging shelf-life of fresh and stored agricultural produces.

Currently, nanofilms are being produced by adding nanoclays or silver nanoparticles in conventional packaging material to improve the tensile properties, stiffness, dimensional stability, and thermal resistance. Such processes enable extension of shelf-life of products by avoiding microbial contamination or delaying in the microbial growth through limiting gaseous exchange along with moisture. Moreover, incorporation of the silver nanoparticles in packaging materials will serve as an antimicrobial (inhibitory) agent and protect the agricultural products from spoilage causing microorganisms.

1.2.9 BIOSENSORS

Sensors are devices that react to physical, chemical, and biological attributes; and subsequently transmit that response into a signal or output. They can allow tracing of contaminants such as microbes, pests, nutrient content, and plant stress due to drought, temperature, insect or pathogen attack, or lack of nutrients. Nanosensors have the capability to allow farmers to utilize inputs more effectively through displaying the nutrient or water status of plants over fine spatial and temporal scales. This allows farmers to use nutrients, water, or pesticides (insecticide, fungicide, or herbicide) only where and when necessary. The significant application of nanotechnology enabled devices is increasing the use of independent sensors linked to global positioning system (GPS) for real-time monitoring. Such devices could be dispersed all over the ground in order to analyze soil conditions and crop growth. Further, these nanoparticles or nanosurfaces can be engineered to induce an electrical or chemical indication of existence of a contaminant including microorganisms. Ultimately, precision farming with the help of smart sensors lead to improve crop yield by giving precise facts, and hence assist farmers to take better decisions.

1.2.10   NANOTECHNOLOGY IN FILTRATION AND IRRIGATION

Nano-enabled water treatment techniques based on membrane filters derived from carbon nanotubes, nanoporous-ceramics, and magnetic nanoparticles, in spite of using chemicals and UV light, are common in traditional water treatment system.[14] Filters prepared from carbon nanotube could be used in removing contaminants and toxicants from potable water. Carbon nanotube fused mesh, that can remove waterborne pathogens, heavy metals like lead, uranium, and arsenic, has been recommended by researchers. Employing nanoceramic filter with positive charge can capture bacteria and viruses with a negative charge. This filtering device eliminates microbial endotoxins, pathogenic viruses, genetic materials, and micro-sized particles.[1] At low magnetic field gradients, the usage of magnetic nanoparticles and magnetic separation is now possible. Nanocrystals, including monodisperse magnetite (Fe_3O_4), have a robust and irreversible interface with arsenic while holding their magnetic features.[36] Simple magnet can be employed to eliminate arsenic from water. Such a treatment can also be employed for irrigation water filtration process. Detoxification or remediation of toxic pollutants is usually done using synthetic clay nanominerals. The water to be filtered is percolated through a column of hydrotalcite. Zinc oxide nanoparticles could be employed to eliminate arsenic by a point-of-source purification device. Other nanomaterials in the remediation process are nanoscale zeolites, metal oxides, carbon nanotubes, fibers, enzymes, various noble metals, and titanium dioxide.

1.2.11   NANOPARTICLE FILTERS TO REMOVE ORGANIC MOLECULES

Nanoencapsulation of pesticides permits appropriate absorption of chemicals in plants.[24] This procedure can also be helpful in carrying DNA and chemicals into the plant tissues to protect the host plant against insect pests.[31] The mode of release of nanoencapsulation consists of dissolution, diffusion, osmotic pressure, and biodegradation with specific pH.[11,32] Nanofertilizers are synthesized to manage the regulation of discharge of nutrients subjected to crop needs. Reports have revealed that nanofertilizers are more efficient than ordinary fertilizers.[17] Nanofertilizers could be applied to minimize the nitrogen loss because of processes such as leaching, emissions, and enduring integration by soil microbes. This could pave the way of selective release linked to time or environmental conditions. Slow release of fertilizers may also increase the soil quality by reducing toxic impacts related to fertilizer.[29] Combined with a smart delivery system, herbicide could be used as and when required thus increasing crop yield, with least health damage to workers in the field. Numerous commercial manufacturers are developing preparations having nanoparticles of 100–250 nm, which are easily soluble in water more efficiently than existing ones (thus, increasing their activity). Other agro-food businesses employ suspensions of water or oil-based nanoscale particles (nanoemulsions) that are identical to suspensions of pesticidal or herbicidal nanoparticles ranging from 200 to 400 nm. Such formulations can be simply integrated in several medias including gels, creams, liquids, and so forth and have wide potential for protective actions, treatment, or conservation of reaped food and agro-products.

1.2.12 NANO-COATINGS AND NANO-FEED ADDITIVES

Self-sanitizing photocatalyst coating used in poultry with nano-titanium dioxide (TiO_2) could be used to oxidize and destroy bacteria in the sunlight and humidity. Poultry feed having nanoparticles that bind to pathogenic bacteria could be helpful in decreasing foodborne pathogens. This unique photocatalytic characteristics of the nano-TiO_2 are activated when the coating is exposed to natural or UV light. In sunlight and humidity, TiO_2 oxidizes and destroys bacteria. Once coated, the surface remains self-sanitized as long as there is enough light to activate the photocatalytic effect; the coating is approved by the Canadian Food Inspection Agency (CFIA). In Denmark, the Chicken and Hen Infection Program (CHIP) involves self-cleaning and disinfection by nanocoatings.[8] The nanoscale smooth surface makes disinfection and cleaning more effective. Surface-modified hydrophobic as well as lipophilic nanosilica can be used as new drugs for healing nuclear polyhedrosis virus (BmNPV), a major problem in silkworm industry. Modified nanoclays (montmorillonite nanocomposite) can ameliorate the harmful impacts of aflatoxin on poultry.[26]

1.2.13   NANOCOMPOSITES AND NANO-BIOCOMPOSITES

Composites prepared from nanosize ceramics or metals lesser than 100 nm can unexpectedly become much stronger than predicted by existing models of materials science. Metals with grain size of 10 nm are seven-folds tougher and harder compared to their usual counterparts having grain sizes of 100 nm. Examples of polymer nanocomposites, combined with metal or metal oxide nanoparticles exploited principally for their antimicrobial action, include nano-zinc oxide and nano-magnesium oxide.[7] Cellulose nanocrystals can integrate into composites because they provide extremely multipurpose chemical functionality.[6] In food packaging, nanocomposites emphasise on the improvement of high barrier attributes against diffusion of oxygen, carbon dioxide, flavor compounds, and water vapors. Nano-clay minerals are found mainly in nature and might be integrated into the packaging films. Bio-nanocomposites are appropriate for packaging applications including starch and cellulose derivatives, poly (lactic) acid, polycaprolactone, polybutylene succinate, and polyhydroxybutyrate.[30] Nanocomposites have broad applications in various fields including agro-food packagings. The use of nanocomposites with new thermal and gas barrier properties can prolong the post-harvest life of foods, and this application could facilitate the transportation and storage of food.[9,28] Water absorption capability is an important property of nanoclay composites that are super absorbents, particularly when used under rain-fed conditions. In this context, the water-holding capacity, water absorbency, and water-retention capacity of zinc-coated nanoclay composite cross-linked polyacrylamide polymer have been developed for the promotion of rain-fed rice crop. Water-holding capacity of the soil with nanoclay composite was 8.5% higher compared to the original soil.[16] Applications of biochar include carbon sequestration, soil amendment, and absorption of several classes of undesirable components from water, soil, or industrial processes.

1.2.14   NANOTECHNOLOGY IN HYDROPONICS

Hydroponics is a technology of producing plants without soil and is extensively used around the globe for growing food crops.[25] The most commonly crops grown hydroponically are: tomatoes, sweet peppers, cucumbers, melons, lettuce, strawberries, herbs, and chilies. In addition, it has been used in the production of fodder and biofuel crops. Scientists have exploited hydroponics in nanotechnology by "growing" metal nanoparticles in living plants.[10,13,23] Nutrient management in agricultural production is more effective in hydroponic than in soil-based production. A nanophosphor-based electroluminescence lighting device has the prospective to decrease energy costs significantly. Such nanotechnology-based light could decrease energy expenses and encourage photosynthesis in indoor and in hydroponic agriculture.[35]

1.2.15   NANOTECHNOLOGY FOR CROP IMPROVEMENT

A variety of nanomaterials, mostly metal-based nanomaterials and carbon-based nanomaterials, have been exploited for their absorption, translocation, accumulation, and impact on growth and development of plants.[19,22] The positive morphological effects involved improved rate of germination; root and shoot length, and vegetative biomass of seedlings in various crops; for example, corn, wheat, ryegrass, alfalfa, soybean, rape, tomato, radish, lettuce, spinach, onion, pumpkin, and cucumber. Augmentation of many physiological parameters such as improved photosynthetic activity and nitrogen metabolism by metal-based nanomaterials has been reported in few crops, including soybean, spinach, and peanut. Magnetic nanoparticles coated with tetra methyl ammonium hydroxide led to an increase in chlorophyll-A level in maize. It has been reported that the use of iron oxide in the pumpkin increased root elongation that was attributed to the iron dissolution.[33] Spent tea (solid waste in tea plantation) could be used for the production of biodiesel, bioethanol, and also hydrocarbon fuel gases.

1.3   CONSTRAINTS

In spite of several prospective uses of nanotechnology in numerous sectors (i.e., health, medical, agriculture, space research technology, etc.), there are certain safety issues. Some key limitations and risks associated with the agricultural application of nanotechnology are:

- Exposure of nanomaterials to humans and accumulation in agro-food chains. The current knowledge on the nanotechnology is still in infancy stage; hence, prediction of the effect of nanoparticles on the human health and environment is not possible.

- Interaction of nanoparticles with the nontarget sites that lead to certain environmental and health issues.
- Higher production costs.
- Developments in agricultural sector are limited because of the small investment in research infrastructure, manpower training, and so forth.
- The public is not aware of the nanotechnology and its applications; and the opinion of the public is generally negative.
- The need of labeling of the nanotechnology products further prevents the innovative applications of this technology in agriculture. The various products of nanotechnology and their commercial use in numerous areas still need to be regulated just to make or guarantee their safe use. Hence, the proper knowledge of these materials and their possible interactions with the human body always need to be examined prior to commercial application.

1.4 CONCLUSIONS AND FUTURE PERSPECTIVES

Nanotechnology applications have the potential to modify the agricultural production by allowing better management and conservation of inputs for crop production. Researchers in nanotechnology can benefit a lot to the society through applications in agriculture and food systems. Introduction of innovative technology always has an ethical responsibility associated with it to be apprehensive to the unforeseen risks that may come along with tremendous positive potential. Public awareness on the advantages and challenges of nanotechnology will lead to better acceptance of this emerging technology. Rapid testing technologies and biosensors related to control pests and cross-contamination of agriculture and food products will lead to applications of nanotechnology in future. Nanotechnology applications in agriculture and food systems are still at the nascent stage; and more applications can be expected in the future. Nanoparticles present an extremely gorgeous stage for a diverse range of biological applications. Agriculture sector should gain the benefit of nanotechnology for human welfare. Nanotechnology can endeavor to deliver and streamline technologies currently in use for used environmental detection, sensing, and remediation.

1.5 SUMMARY

Nanotechnology has immense prospective to transform diverse segments of agricultural and food science with contemporary tools for the diagnosis and cure of various diseases, augmenting the capability of plants to absorb nutrients, and so forth. Smart sensors and transport arrangements will aid the agricultural industry to combat viruses and crop pathogens. In the immediate future, nano-based catalysts will enhance the competence of pesticides and herbicides, permitting the use of lower doses. It will also safeguard the environment indirectly by employing renewable energy supplies, filters, or catalysts to reduce pollution and clean up existing pollutants.

KEYWORDS

- Biofertilizers
- ellulose nanocrystals
- nano-based catalysts
- nanoclay composite
- nanocomposites
- nanofilters
- nanotechnology
- pesticides
- pollutant
- silica nanoparticles
- silver nanoparticles

REFERENCES

1. Argonide, *NanoCeram filters, 2005*. Argonide Corporation. http://sbir.nasa.gov/SBIR/successes/ss/9-072text.html; (accessed Dec 10, 2013).
2. Beyrouthya, M. E.; Azzia, D. E. Nanotechnologies: Novel Solutions for Sustainable Agriculture. *Adv. Crop. Sci. Tech.* **2014**, *2*, 3.
3. Biswal, S. K.; Nayak, A. K.; Parida, U. K.; Nayak, P. L. Applications of Nanotechnology in Agriculture and Food Sciences. *IJSID*, **2012**, *2*(1), 21–36.
4. Blackmore, S. Precision Farming: An Introduction. *Outlook Agric. J.* **1994**, *23*, 275–280.
5. Brock, D. A.; Douglas, T. E.; Queller, D. C.; Strassmann, J. E. Primitive Agriculture in a Social Amoeba. *Nature* **2011**, *469*(7330), 393–396.
6. Cao, X.; Habibi, Y.; Magalhães, W. L. E.; Rojas, O. J.; Lucia, L. A. Cellulose Nanocrystals-Based Nanocomposites: Fruits of a Novel Biomass Research and Teaching Platform. *Current Sci.* **2011**, *100*(8), 1172–1176.

7. Chaudhry, Q.; Castle, L.; Watkins, R. Eds. *Nanotechnologies in Foods*. Royal Society of Chemistry: Cambridge, U.K., 2010.

8. Clemants, M. Pullet Production Gets Silver Lining. *Poultry International, 2009*. http://www.wattagnet.com/Poultry_International/4166.html (accessed Oct 28, 2016).

9. Cortes-Lobos, R. *Nanotechnology Research in the US Agri-Food Sectoral System of Innovation: Towards Sustainable Development*. M.Sc. Thesis for Georgia Institute of Technology: Atlanta, GA, 2013.

10. Dimkpa, C. O.; McLean, J. E.; Martineau, N.; Britt, D. W.; Haverkamp, R.; Anderson, A. J. Silver Nanoparticles Disrupt Wheat (*Triticum Aestivum* L.) Growth in a Sand Matrix. *Environ. Sci. Technol.* **2013**, *47*(2), 1082–1090.

11. Ding, W. K.; Shah, N. P. Effect of Various Encapsulating Materials on the Stability of Probiotic Bacteria. *J. Food Sci.* **2009**, *74*(2), M100–M107.

12. Ditta, A. How Helpful is Nanotechnology in Agriculture? *Adv. Nat. Sci.: Nanosci. Nanotechnol.* **2012**, *3*(3), 033002.

13. Giordani, T.; Fabrizi, A.; Guidi, L. Response of Tomato Plants Exposed to Treatment with Nanoparticles. *Environmental Quality*, **2012**, *8*, 27–38.

14. Hillie, T.; Hlophe, M. Nanotechnology and the Challenge of Clean Water. *Nat. Nanotechnol.* **2007**, *2*, 663–664.

15. Hiregoudar, S. Application of Nanotechnology in Enhancing Quality of Agricultural Produce, 2014. http://www.isssonline.in/isss-2014/agri.html (accessed Oct 28, 2016).

16. Jatav, G. K.; Mukhopadhyay, R.; De, N. Characterization of Swelling Behavior of Nano-Clay Composite. *Int. J. Innov. Res. Sci. Eng. Technol.* **2013**, *2*(5), 1560–1563.

17. Liu, F.; Wen, L. X.; Li, Z. Z.; Yu, W.; Sun, H. Y.; Chen, J. F. Porous Hollow Silica Nanoparticles as Controlled Delivery System for Water Soluble Pesticide. *Mat. Res. Bull.* **2006**, *41*, 2268–2275.

18. NAAS. *Nanotechnology in Agriculture: Scope and Current Relevance*. Policy Paper No. 63; 2013; Nat. Acad. Agri. Sci. (NAAS), New Delhi, p 20. http://naasindia.org/Policy%20Papers/Policy%2063.pdf.

19. Nair, R.; Varghese, S. H.; Nair, B. G.; Maekawa, T.; Yoshida, Y.; Kumar, D. S. Nanoparticulate Material Delivery to Plants. *Plant Sci.* **2010**, *179*(3), 154–163.

20. Prasad, R.; Bagde, U. S.; Varma, A. Intellectual Property Rights and Agricultural Biotechnology: An Overview. *Afr. J. Biotechnol.* **2012**, *11*(73), 13746–13752.

21. Rickman, D. Precision Agriculture: Changing the Face of Farming. *Geotimes News Note*, 2001.

22. Rico, C. M.; Majumdar, S.; Duarte-Gardea, M.; Peralta-Videa, J. R.; Gardea-Torresdey, J. L. Interaction of Nanoparticles with Edible Plants and Their Possible Implications in the Food Chain. *J Agric. Food Chem.* **2011**, *59*(8), 3485–3498.

23. Schwabe, F.; Schulin, R.; Limbach, L. K.; Stark, W.; Bürge, D.; Nowack, B. Influence of Two Types of Organic Matter on Interaction of CeO_2 Nanoparticles with Plants in Hydroponic Culture. *Chemosphere* **2013**, *91*(4), 512–520.

24. Scrinis, G.; Lyons, K. The Emerging Nano-Corporate Paradigm: Nanotechnology and the Transformation of Nature, Food and Agro-Food Systems. *Int. J. Sociol. Food Agric.* **2007**, *15*, 22–44.

25. Seaman, C.; Bricklebank, N. Soil-Free Farming. *Chem. Ind. Mag.* **2011**, *6*, 19–21.

26. Shi, Y. H.; Xu, Z. R.; Feng, J. L.; Wang, C. Z. Efficacy of Modified Montmorillonite Nanocomposite to Reduce the Toxicity of Aflatoxin in Broiler Chicks. *Anim. Feed Sci. Technol.* **2006,** *129,* 138–148.

27. Soutter, W. *Nanotechnology in Agriculture, 2014.* http://www.azonano.com/article. aspx?ArticleID=3141 (accessed Oct 28, 2016).

28. Sozer, N.; Kokini, J. L. Nanotechnology and its Applications in the Food Sector. *Trends Biotechnol.* **2009,** *27*(2), 82–89.

29. Suman, P. R.; Jain, V. K.; Varma, A. Role of Nanomaterials in Symbiotic Fungus Growth Enhancement. *Curr. Sci.* **2010,** *99,* 1189–1191.

30. Tiwari, A. Ed. *Recent Developments in Bio-Nanocomposites for Biomedical Applications,* 1st ed.; Nova Science Publishers, Inc.: Hauppauge, N.Y., 2010.

31. Torney, F. Nanoparticle Mediated Plant Transformation. Emerging Technologies in Plant Science Research. In *Interdepartmental Plant Physiology Major Fall Seminar Series*; Physics, 2009; p 696.

32. Vidhyalakshmi, R.; Bhakyaraj, R.; Subhasree, R. S. Encapsulation the Future of Probiotics-A Review. *Adv. Biol. Res.* **2009,** *3*(3–4), 96–103.

33. Wang, H.; Kou, X.; Pei, Z.; Xiao, J. Q.; Shan, X.; Xing, B. Physiological Effects of Magnetite (Fe_3O_4) Nanoparticles on Perennial Ryegrass (*Lolium perenne* L.) and Pumpkin (*Cucurbita mixta*) Plants. *Nanotoxicology* **2011,** *5*(1), 30–42.

34. Welch, R. M.; Graham, R. D. A new paradigm for World Agriculture: Meeting Human Needs, Productive, Sustainable, and Nutritious. *Field Crops Res.* **1999,** *60,* 1–10.

35. Witanachchi, S.; Merlak, M.; Mahawela, P. Nanotechnology Solutions to Greenhouse and Urban Agriculture. *Technol. Innov.* **2012,** *14*(2), 209–217.

36. Yavuz, C. T.; Mayo, J. T.; Yu, W. W. Low-Field Magnetic Separation of Monodisperse Fe_3O_4 Nanocrystals. *Science* **2006,** *314*(5801), 964–967.

CHAPTER 2

NANOTECHNOLOGY APPLICATIONS IN AGRICULTURAL AND BIOLOGICAL ENGINEERING

DEEPIKA CHOUDHARY[1] AND SUDESH KUMAR[2,*]

[1]*Department of Chemistry, Banasthali University, 514 Gautam Budh, Banasthali, Rajasthan 304022, India*

[2]*Department of Chemistry, Banasthali University, 514 Gautam Budh, Banasthali, Rajasthan 304022, India*

*Corresponding author. E-mail: sudeshneyol@gmail.com

CONTENTS

2.1   INTRODUCTION

Nowadays, the field of agricultural science is diversifying worldwide. The agriculture is the main source for more than 60% of the population for their living.[27] To feed the increasing population day by day, the population worldwide is demanding agricultural products at affordable prices. To improve the crop yield, it is essential to utilize most recent innovations.

To increase the crop yield and redesign agribusiness, nanotechnology and nano-biotechnology have enormous possibilities. Currently, nanotechnology in the agriculture focuses on the use of nanoparticles (NPs) with notable properties to sustain yield and effectiveness of household animals.[54,95] The development of nanotechnology has been able to build up nourishment quality, nutrition generation, plant protection, identification of ailments in plants and animals, observation of plant development, and diminish agricultural waste.[7,29,43,56,57,84,105] The potential applications of nanotechnology in agriculture includes: enhancing plant growth and crop yield, sensors for checking soil quality, pesticides, and nanofertilizers. To increase the capability and sustainability of cultivating practices in agriculture, nanomaterials are being used through embedding minor inputs; in addition to this it also produces small amount of waste than traditional methodologies and sustainability. In agribusiness field, the usage of conventional fertilizers for a long time and at an enhanced rate has been considered major environmental issue all over the world. The utilization of phosphorus (P) and nitrogen (N) fertilizers has turned into prime issue globally leading to eutrophication in coastal ecosystems as well as freshwater bodies.[24,25] Therefore, there are major thrusts to develop bio-fertilization by involvement of nanomaterials (NMs).

The main perspective of sustainable agriculture and implication of present nanotechnology in agriculture[20] are regarded as important factors to feed and nourish world's quickly developing population.[61] Nanoencapsulated nutrients or nanofertilizers possess tendency to dispense chemical fertilizers and nutrients on the basis that control target activity and plant growth.[26] Nanotechnology and nanoscale science can possibly change the food systems and agriculture.[79] Nanotechnology provides challenging possibilities in horticultural upheaval, better bioavailability, high reactivity, and surface and bioactivity impacts on NPs.[47,104] Nanotechnology alludes to a nanoscale innovation that has promising applications in everyday life. This technology accentuates the ramifications of individual atoms, molecules, or submicron measurements regarding their applications to physical, chemical, and biological systems, and in the end their reconciliation into bigger complex frameworks. Nanoscience is a an emerging science, which contains particles, macromolecules, quantum dabs/dots, and macromolecular assemblies.[4] Nanotechnology pacts with structures that mainly range from 1 to 100 nm and includes devices or materials within that size.[19–27] Nanoparticle size can affect physicochemical stability, biological activity, and mainly characteristics of encapsulation and release of active compounds.[45,53] At the nanoscale level, matter shows completely different properties than the bulk material. By one mean or another in nanotechnology, nuclear and atomic means can strengthen determinant components by different impacts on importance, which are at larger scale.[47]

In 1970s, Green Revolution focused on four essential components that resulted in generation framework viz. semi-overshadow large yielding assortments of wheat[93] and rice, broad water utilization system, agrochemicals, and manures. Therefore, they brought about awesome increments in the farming generation. However, the agrarian generation is now encountering a level that has unfavorably influenced the vocation base of the cultivation. Nanoscale science and nanotechnologies can possibly alter horticulture and nourishment frameworks, additionally, creating enhanced frameworks for checking ecological conditions and conveying supplements or pesticides. Nanotechnology can enhance our comprehension of the science of various products and hence conceivably upgrade yields or nourishing qualities. Nanoinnovation procedures or instruments are utilized amid development, generation, handling, or bundling of the sustenance.[31,34]

In recent years, nanotechnology has been progressively connected to the improvement of novel antimicrobials for the administration of pathogenic microscopic organisms influencing crop yield, people, and animals. Specifically, huge improvement in nanomaterials amalgamation, for example, polymeric, carbon-based, and metallic, has pulled in specialists' consideration toward applications in overseeing plant maladies brought on by microorganisms. Toxicity considerations, including negative natural impacts, have likewise prompted the overhaul of nanomaterials by surface alteration and tuning the shape with size, prompting expanded antimicrobial action, and diminished environmental lethality.[37,82] Nanotechnology can possibly upset the agricultural sector and sustenance with novel devices, for example, sub-atomic treatment of diseases, fast ailment discovery, improving the quality ingestion of plant supplements, and so forth. Maintainable escalation is an idea identified with a creation framework intending to expand the yield without any unfriendly ecological effect while developing the same farming regionThis worldview gives a structure to assess the determination of the best mix of ways to deal with agrarian creation by looking over the impact of social, monetary and the present biophysical conditions.[38] Nanomaterials in the utilization of lipid, inorganic and polymeric NPs are examples of polymerization, ionic gelatin, emulsification, oxy-reduction, and so forth.[90]

Nano originates from the Greek word for diminutive person and NPs are portrayed as those with the molecule size of hundred nano meters.[85] However, nanotechnology is not only known for their dwarf nature. But it also shows the excellence structural property at the nuclear or sub-atomic level and expanded mechanical quality. These outcomes in materials and frameworks regularly show novel and fundamentally changed physical, compound, and organic properties because of their size and structure.[16,67,72,101,117] The current challenges of sustainability, food security and environmental change are drawing analyst's attention in of nanotechnology, due to exceptional chemical and physical property of nanoscale particles widely used in various fields such as medicine, biotechnology, electronics, material science and energy sectors, and so forth.[83,113,117] Nanotechnology goes for accomplishing for control of matter what PCs accomplished for our control of data.[108,112] Nanotechnology has offered advantages over the agro-nourishment framework like expanding efficiency. Environmental benefits include diminishing deposit and contamination and consequently more nutritious and safe nourishment. Nanotechnology is being considered as a potential sector to meet the needs of growing population, fighting yearning difficulties of environmental change and other biological aggravations.[17,87]

The adequate handling and processing techniques have been developed in agricultural practices, to fulfill the demand of growing consumers and for reliable supply of safe and high quality nutritious food products.[13,45,66,74] A various kind of specific NPs have been developed to enhance soil fertility and for even distribution of water. In agriculture, scientists have done an endless commitment by using innovative technology in farming and sustenance creation frameworks for effective process of sensing, bioremediation, and environment detection.

The primary report on utilization of nanotechnology in agribusiness was distributed in 2003 by the U.S. Bureau of Agriculture. Nanotechnology is presently one of the needs of the nation, which has been underscored as a rule approach and Fifth 5 Year Plan of India. With respect to the part of farming has in the national economy and confronting obstructions to build profitability from one hand, nanotechnology's potential and capacity to lessen or wipe out a large portion of these issues from one viewpoint, improvement of this innovation in the horticultural segment in the National Innovation System is vital in all accounts.[3,22,51,80] Later on, nanotechnology for farming applications include accomplishments of motorization and computerization. Successful fusion of emerging technological innovations in agriculture requires the instruction of people with great learning the needs of the agribusiness and who can incorporate these necessities into the configuration of creative gadgets and procedures.[13,35]

This chapter discusses potential applications of nanotechnology in agricultural and biological engineering. It also augments knowledge to convey the agriculture applications, potential of nanomaterials in agriculture management, pest control, fertilizers, clay and disease treatment and plant protection, and production.

2.2   APPLICATIONS OF NANOTECHNOLOGY IN AGRICULTURE

- Improve in agriculture production.
- Food security and food processing materials.
- Pollutant remediation.
- Improving feeding efficiency and nutrition of farm animals.
- Climate change.
- Development of tools for disease detection and treatment.
- Monitoring crop growth: nanocapsules and nanosensors.
- Pest control.

2.2.1   IMPROVEMENTS IN AGRICULTURE PRODUCTION

Nanotechnology can enhance crops yield as well as increase the value of harvests or normal remediation. Molecule cultivating is a process of NPs yielding modern practice of creating plants as has been reported in various soil samples. Mechanically, NPs can be separated from different plant tissues after harvesting. This application provides the opportunity to remediate wastes and make them reusable in different fields like, food, cosmetics or pharmaceutical products.[20,83] Investigators have reported eco-friendly methods to prepare silver and gold NPs[71] by different plants such as: *Medicago sativa, Arachis hypogaea, Pennisetum glaucum, Cyamopsis tetragonolobus, Zea mays, Sorghum vulgare, Vigna radiata,* and *Brassica juncea* or extricates from *M. sativa, B. juncea, Allium sativum L.,* or *Memecylon edule*. Nanoparticles can acquire various shapes and sizes depending upon the temperature and species of plant or tissue in which they are inserted. In environmental-friendly use, applications of nanotechnology have huge power to convert rural generation for providing best experimental conditions for plant production. It can improve agriculture by utilizing:

- Nanoporous zeolites for controlled discharge and effective measure of water manure and so forth.
- Nanocapsules for conveying of herbicide, vector and overseeing of vermin.
- Nanosensors are used identifying aquatic poisons and pests.
- Biopolymers nanoscale (e.g., proteins, sugars) NPs with couple of properties, for example, minimum effect upon human well-being and earth might be utilized as a part of sterilization and reusing of substantial materials.
- Nanostructured metals can be investigated in decay of unsafe organics at room temperature.
- Smart particles can be helpful in successful ecological checking and purification procedures.
- Nanoparticles as a new photo catalyst.[101]

In agriculture, new apparatuses for sub-atomic and cell science for identification, quantification of individual genes, separation, and molecules.[58] Nanotechnology can possibly convey qualities to particular destinations at cell volume and revamp the DNA molecule similar life form for articulation fancied appearance, in this way, avoiding tedious procedure to exchange the quality of foreign pathogen. Nanotechnology has likewise demonstrated capacity in changing hereditary combination of the plants helping in better exchange. Rather than utilizing certain substance mixes, physical mutagens and energy management system (EMS), X-ray beam, gamma beam, and so forth for routine actuated change, nanotechnology has demonstrated another measurement in transformation research. Researchers had changed the shade by utilizing nanotechnology of the stems and leaves of Khao Kam from purple to green and the grain was changed to white. The exploration includes instructing a nanosized hole that concludes the divider as well as layer cell of rice in order to embed an iota of nitrogen; utilizing a molecule shaft as well as particle of nitrogen is shot by the hole to empower modification DNA of rice. Recently, living organism derived changes is assigned by "molecularly modified organism (MMO)."[35]

2.2.2   FOOD SECURITY AND FOOD PROCESSING MATERIALS

The growing population is, likewise, confronting natural dangers including environmental change that would influence nourishment profitability. It is basic to guarantee heightening of agriculture, combined with productive sustenance taking care of, preparing, and distribution. Nanotechnology promises to change the entire way of life—from creation to preparing, storage, and improvement of inventive materials, items, applications.[92] Nanotechnology has given new answers for the issues of nourishment science in plants and also gives new ways for dealing with the choice in crude materials, and handling the materials for upgrade of various products from plants. Food applications of nanotechnology demands preservation, smart packaging as well as production of healthy foods. Expanding the idea that interest sustenance, possibility of intelligent nourishment permits purchasers to change sustenance depending on their own particular needs. The basic idea is that a huge number of nanocapsules enclosing flavor or healthy components, (for example, vitamins), would stay torpid in the sustenance and might be discharged only when activated by the consumer.

The primary standard behind the advancement of nanosized fixings and added substances has all the earmarks of being coordinated toward upgraded bioavailability of nanosized substances and uptake in the body, however, different advantages are the change in consistency, soundness, taste as well as surface. Likewise, nanopackaging was intended to empower various materials cooperated with nourishment; antimicrobials emanating, cell reinforcements, and different data sources. For example, bundling nanosensors, which are designed started as off chance that it has been tainted by pathogens. Various kinds of metal oxides in nano form such as nano copper oxide, nano magnesium oxide, and nano titanium dioxide have been used in food packaging. Carbon nanotubes are elucidated for antimicrobial food packaging use in future.[18] The use of new advances in the nourishment business has been exceedingly inquired to get benefits as far as health, security and items with better quality are concerned.[78,110]

The nano-precipitation strategy is a decent alternative of advancement as nanospheres of NPs take just a brief timeframe, and just a little measure of crude matter is necessary as it uses less energy.[1,10,12,28,30] Both governments and advancement offices have accordingly perceived the need to focus on the poor agriculturists and developed a methodology to lighten the issues of nourishment uncertainty and destitution confronting these regions.[107]

Nourishment security has been the greatest concern among all sectors. The first and most essential need of each human is nourishment, and sustenance supply for people connected with farming specifically and by implication. Agricultural sector growth with the reference for objectives seems essential for developing countries. Presently, following green insurgency for quite a while and decrease in the farming items proportional to the total populace development, the need of utilizing new innovations in the horticulture business is evident like never before. Modern technologies, for example, bio and nanotechnologies can assume a critical role in expanding generation and enhancing the nature of nourishment delivered by agriculturists.[107]

Nanofoods portray the nourishment, which has been developed, delivered, handled, or bundled utilizing nanotechnology systems or instruments. Nanotechnologies economically utilize a part of the sustenance business, bundling and capacity applications. Evaluations of financially accessible nano-oods change broadly in the range of 150–600 nm in nanofoods and 400–500 nm in nanofood applications of bundling. Generation of useful substances (e.g., soda pops, frozen yogurt, chips, and chocolates) are advertised as nourishments for decreasing sugar or calorie and are also expanding as vitamins, protein, or fiber substance. Improvement of nourishments is equipped for altering their shading, enhancement of healthful properties as indicated by a man's dietary needs, hypersensitivities, or taste inclinations, generation of more grounded flavorings, colorings and nutritious added substances and bringing down expenses of fixings. Expanding the time span is by utilizing bundling substances, those discharge control air, antimicrobials and dampness trade the earth, which is possible because of nanotechnology methods.

Nowadays, various nanofood items are being developed. Nanofoods are having nanoscale fixings in addition of the substances that are presently accessible on the racks of grocery store. Nestle and Unilever are leaders to create emulsion-based nanosized frozen yogurt with less fat substance and yet having greasy surface and flavor. Nanoencapsulated fixings consisting of unsaturated fats and vitamins are currently exported economically in preparing and preservation of meat, drinks, cheese, and other foods like cured meat production and industrial sausage. Many agencies are required to improve and stabilize color and flavor of a particular food. Aquanova, German company, developed a carrier system which was purely based on nanotechnology to encapsulate active ingredients by utilizing 30 nm micelles like fatty acid and vitamins C and E, to be used as preservatives and commercialized as NovaSOL. This offered speedier preparation of meat, less expensive fixings, and higher shading stability. Nanoparticles are added to numerous nourishments to enhance stream properties, shading and solidness amid handling, or to expand time span of usability. Moreover, the materials of alumina–silicate were usually utilized as anticaking specialists as a part of powdered or granular handled nourishments, while titanium dioxide is a typical added substance for sustenance brightening and whitening, and are utilized in a dessert shop, and for a few cheeses and sauces.[19,86]

The Institute of Medicine of U.S. National Academy of Sciences characterizes foods like sustenances which are given a medical advantage past customary supplements present in it. Nanoencapsulation includes a dynamic fixing in a nano-estimated container. Dairy items, oats, breads, and refreshments are presently braced with minerals, vitamins, bioactive peptides, probiotics, cancer prevention agents, and plant sterols. Recently, dynamic fixings having vitamins, additives, and chemicals are being added to sustenances in microscale cases. For example, a significant number of the usually utilized Omega-3 sustenance added substances are 140–180 μm encapsulated in fish oils, and are utilized by Nu-Mega Driphorm to strengthen Australian bread. Aquanova, NovaSOL scope of nano-exemplified bioactive fixings, incorporates vitamins, co-catalysts, isoflavones, flavonoids, carotenoids, protecting specialists, nourishment shading materials and other bioactive substances. These items present an extensive variety of nourishment added substances and in refreshments added substances, for example, Solu-tm-E200 showcased using BASF, which is a vitamin E nano-arrangement, particularly defined for clear drinks like game drinks and improved water.

Basic viability of these fixings relies on improving their bioavailability as well as safeguarding. Nanoencapsulating dynamic fixings convey more prominent bioavailability, enhanced solvency, and expanded intensity contrasted with the substances in bigger or smaller scale exemplified shape. The more prominent intensity of NPs added substances may well lessen the amounts of the required added substances, furthermore, their more noteworthy compound reactivity.[73] The real potential applications of nanocircle/microsphere framework include heated merchandise, refrigerated/solidified players, tortillas and level breads, handled meat items, occasional sweet shop, strength items, biting gums, dessert blends, and wholesome nourishments.

There are numerous organizations in the sustenance utilizing nanotechnology. Nutralease Ltd. Company has created novel transporters for nutraceuticals to be fused in sustenance frameworks or makeup details, expanding the bioavailability of an item. A portion of the nutraceuticals is joined in the transporters incorporate, beta-carotenes, phytosterols, and lycopene. These items are utilized as a part of solid nourishments, particularly to keep the collection of cholesterol. The huge motivation behind nano-packaging is to set the longer time span of usability by enhancing the obstruction capacity of sustenance bundling to decrease gas and dampness trade and UV light introduction. For instance, DuPont has declared the arrival of a nano-titanium dioxide plastic added substance to be specific "DuPont light stabilizer-210," that could decrease the UV-harm of foods in transparent packaging. By 2003, more than 90% of nano-packaging depended on nanocomposites, in which nanomaterials were utilized to enhance the obstruction capacity of plastic wrapping for nourishments, and plastic jugs for brew, soda pops, and squeezes. Nano-packaging can likewise be intended to discharge antimicrobials, cancer prevention agents, catalysts, flavors, and nutraceuticals to expand time span of usability.

The UK research institute has recognized protected and compelling antimicrobial NPs for nourishment bundling, a disclosure that could alter sustenance bundling later on. Researchers reported that zinc oxide and magnesium oxide NPs have capability to kill microorganisms. Dr. El Amin with El Obeid agricultural research station at El Obeid, Sudan reported that energizing new nanotechnology items for sustenance bundling are in the pipeline or have now entered the market. It was included that in the sustenance bundling segment, nanomaterials are being produced with improved mechanical and warm properties to guarantee better insurance of nourishments from outside mechanical, warm, concoction, or microbiological impacts. This would enrich packaged foods with an extra level of well-being and usefulness. This technology would potentially increase the shelf life of foods. Dr. El Amin with El Obeid agricultural research station have reviewed that rising shopper's interest for enhanced quality, augmented time span of usability, and naturally cordial items alongside improved directions in the EU is pushing for more development in the food packaging sector. He mentioned that the sustenance bundling makers are reacting to the purchaser and administrative patterns by not just concentrating on creating financial and successful bundles for securing the nourishment items, but also on the tasteful estimation of the packages. Clients today expect significantly more from packaging for securing the quality, freshness, and well-being of nourishments.

An experimental and modern research at the Norwegian Institute of Technology is utilizing nanotechnology to make minor particles in the film, to enhance the transportation of some gasses through the plastic film to pump out dirty air, for example, carbon dioxide. It is trusted that the idea could be utilized to shut out unsafe gasses that abbreviate the time span of usability of the foods. The researchers are investigating whether the film could likewise give obstruction assurance and prevent gases, for example, oxygen and ethylene from falling apart sustenances. Waxy coating is used widely for apples and cheeses. As of late, nanotechnology has empowered the advancement of nanoscale consumable coatings as thin as 5 nm broad. These edible nano-coatings could be utilized on meats, cheddar, organic products, vegetables, ice cream parlor, bread kitchen merchandise, and fast foods. They could give an obstruction to dampness and gas trade, go about as a vehicle to convey flavors, cell reinforcements, compounds and hostile to cooking specialists and could likewise build the timeframe of realistic usability of made sustenances, even after the packaging is opened.

The U.S. Organization Sono-Tec Corporation claimed in mid-2007 that it has built a consumable antibacterial nano-covering, which can be connected straightforwardly to bread kitchen products; it is presently testing this procedure with its customers. Another pattern in this regard is the synthetic discharge nanobundling. This procedure empowers nourishment bundling to connect with the food; the exchange can be handled in both directions. Packaging can discharge nanoscale antimicrobials, cell reinforcements, flavors, aromas, or nutraceuticals into the nourishment or refreshments to extend its timeframe of realistic usability or to enhance its taste or smell. In numerous occurrences, compound discharge bundling can additionally fuse observation components. On the other hand, nanobundling utilizing carbon nanotubes are being produced with the capacity to pump out carbon dioxide that can cause food or beverage deterioration. Nano-packaging that retains undesirable flavors is likewise at a developed stage. The synthetic discharge packaging is likewise intended to discharge biocides in microbial populace, mugginess, or other evolving conditions. Other packaging and food contact materials fuse antimicrobial nanomaterials that are planned not to be discharged, so that the bundling itself goes about as an antimicrobial operator. These products commonly use silver NPs, although some use nanosize zinc oxide or chlorine dioxide. Packaging furnished with nanosensors is additionally intended to track either the interior or outer states of nourishment items, pellets, and holders, all through the inventory network, for example; such packaging can monitor temperature or humidity over time and then provide relevant information of these conditions. Some of these nanosensors are in progress, such as the Georgia Tech utilized carbon nanotube as biosensor to identify microorganisms, harmful substances, and waste of nourishments or drinks.

OPALFILM joining 50 nm carbon dark NPs from Opel Company is being used as biosensor that can change shading in response to food spoilage. Titanium dioxide NPs based oxygen-sensing inks were used as tamperproofing. Mechanical nanotech (OTC: INTK) has reported the effective use of Nanosulate defensive covering to give warm protection and erosion assurance of dairy handling gear-based plastics for sustenance bundling. The analysts have connected thin films of less than 20 nm within the surface of food packaging. However, there are some major concerns including manufactured nanomaterials might be released into the environment from waste streams or during recycling. Recognition of small measures of a compound contaminant, infection, or microscopic organisms in nourishment framework is another potential utilization of nanotechnology.

The energizing possibility of consolidating science and nanoscale innovation into sensors is promising as it will bring about more well-being for the nourishment preparing framework. Nanosensors that are created by specialists at both Purdue and Clemson Universities utilize NPs, which can either be carefully fit to flavors, hues or then can be fabricated out of magnetic materials. These NPs can then specifically assault any food pathogen. Additionally, these sensors utilize either infrared light or attractive materials. The benefit of such a framework is that possibly a huge number of NPs can be put on a solitary nanosensor to quickly and precisely distinguish the nearness of any number of various microorganisms and pathogens. These nanosensors can obtain entrance into the modest hole where the pathogens frequently stow away. The use of nanotechnologies on the discovery of pathogenic living beings in nourishment and the improvement of nanofood well-being is likewise learned at the Bio-explanatory Microsystems and Biosensors Research Center at Cornell University. Here, research studies focus on the advancement of quick and versatile biosensors for the location of pathogens and sustenances. This framework focuses on the exceptionally quick identification of pathogens in routine drinking water, sustenance investigation, natural testing, and in clinical diagnostics.

The silver NPs can enhance the security of the world's sustenance supply, as indicated by an exploration research at Iowa State University, USA. Silver NPs cannot currently be added directly to foods as little is known about their adverse effects on human health and their impact on ecological systems. In any case, research program looking at silver NPs focus on how these act as antimicrobial operators in nourishments, the objective of creating sustenance known applications, for example, organism safe textures and non-biofouling surfaces in dairy products. It may prompt new methodologies for invading pathogens as well as upgrading the sustenance well-being. The potential for such sustenances to posture new well-being dangers must be researched with a specific end goal to figure out whether it is related or not related to food safety standards. New earnest requirement for administrative systems equipped for dealing with any dangers is connected with nanofood and the utilization of nanotechnologies in nourishment industry.

Governments should likewise respond to nanotechnologies more on extensive social, monetary, common freedoms, and moral issues. To assure control of developed modern technologies in the significant area of food and horticulture, open contribution in nanotechnology basic leadership is essential. The fixings in NPs must experience a full security appraisal by the significant investigative admonitory relationship before these are allowed to be utilized as a part of the food products. Additionally, the fixings arrangements of sustenance items ought to reveal the way that NP materials have been incorporated.[5]

2.2.3 POLLUTANT REMEDIATION

Nanoscale particles speak to another era of natural remediation innovations that could give savvy answer for probably the most difficult ecological cleanup issues. Particles of nanoscale iron have expansive surface zones with high surface reactivity, providing tremendous adaptability to interior utilization. Research has demonstrated that NPs are extremely viable to change and wide assortment of detoxification regular natural contaminants, for example, natural solvents (chlorinated), polychlorinated biphenyls (PCBs), and organochlorine pesticides. Altered NPs of iron are being orchestrated for improving the quality and productivity for remediation. Mainly ecological science of zero valent iron or metal is broadly recorded. Current research laboratory mainly established nanoscale particles as reductants, which are the most wide variety catalyst of natural environmental contaminants involving organic compounds that are chlorinated and metal ions. Fast and finish chlorinated contaminants by the method of de-chlorination are accomplished inside soil-water slurries and also in water. Ethane is genuine among all tests. More than 99% ejection was refined with nanosized squeeze atom within 24 h. Couple of pesticides, which are consistent high-affect circumstances, are expeditiously degraded in lessening conditions.

Consumption method utilizes zero-valent iron as substance reductant. At high-impact circumstances standard electron acceptor oxygen under anaerobic environment, electron discharge from the response of ZVI with water is coupled with response to chlorinated as well as nitro-aromatic mixes. Usage of "attractive microscopic organisms" appears to be helpful for metallic particles and substantial metal expulsion from watery arrangements such as Hg, Ag, Pb, Zn, Cu, Mn, Sb, and As, Fe, Pd, Pt, Ru, and Al. For example, press sulfide, substantial metal encourages bacterial cell dividers, preparing the microorganisms adequately charged for expulsion of suspension to attractive partition technique. Scrutinization has demonstrated that specific microorganisms could create press sulfonide, which would work as adsorbent for a few metallic particles. A new thought was generated for arranging mesoporous alluring nanocomposites. These nanocomposites are utilized for removal of perilous administrator's show in the earth. This new methodology of nuclear arrangements uses mesoporous silica to coat NPs of magnetite.[118]

2.2.4 IMPROVING FEEDING EFFICIENCY AND NUTRITION OF FARM ANIMALS

A standout amongst the most basic commitments to animal era is feedstock. Low feeding adequacy achieves advance of high arrivals of waste, overpowering environmental weight, high era cost, and equaling diverse livelihoods of the grains, biomass, and other reinforce materials. Nanotechnology may altogether enhance the supplement profiles and adequacy of minor supplement conveyance of nourishes. Most creature bolsters are not ideal health wise, particularly in developing countries. Including supplemental nutrients is a viable way to deal with the productivity of protein blend and the use of minor supplements. Other digestive aids, for example, cellulosic compounds can encourage better usage of the vitality in plant-based materials. Besides, minor supplements and bioactives can enhance general strength of creatures so that an ideal physiological state can be accomplished and kept.

An assortment of nanoscale conveyance frameworks have been explored for food applications. They incorporate micelles, liposome, biopolymeric NPs, nanoemulsions, protein-sugar nanoscale buildings, dendrimers, strong nano lipid particles and many others. These frameworks on the whole have demonstrated various favorable circumstances including better dependability against natural hassles and preparing impacts, high ingestion and bioavailability, better solvency and scatter capacity in watery based frameworks (nourishment and encourage), and controlled discharge energy.[79] Self gathered and thermodynamically stable structures require little vitality in preparing subsequently tending to issues identified with supportability. Nanoscale conveyance can be utilized to enhance the dietary profiles of bolstering effectiveness.

The nanoscale conveyance frameworks can be intended for veterinary medication conveyance, which ensures the medication in GI tract, and takes into account discharge at the wanted area and rate for ideal impact. These favorable circumstances enhance the productivity by which creatures use supplement assets, decrease material and monetary weight of the makers, and enhance item quality and generation yield. Like sustenance applications, the plan of a suitable nanoscale conveyance framework will require a full thought of the adequacy of its planned use while keeping any unfavorable impacts or unintended results. The nanoscale particles ought to be liable to a thorough hazard appraisal to guarantee dependable and safe advancement and organization in the items.[81]

2.2.5 CLIMATE CHANGE

Individuals depend intensely on Earth's common capital—its organic assets and biological systems—to live and thrive. Earth's natural assets incorporate types of plants, creatures, and microorganisms, and they give a critical portion of humankind's sustenance, agrarian seeds, pharmaceutical intermediates, and wood items. At present, fossil fills up roughly 80 % of the vitality utilized around the world. The world will keep on burning noteworthy measures of fossil fills within a reasonable time-frame. Along these lines, carbon catch and capacity (CCS) is rising as practical short to medium term elective in diminishing between measures of anthropogenic CO_2 discharged into the air. Earth likewise has an assortment of biological systems such as wetlands, rainforests, seas, coral reefs, and icy masses) that give basic administrations: (1) water stockpiling and discharge; (2) CO_2 ingestion and capacity; (3) supplement stockpiling and reusing; and (4) contamination take-up and breakdown.

Conservation of the Earth's biodiversity in various biological communities is basic to human life and success. In recent decades, an agreement developed that expanding outflows of carbon dioxide (CO_2) of the burning fossil powers (like petroleum and petroleum) are key drivers of worldwide environmental determination. The customary vitality sources like lights of oil, diesel generators and wood stoves can be used with renewable off lattice power (sun oriented, wind), enhanced cooking, gadgets keep running on battery with off network power charging. This ought to be joined with ability for assembling/amassing, keeping up renewable-based hardware and reasonable sun-powered cells would give power to rustic zones.

Nanotechnology has been utilized as a part of sun-based application of photovoltaic (PV) cells to enhance effectiveness. Nanomaterials can be utilized as a part of sun-powered cells, for example, natural, thin film, color sharpened and half and half ones. Hybrid sun powered cells utilizing layer of polymer secured with a thin film of nanocrystal Pb; Se has been produced to expand proficiency and shield the polymeric layer from UV radiation. Sunlight based controlled electrochemical or photograph synergist frameworks, which deliver hydrogen by means of water part utilizing natural toxins as conciliatory electron contributors, give a conceivable answer for achieving two goals: era of vitality and creation of clean water. Nanotechnology applications in the vitality and environment areas, for example, expanded utilization of renewable vitality sources, remediation of contaminated water and soils contribute towards giving an enhanced situation to farming exercises reported that one billion individuals are at hazard since they do not have admittance to consumable water and another 2.6 billion need access to clean water. Seawater and harsh water from saline aquifers constitute about 97 % of the water on Earth. Absence of safe water and satisfactory sanitation conditions adversely affects a well-being as far as waterborne illnesses, looseness of the bowels, typhoid, that are transmitted through drinking degraded water, and water-washed ailment like skin and eye diseases, which happen because of absence of water for washing and individual cleanliness. There are around 4 billion diarrheal cases every year out of which 1.8 million persons die.

Degraded surface water sources influence inland fisheries, which is a noteworthy nourishment source in a few sections of the world. Approximately 70–90 % of the water utilized as part of agribusiness and industry and for human utilization comes back to the earth as wastewater. Different substance gatherings can likewise be added to NPs to enhance their specificity in expelling certain toxins. Carbon nanotube channels can viably expel microscopic organisms and infections from water. Gram-positive and -negative microbes can be eliminated by NPs of silver mixes and magnesium oxide, which disturb bacterial cell films.

Magnetic NPs can be utilized to filter water to expel arsenic. Nano-empowered water treatment procedures consolidating carbon nanotubes, nanoporous pottery, and attractive NPs can be utilized to expel pollutions from drinking water and could conceivably evacuate microorganisms, infections, water-borne pathogens, lead, uranium, and arsenic, among different contaminants. The nanostructured films can be delivered from nanomaterials, such as carbon nanotubes, NPs and dendrimers, nanoreactive layers from metal NPs, and different nanomaterials;[8] and polysulfonate ultrafiltration films impregnated with silver NPs were discovered successful against *E.coli* K12 and *P. mendocina* microscopic organism's strains and demonstrated a critical change in infection expulsion. Nanoscale zero-valent iron and different nanomaterials, such as nanoscale zeolites, metal oxides, carbon nanotubes, can be utilized to remediate poisons in soil or groundwater. Nanotechnology and especially NPs can be used to tide up soil tainted with overwhelming metals. The molecule stream alongside the ground water is substantially less costly than uncovering the dirt to treat it. Amongst the most important areas for increasing the productivity of crops is precision farming by applying contributions to vitally required amount and in required time.

Minor sensors and observing frameworks empowered by nanoinnovation will affect future exactness cultivating techniques. Precision cultivation has been a long-felt objective to amplify yield like trim yields while minimizing input of manures, pesticides, herbicides, and so forth through observing natural factors to diminish the rural waste and, in this way, keep ecological contamination to a base. Nanotechnology-empowered gadgets utilize self-sufficient sensors connected into a GPS framework for continuous checking. These nanosensors could be disseminated throughout the field where they can screen soil conditions and yield development. Remote sensors are being used in specific parts of the USA and Australia. For instance, in one of the Californian vineyards, Pick-berry, in Sonoma County has introduced Wi-Fi frameworks with the assistance of Accenture. The underlying expense of setting up such a framework is defended by the way that it empowers the best grapes to be developed, thus, creating better wines, which command a premium cost. The union of biotechnology and nanotechnology in sensors will make equipment of expanded affectability, permitting a prior reaction to ecological changes.

Environmental change has risen as a standout amongst the most genuine ecological worries of our times. Warming of the atmosphere framework is unequivocal, it could be seen from the increments in worldwide normal air and sea temperatures, broad melting of snow and ice, and rising worldwide mean ocean level. The vast majority of the warming that has happened in the course of the most recent 50 years is prone to have been created by human exercises. Other than atmosphere compelling, human exercises (such as smoldering of fossil fuels, horticulture) and land-utilization changes (such as deforestation, creature farming) have caused ozone exhaustion due to mist concentrates. To address adverse effects of climatic change, worldwide nursery gas outflows must be reduced significantly. From time to time single division or innovation can address the whole relief challenge. All segments including structures, industry, vitality creation, farming, transport, ranger service, and waste administration could add to the general moderation endeavors, for example through more prominent vitality effectiveness.

Numerous technologies and procedures, which emanate less greenhouse gasses, are now or will be economically accessible in the coming decades. Today, nanotechnology is having immediate effect, its application into bigger frameworks, for example, the hydrogen-based economy, sun-based power innovation or cutting edge batteries, possibly could profoundly affect vitality utilization and nursery gas outflows. Conceivable zones distinguished to mediate through nanotechnology to diminish greenhouse gas outflows are:

- The improvement of hydrogen powered vehicles.
- Upgraded and less expensive photovoltaics or solar power innovation.
- New era of batteries and super capacitors.
- Improved protection of structures.
- Fuel-added substances to upgrade the energy effectiveness of engine vehicles.

These innovations are being created contemporarily to moderate the dependence on fossil fuels and subsequently start a way toward decoupling carbon dioxide emissions from energy. Furthermore, these advances are likely to have positive effects in diminishing the convergences of NO_x and SO_x in the air by decreasing the amount of fossil fills utilized as a part of the era of power. For power era, hydrogen energy component is an effective, non-contaminating source.

Hydrogen Solar Ltd. of United Kingdom has developed Tandem Cell™ innovation for the era of hydrogen fuel utilizing sun-powered vitality with zero carbon emanations. Other than hydrogen fuel cell, yet another innovation which changes over solar energy, renewable, boundless wellspring of outflow free, to power is photovoltaic advances. Nanotechnology is generally utilized as a part of current research and development in photovoltaics. Part of the primary research include: NP silicon frameworks; utilization of non-silicon materials, for example, calcopyrites to grow thin film innovation; atomic natural sun-based cells; natural polymer photovoltaic frameworks and III–V nitride sunlight-based cells. A few distinct sorts of photovoltaic boards accessible in the market are exceedingly costly and have restricted time of lifetime. Endeavors are being made to bypass this issue through nanomechanical approach. One such approach is the purported crystalline silicon on glass and utilization of substitute materials, for example, cadmium telluride. Next imperative zone, which could reduce the environmental change, is vitality stockpiling.

The cutting edge batteries, more significant to environmental change, will be more reasonable for use in electric autos and different vehicles, is being endeavored utilizing nanotechnology. The automobile industries (Nissan, Mitsubishi, and Sanyo) are included in the advancement of cutting edge batteries like lithium particle and nickel metal hydride batteries having more limits of those officially utilized as a part of hybrid electric vehicles. Nissan has built up another overlaid lithium-particle battery for electric vehicles. As indicated by Nissan, it is the same size as an ordinary auto battery however has multiplied the limit (140 Wh/kg) and 1.5 times more power even after 100,000 km utilization. The outcome is twofold the driving separation, accomplished with no expansion in battery load.[81]

2.2.6 DEVELOPMENT TOOLS FOR DISEASE CONTROL

2.2.6.1 NANOFERTILIZERS AND THEIR ROLE

Composts have axial part in improving the sustenance generation. It is realized that crop yields have started to melancholy as an aftereffect of imbalanced preparation and decline in soil natural fertility. Additionally, luxurious uses of nitrogen and phosphorus fertilizers influence the groundwater prompting eutrophication in oceanic biological communities. The manure utilization proficiency is around 20–50% for nitrogen and 10–25 % for phosphorus composts and it suggests that sustenance generation must be substantially more effective.[84,109]

According to Royal Society,[21] nanotechnology is dynamically moving far from the exploratory into the reasonable range. For example, nanotechnology has given the practicality of misusing nanoscale or nanostructured materials as compost bearers or controlled-discharge vectors for working of supposed "brilliant manure" as new offices to improve supplement utilization proficiency and less expenses of natural assurance.[11] Encapsulation of composts inside a NP is one of these new modalities, which are done in three ways, for example, the supplement can be epitomized inside nanoporous materials, covered with thin polymer film, or conveyed as molecule or emulsions of nanoscale measurements.[94]

Moreover, nanofertilizers will consolidate nanodevices keeping in mind the end goal to synchronize the manure N and—P with their take-up by-products, thus, preventing undesirable supplement misfortunes to soil, water, and air.[88] Savvy composts may become reality through changed plan of ordinary items utilizing nanotechnology.[26] The nano-organized plan may allow compost cleverly control the discharge speed of supplements to coordinate the take-up example of product. Solvency and scattering of mineral micronutrients may cause controlled discharge detailing.[75,92]

Nanosized definition of mineral micronutrients may enhance dissolvability and scattering of insoluble supplements in soil, decrease soil assimilation and obsession and increase the bioavailability prompting expanded nutrient take-up. Nanostructured modality may expand compost productivity and take-up proportion of dirt supplements, and save fertilizer source. Controlled discharge modes have properties of both discharge rate and discharge. Supplements for water-solvent composts may be absolutely controlled through epitome in envelope types of semipermeable films covered by gum polymer, waxes, and sulfur. Effective span of supplement discharge has attractive property of nano-organized detailing; it can develop viable length of supplement supply of manures into soil. Nano-organized detailing can lessen misfortune rate of compost supplements into soil by filtering and/or spilling.[66]

2.2.6.2 DETECTION OF DISEASES

Infections are one of the real constraining variables for harvest efficiency.[119] Pesticides are generally utilized as a part of agribusiness to enhance yield and proficiency. Nanotechnology can be connected for sheltered and productive uses of pesticide. It keeps the unfavorable impacts of pesticides on target and nontarget living being. In addition, the advancement of nanometric-scale materials, whose properties contrast significantly from the comparing mass materials, are utilized for the controlled arrival of operators for vermin control, and additionally, plant supplements. Through exemplification of bug sprays, fungicides, or nematicides with NPs may give more successful arrangement towards nuisances and it may cause less aggregation in soil.

With the assistance of nanotechnology, small amount of pesticide is utilized that can give much better efficiency. In the present phase of innovative work,[32] nanosized agrochemicals are for the most part nano-reformulations of existing pesticides and fungicides,[33,41] which would expand the evident dissolvability of inadequately solvent dynamic fixings, to discharge the dynamic fixing in a moderate way to ensure against premature degradation.[9]

The disease management is an important factor that contributes to better crop productivity; and the issue with disease management involves detection of correct phase of prevention. Pesticides are used as a preventive measure that results in residual toxicity and environmental hazards. Although application of pesticides on the diseased plant may lead to certain degree in reduction of crop yield, yet various kinds of disease infections in crop (e.g., fungal, bacterial, and viral) can be adequately kept under control. Among all kinds of diseases, viral infection is difficult to cure because it is spread by vectors. To prevent the viral infection, different types of nano-based viral diagnostic kits have been developed, which includes multiplexed diagnostic kit in order to detect the exact strain of virus. These nano-based diagnostic kits not only enhance the speed of detection but also increase the power of detection.[70]

Nanotechnology has the capacity to alter hereditary constitution, thus, helping in change of crop plants.[111] The improvement of nanometric-scale materials are used to control diseases, release agents of pests, and preserve the plant nutritious product. Lower quantity of pesticide in nanotechnology is needed to cure the disease. At present, the most encouraging innovation for protection the of host plants against insect and pests is nanoencapsulation. Nanoencapsulation has been proved to reduce chemical discharge into the soil and to improve crop production efficiency.[49] Nanosized agrochemicals are nano-reformulations of accessible pesticides and fungicides that would moderately discharge the dynamic fixing to protect against inopportune corruption.[44,96]

2.2.7   MONITORING CROP GROWTH

Today, use of farming composts, pesticides, anti-infection agents, and supplements is ordinarily by spray or splash application to soil or plants, or through infusion frameworks to creatures.[7,115] Nanotechnology enhances their execution and acceptability by expanding viability, protection, persistent obedience, and in addition eventually lessening social insurance costs.[6] Nanoscale devices have the capacity to distinguish and treat a contamination, supplement deficiency, or other well-being issue, much sooner than side effects are observable at full scale.[106]

2.2.7.1   NANOCAPSULES

With potential applications over the natural way of life (in pesticides, immunizations, veterinary solution, and nutritiously improved sustenance), the nano and smaller scale formulations are being created and licensed by agribusiness and nourishment companies, like Monsanto, Syngenta, and Kraft. Pesticides containing nanoscale products are available now in the market, and a number of the world's driving agrochemical companies are experimenting on the advancement of new nanoscale pesticides like BASF of Germany, which ranks fourth among agrochemical corporations in the world and is also conducting basic research and has applied for a patent on a nano-pesticide formulation. Therefore, there is a huge potential of usefulness in the formulation of nano-pesticides that involve an active ingredient with particle size in the range of 10–150 nm.

In case of nano-formulation, the toxic pesticide dissolves more easily in water and stays more stable, and has tendency to kill different types of diseases that is related to herbicide, insecticide, or fungicide. Bayer Crop Science of Germany has patented an agro-pesticide in the form of a nanoemulsion that consists of active ingredient in the range of 10–400 nm. The organization alludes to the development of a microemulsion concentrate with several advantages such as: more quick and solid action and augmented long-term activity. Primo MAXX Plant Growth Regulator and Banner MAXX fungicide of Bayer Crop Science, Syngenta have nano-formulations with particle size of 100 nm that has tendency to prevent clogging in spray tank filters; these are completely soluble in water; and does not settle in the spray tank. Their effectiveness may last more than 1 year. Furthermore, the fungicides are absorbed into the plant's system and cannot be washed off by rain or irrigation.

2.2.7.2   NANOSENSORS

The literature on multidiscipline research papers on biosensors in agriculture has been lacking.[52] Utilization of nanoscale materials for electrochemical biosensors has been increasing exponentially because of high affectability and fast response time. In these applications, powerful immobilization of biomolecules without adjusting bioactivity is the key in development of stable and organized electrode material for biosensor.[68] The created biosensor framework is a perfect instrument for Internet observing of organophosphate pesticides and nerve specialists. Bioanalytical nanosensors are being used to identify and measure micro quantities of contaminants in horticulture and nourishment frameworks such as infectious microorganisms, poisonous bio-dangerous substances, and so forth. Most research on these poisons is still directed to utilize traditional techniques; notwithstanding, biosensor strategies are presently being created as screening instruments for use in field examination.[98]

Nanosensor frameworks have been produced for the checking of natural conditions, and also the communications amongst pathogens and plants.[52] In farming fields, nano-biosensor helps in breaking down physiochemical and organic changes in soil and transmits these adjustments as signals, thus, helps locating organisms and pests. It furnishes information on the level of supplements and water status in horticultural fields. Consequently, the utilization of nano-biosensors in agriculture can improve profitability by providing precise data on the field and harvest conditions. The fundamental method to apply nanotechnology in horticulture framework is to build efficiency along with the utilization of lesser quantities of composts, pesticides, herbicides with a specific end goal to keep away from negative results on yields and people.

Nanofertilizers play a vital part of increasing crop production.[40] It enhances the various parameters like root–shoot length, germination rate, biomass of seedlings and many other physiological parameters, like upgraded photosynthetic movement and nitrogen digestion system also additionally expanded in vegetable products with the assistance of metal-based nanomaterials in various plants, such as, corn, wheat, horse feed, soybean, tomato, radish, lettuce, spinach, onion, pumpkin, cucumber, and so forth.[89,103,108,116]

2.2.8   PEST CONTROL

Pests, including insects, mites, nematodes, and pathogens are real restricting elements in crop production. The utilization of pesticides has brought improvement of vermin and illness resistance, aggregating deposits in ecological contamination. Utilization of nanotechnology in harvest insurance holds a huge guarantee in administration of creepy-crawlies and pathogens, by control and focus on conveyance of agrochemicals furthermore, by giving analytic apparatuses to early recognition. Nanoparticles are of <100 nm in size with more charge and bigger surface area thus, higher strength and solubility.[23]

Vermicomposting is the strategy by which worms are used to change natural materials normally squanders into a humus-like material known as vermicompost. The goal is to handle the material as quick and adequately as sensibly normal.[55,76] Vermicomposting is a fundamental biotechnological technique of treating the dirt, in which certain sorts of worms are used to update the strategy of waste change and improve the finished results. Vermicomposting[2,104,36] is quicker than treating the dirt fertilizing the soil so that the material experiences the night crawler gut, by which propped with irritation repellence properties too. The subsequent manure is rich in microbial action for plant development. And! night crawlers, through a kind of natural speculative chemistry, are prepared to transform garbage (waste) into gold (harmless materials).[50]

2.2.8.1 IMPORTANCE OF VERMICOMPOST

- Source of plant nutrients.
- For increasing plant growth promoting activity.
- Development in crop growth and yield.
- Decrease in soil C: N proportion.
- Improved soil physical, chemical, and biological properties.
- Vermicompost is generally advanced than fertilizers in a number of significant manners.
- Vermicompost has better quality than most of compost as inoculants in the production of compost teas.
- Uses of worms on farms, including value as a high-quality animal feed.
- By using vermicomposting and vermiculture, organic farmers have sources of supplemental income.
- Vermicompost is free of pathogens and toxic chemicals.

The primary burdens of pesticides are advancement of pathogen and irritation resistance, diminishes nitrogen obsession, decreases soil biodiversity, adds to bioaccumulation of pesticides, pollinator decay and devastates environment for winged creatures.[116] Along these lines, utilization of NPs resolves these issues to a greater degree: its application with herbicides diminishes the measure of herbicides prerequisite for weed annihilation. With the dynamic fixing and conveyance frameworks, herbicides are discharged in the dirt as indicated by the soil condition.[39]

Ag NPs have pesticides action against pathogenic organisms and affect conidial germination of class Raffaele which causes mortality of oak trees.[46,74] Conventional strategies to control the pathogens affect both nature and economy of farmers. The studies directed under non-sterile conditions make it clear that the expansion in product development/yield is the consequence of diminished sickness. This is conceivable from the counter pathogenic movement of the NP itself. Nano-manures are NMs, which can give supplements to plants or they can help to enlarge the exercises of ordinary composts. Swap of nano-manures for conventional compost is useful as it discharges supplements into the dirt relentlessly and controlled thus, avoiding water contamination.[14,69]

Hydroxyapatite ($Ca_5 (PO_4)_3OH$) NPs of 16 nm in size showed treating impacts on soybean. The number of reports confirms that utilization of nanofertilizers reflects beneficial results to increase crop yield and to reduce natural contamination. The use of NPs expanded the development rate and seed germination by 33 and 20 %, respectively compared with general compost. The results showed that the foundations of soybean can retain hydroxyapatite NPs as a powerful supplement source. Soil revised with metallic Cu NPs fundamentally expanded 15 day lettuce seedling development by 40%.[62] Research studies on the qualities of NPs reported that NPs can enter plant cells and transport DNA and chemicals into the cells.[42,97,114] These studies prove that NPs can likewise convey supplements to the plants as manures. The nano-natural iron chelated composts exhibited high ingestion, increment in photosynthesis, and development in leaf surface region.[99] Moreover, NMs have awesome effect on the dirt, as nanofertilizers can diminish the lethality of the dirt and lessening the recurrence of compost application.[63]

In NMs, supplements can be exemplified by NMs, covered with a thin defensive film or conveyed as emulsions or NPs.[100] Nano and sub-nano composites can control the arrival of supplements from the compost case.[91] Urea adjusted hydroxyapatite NP-typified *Gliricidia sepium* nanocomposite displayed a moderate and managed arrival of nitrogen after some time at 3 diverse pH values.[26] Manikandan et al.[65] reported that nanoporous zeolites on N compost may be utilized as substitute procedure to upgrade the adequacy of N utilized as a part of harvest creation framework. Nanofertilizers because of their trademark highlights have incredible part in supportable horticulture.[119] The utilization of nanofertilizers prompts an enlarged viability of the miniaturized scale and full scale components, lessens the lethality of the dirt, and decreases the rate of use of traditional manures.[65]

2.3   SUMMARY

Nanotechnology is an emerging science and technology according to European Commission. The NPs have size less than 100 nm diameters with different size-dependent properties. Nanoparticles have wide applications in medicine, nanocomposites, biotechnology, electronics, material science, energy sector, environmental challenge, nano-filter, and in improvement agricultural production. The synthesis of NPs is done by physical, chemical, biological, and hybrid methods.

Nanotechnology offers applications in pollutant remediation, agricultural production, food securing, animal feed, climate change, and food processing materials. Nanotechnology offers new applications for agricultural research and development tools for diseases molecular treatment, rapid disease detection, enhancing absorption of nutrients and so forth. In agriculture, nanotechnology is used as nanocapsules, nanosensors for monitoring crop growth and pest control of animal or plant diseases identification. This chapter augments the knowledge to convey the agriculture applications, potential of nanomaterials in agriculture management, pest control, fertilizers, clay and diseases treatment and plant protection, production.

KEYWORDS

- Ag NPs
- bioavailability
- bionanotechnology
- nanofertilizers
- nanomaterials
- nanoparticles
- nanosensors
- nanotechnology
- quantum dots

REFRENCES

1. Abbas, K. A.; Saleh, A. M.; Mohamed, A.; Mohamad-Azhan, N. The Recent Advances in the Nanotechnology and its Applications in Food Processing: A Review. *J. Food Agric. Environ.* **2009**, *7*(3–4), 14—17.
2. Adhikary, S. Vermicompost, the Story of Organic Gold: A Review. *Ag. Sci.* **2012**, *3*(7), 905–917.
3. Agrawal S.; Rathore P. Nanotechnology Pros and Cons to Agriculture: a Review. *Int. J. Curr. Microbiol. App. Sci.* **2014**, *3*(3), 43–55.
4. Ahmadi, H.; Rezaei, R.; Kheiri, S. Factor Analysis of Barriers and Problems Affecting the Development of Nanotechnology in Agriculture. *Ann. Biol. Res.* **2013**, *4*(1), 131–134.
5. Ahmed, N.; Helal, S. Nanotechnology in Agriculture: A Review. *Agric. For.* **2013**, *59*(1), 117–142.

6. Alfadul, S. M.; Elneshwy, A. A. Use of Nanotechnology in Food Processing, Packaging and Safety—Review. *Afr. J. food Agric. Nutr. Dev.* **2010**, *10*(6), 2719–2739.

7. Anwunobi, A. P.; Emeje, M. O. Recent Application of Natural Polymers in Nanodrug Delivery. *J. Nanomed. Nanotechnol.* **2011**, *4*, 1–6.

8. Arif, N.; Yadav, V.; Singh, S.; Mishra, R. K.; Sharma, S.; Dubey, N. K.; Tripathi, D. K.; Chauhan, D. K. Current Trends of Engineered Nanoparticles (ENPs) in Sustainable Agriculture: An Overview. *Environ. Anal. Toxicol.* **2016**, *6*(5), 1–5.

9. Arivalagan, K.; Ravichandran, S.; Rangasamy, K. Nanomaterials and its Potential Applications. *Int. J. ChemTech Res.* **2011**, *3*, 534–538.

10. Arya, H.; Kaul, Z.; Wadhwa, R.; Taira, K.; Hirano, T.; Kaul, S. C. Quantum dots in bio-imaging: revolution by the small. *Biochem. Biophys. Res. Commun.* **2005**, *329*(4), 1173–1177.

11. Azeredo, H. M. C. Nanocomposites for Food Packaging Applications. *Food Res. Int.* **2009**, *42*, 1240–1253.

12. Baruah, S.; Dutta, J. Nanotechnology applications in Pollution Sensing and Degradation in Agriculture: a Review. *Environ. Chem Lett. J.* **2009**, *7*, 191–204.

13. Berekaa, M. M. Nanotechnology in Food Industry; Advances in Food Processing, Packaging and Food Safety. *Int. J. Curr. Microbiol. Appl. Sci.* **2015**, *4*(5), 345–357.

14. Beyrouthya, M. E.; Desiree, E. A. Nanotechnologies: Novel Solutions for Sustainable Agriculture. *Adv. Crop Sci. Technol.* **2014**, *2*(3), 1–2.

15. Bhattacharyya, A.; Bhaumik, A.; Rani, P. U.; Mandal, S.; Epidi, T. T. Nanoparticles—a Recent Approach to Insect Pest Control. *Afr. J. Biotechnol.* **2010**, *9*(24), 3489–3493.

16. Biswa, S. K.; Nayak, A. K.; Parida, U. K.; Nayak, P. L. Applications of Nanotechnology in Agriculture and Food Sciences. *Int. J. Sci. Innovations Discoveries* **2012**, *2*(1), 21–36.

17. Bouwmeester, H. Review of Health Safety Aspects of Nanotechnologies in Food Production. *Regul. Toxicol. Pharmacol.* **2009**, *53*(1), 52–62.

18. Chakraborty, S.; Chowdhury, A. Nanotechnology—the Mantra of Present and Future in Agriculture and Food. *Agriculture,* **2015**, *2*, 401–412.

19. Chaudhry, Q.; Scotter, M.; Blackburn, J.; Ross, B.; Boxball, A.; Castle, L.; Aitken, R.; Watkins, R. Applications and Implications of Nanotechnologies for the Food Sector. *Food Addit. Contam.* **2008**, *25*(3), 241–258.

20. Chellaram, C.; Murugaboopathi, G.; John, A. A.; Kumar, R. S.; Ganesan, S.; Krithika, S.; Priya, G. Significance of Nanotechnology in Food Industry. *APCBEE Procedia,* **2014**, *8*, 109–113.

21. Chen, H.; Yada, R. Nanotechnologies in Agriculture: New Tools for Sustainable Development. *Trends Food Sci. Technol.* **2011**, *22*, 585–594.

22. Chinnamuthu, C. R.; Boopathi, P. M. Nanotechnology and Agro Ecosystem. *Madras Agric. J.* **2009**, *96*(1–6), 17–31.

23. Chizari, M.; Khayyam-Nekouei, S. M.; Tabatabaei, M. Investigating the Researchers Attitude and the Obstacle Hampering Nanotechnology Development in the Agriculture. *J. Sci. Tec.* **2012**, *14*, 493–503.

24. Chowdappa, P.; Gowda, S. K. Nanotechnology in Crop Protection: Status and Scope. *Pest. Hortic. Ecosyst.* **2013**, *19*(2), 131–151.

25. Conley, D. J.; Paerl, H. W.; Howarth, R. W.; Boesch, D. F.; Seitzinger, S. P. Ecology-controlling Eutrophication: Nitrogen and Phosphorus. *Science,* **2009**, *3*(3), 1014–1015.

26. Correl, D. L. The Role of Phosphorus in the Eutrophication of Receiving Waters: a Review. *J. Environ. Qual.* **1998**, *27*, 261–266.

27. DeRosa, M. R.; Monreal, C.; Schnitzer, M.; Walsh, R.; Sultan, Y. Nanotechnology in Fertilizers. *Nat. Nanotechnol. J.* **2010**, *5*(2), 1–91.

28. Dhewa, T. Nanotechnology Applications in Agriculture: An Update. *Octa J. Environ. Res.* **2015**, *3*(2), 204–211.

29. Dingman, J. Nanotechnology: it's Impact on Food Safety. *J. Environ. Health,* **2008**, *70*(6), 47–50.

30. Ditta, A. How Helpful is Nanotechnology in Agriculture? *Nanosci. Nanotechnol.* **2012**, *3*(3), 298–302.

31. Duncan, T. V. Applications of Nanotechnology in Food Packaging and Food Safety: Barrier Materials, Antimicrobials and Sensors. *J. Colloid Interface Sci.* **2011**, *363*(1), 1–24.

32. Fakruddin, P.; Chakraborty, A. Nanotechnology in Agriculture. *Innovative Farming,* **2016**, *1*(1), 18–20.

33. Fan, C.; Wang, S.; Hong, J. W.; Bazan, G. C.; Plaxco, K. W.; Heeger, A. J. Beyond Super Quenching: Hyper-Efficient Energy Transfer from Conjugated Polymers to Gold Nanoparticles. *Proc. Nati. Acad. Sci.* **2003**, *100*(11), 6297–6301.

34. Feng, Y.; Cui, X.; Shiying, H.; Dong, G.; Chen, M.; Wang, J.; Lin, X. The Role of Metal Nanoparticles in Influencing Arbuscular Mycorrhizal Fungi Effects on Plant Growth. *Environ. Scie. Technol.* **2013**, *47*, 9496–9504.

35. Fessi, H. P. F. D.; Puisieux, F.; Devissaguet, J. P.; Ammoury, N.; Benita, S. Nanocapsule Formation by Interfacial Polymer Deposition Following Solvent Displacement. *Int. J. Pharm.* **1989**, *55*(1), R1–R4.

36. Fraceto, F. L.; Grillo, R.; Medeiros, G. A.; Scognamiglio, V.; Rea, G.; Bartolucci, C. Nanotechnology in Agriculture: Which Innovation Potential Does it Have? *Front. Environ. Sci.* **2016**, *4*(20), 1–5.

37. Gandhi, M.; Sangwan, V.; Kapoor, K.K.; Dilbaghi, N. Composting of Household Wastes with and Without Earthworms. *Environ. Ecol.* **1997**, *15*(1), 432–434.

38. Garcia, M.; Forbe T.; Gonzalez, E. Potential Applications of Nanotechnology in the Agro-Food Sector. *Aliment. Campinas,* **2010**, *30*(3), 573–581.

39. Garnett, T.; Charles, H.; Godfray, J. Sustainable Intensification in Agriculture. *Ecol. Soc. Am.* **2012**, *1*, 1–51.

40. Ghormade, V.; Deshpande, M.V. Paknikar, K.M. Perspectives for Nano-Biotechnology Enabled Protection and Nutrition of Plants. *Biotechnol Adv.* **2011**, *29*(1), 792–803.

41. Giraldo JP, et al. Plant Nanobionics Approach to Augment Photosynthesis and Biochemical Sensing. *Nat. Mater.* **2014**, *13*(1), 400–408.

42. Goluch, E. D.; Nam, J. M.; Georganopoulou, D. G.; Chiesl, T. N.; Shaikh, K. A.; Ryu, K. S.; Barron, A. E.; Mirkin, C. A.; Liu, C. A Biobarcode Assay for on-chip Attomolar-Sensitivity Protein Detection. *Lab on a Chip,* **2006**, *6*(10), 1293–1299.

43. Goswami, A.; Roy, I.; Sengupta, S.; Debnath, N. Novel Applications of Solid and Liquid Formulations of Nanoparticles Against Insect Pests and Pathogens. *Thin Solid Films,* **2010**, *519*(3), 1252–1257.

44. Govorov, A.; Carmeli, I. Hybrid Structures Composed of Photosynthetic System and Metal Nanoparticles: Plasmon Enhancement effect. *Nano Lett.* **2007**, *7*(3), 620–625.

45. Green, J. M.; Beestman, G.B. Recently Patented and Commercialized Formulation and Adjuvant Technology. *Crop Prot.* **2007**, *26*, 320–327.
46. Gregorio, C.; Bareras-Urbina, Ramirez-Wong, B.; Lopez-Ahumada, G.A.; Burruel-Ibarra, S. A.; Martinez-Cruz, O.; Tapia-Hernandez, J. A.; Rodriguez Felix, F. Nano- and Micro-Particles by Nanoprecipitation: Possible Application in the Food and Agricultural Industries. *Int. J. Food Prop.* **2016**, *19*, 1912–1923.
47. Gruere, G.; Clare, N.; Linda, A. Agricultural, Food and Water Nanotechnologies for the Poor Opportunities, Constraints and Role of the Consultative Groups on International Agricultural Research. *J. Int. Food Policy Res. Inst.* **2011**, *1*(1), 1–35.
48. Gutierrez, F. J.; Mussons, M. L.; Gaton, P.; Rojo, R. Nanotechnology and Food Industry. *Sci. Health Soci. Aspects Food Ind.* **2012**, *1*, 95–121.
49. Hobbs, P. R.; Sayre, K.; Gupta, R. The Role of Conservation Agriculture in Sustainable Agriculture. *Philos. Trans. R. Soc. London Ser. B Biolo Sci.* **2008**, *363*, 543–555.
50. Jampilek, J.; Karlova, K. Application of Nanotechnology in Agriculture and Food Industry, its Prospects and Risks. *J. Soc. Ecol. Chem. Eng.* **2015**, *2*(3), 321–361.
51. Jayanta, M. Vermicompost, a Best Superlative for Organic Farming: a Review. *J. Adv. Stu. Ag. Bio. Env. Sci.* **2015**, *2*(3), 38–46.
52. Jha, D.; Behar, N.; Sharma, S. N.; Chandel, G.; Sharma, D. K.; Pandey, M. P. Nanotechnology: Prospects of Agricultural Advancement. *Nano Vision* **2011**, *1*(2), 54–110.
53. John, J. H.; Yaung, G.; Mielke, R. E.; Horst, A. M.; Moritz, S. C.; Espinosa, K.; Gelb, J.; Walker, S. L.; Nisbet, R. M.; Youn-Joo, A.; Schimel, J. P.; Palme, R. J.; Hernandez-Viezcas, J. A.; Zhao, L.; Gardea-Torresdey, J. L.; Holden, P. A. Soybean Susceptibility to Manufactured Nanomaterials with Evidence for Food Quality and Soil Fertility Interruption. *Curr. Issue.* **2012**, *109*, 2451–2456.
54. Joye, I. J.; Davidov-Pardo, G.; McClements, D. J. Nanotechnology for Increased Micronutrient Bioavailability. *Trends in Food Sci. Technol.* **2014**, *40*(2), 168–182.
55. Judy, J. D.; Unrine, J. M.; Bertsch, P. M. Evidence for Biomagnification of Gold Nanoparticles Within a Terrestrial Food Chain. *Environ. Sci. Technol.* **2011**, *45*, 776–781.
56. Kah, M.; Beulke, S.; Tiede, K.; Hofmann, T. Nanopesticides: State of Knowledge, Environmental Fate, and Exposure Modeling. *Crit. Rev. Environ. Sci. Technol.* **2013**, *39*(1), 1823–1867.
57. Khan, M. R.; Rizvi, T. F. Nanotechnology: Scope and Application in Plant Disease Management. *Plant Pathol. J.* **2014**, *13*(3), 214–231.
58. Khodakovskaya, M.; Dervishi, E.; Mahmood, M. Carbon Nanotubes are Able to Penetrate Plant Seed Coat and Dramatically Affect Seed Germination and Plant Growth, *ACS Nano.* **2009**, *3*, 3221–3227.
59. Khot, L. R.; Sankaran, S.; Maja, J. M.; Ehsani, R.; Schuster, E. W. Applications of Nanomaterials in Agricultural Production and Crop Protection: A Review. *Crop Prot.* **2012**, *35*, 64–70.
60. Kottegoda, N.; Munaweera, I.; Madusanka, N.; Karunaratn, V. Green Slow Release Fertilizer Composition Based on Urea Modified Hydroxyapatite Nanoparticles Encapsulated Wood. *Curr. Sci.* **2011**, *101*(1), 73–78.
61. Lahir Y. K. Role and Adverse Effects of Nanomaterials in Food Technology. *J. Toxicol. Health,* **2015**, *2*(2), 2056–2699.
62. Lal, R. Promise and limitations of soils to minimize climate change, *J. Soil Water Conserv.* **2008**, *63*, 113A–118A.

63. Lamsal, K.; Kim, S. W.; Jung, J. H.; Kim, Y.S.; Kim, K. S.; Lee, Y. S. Inhibition effects of silver nanoparticles against powdery mildews on cucumber and pumpkin. *Mycobiology*, **2011**, *39*(1), 26–32.

64. Liu, R.; Lal, R. Synthetic apatite nanoparticles as a phosphorus fertilizer for soybean (*Glycine max*). *Sci. Rep,* **2014**, *4*(1), 5686–5691.

65. Liu, X.; Feng, Z.; Zhang, S.; Zhang, J.; Xiao, Q.; Wang, Y. Preparation and testing of cementing nano-subnano composites of slow or controlled release of fertilizers. *Sci. Agr. Sin*, **2006**, *39*, 1598–1604.

66. Manikandan, A.; Subramanian, K. S. Fabrication and Characterization of Nanoporous Zeolites Based N Fertilizer. *Afr. J. Agric. Res.* **2014**, *9*(2), 276–284.

67. Manjunatha, S. B.; Biradar, D. P.; Aladakatti, Y. R. Nanotechnology and its applications in agriculture: A review. *J. Farm Sci.* **2016**, *29*(1), 1–13.

68. Masciangioli, T.; Zhang, W. X. Nanotechnology and the Environment: Applications and implications. *Occas. Pap. Ser.* **2003**, *7*(1), 1–14.

69. Melendi, P.; Fernandez-Pacheco, R.; Coronado, M. J.; Corredor, E.; Testillano, P. S.; Risue, M. C.; Marquina, C.; Ibarra, M. R.; Rubiales, D.; Perez-de-Luque, A. Nanoparticles as Smart Treatment Delivery Systems in Plants: Assessment of Different Techniques of Microscopy for Their Visualization in Plant Tissues. *Ann. Bot.* **2007**, *101*(1), 187–195.

70. Moaveni, P.; Kheiri, T. Eds. *Second International Conference on Agricultural and Animal Science*; IACSIT Press: Singapore, **2011**; Vol. *22*, pp 160–163.

71. Mohammad, A.; Khiyamia, H. K.; Yasser, M.A.; Mousa, A. A.; Kamel, A.; Abd-Elsalamd, E. Plant Pathogen Nanodiagnostic Techniques: Forthcoming Changes? *Biotechnol. Biotechnol. Equip.* **2014**, *28*(5), 775–785.

72. Morones, J. R.; Elechiguerra, J. L.; Camacho, A.; Holt, K.; Kouri, J. B.; Ramirez, J. T.; Yacaman, M. Bactericidal Effect of Silver Nanoparticles. *Nanobiotechnology* **2005**, *16*, 2346–2353.

73. Mousavi, S. R.; Rezaei, M. Nanotechnology in Agriculture and Food Production . *J. App. Environ. Biol. Sci.* **2011**, *1*(10), 414–419.

74. Mozafari, M.; Flanagan, J.; Matia, L.; Merino, M.; Awati, A.; Omr,i A.; Suntres, Z.; Singh, H. Recent Trends in the Lipid-Based Nanoencapsulation of Antioxidants and Their Role in Foods. *J. sci. Food Agric.* **2006**, *86*, 2038–2045.

75. Mukhopadyay, S. Nanotechnology in Agriculture: Prospects and Constraints. *Nanotechnol. Sci. Appl.* **2014**, *7*, 63–71.

76. Naderi, M. R.; Shahraki, D. A. Nanofertilizers and Their Roles in Sustainable Agriculture. *Inte. J. Agric. Crop Sci.* **2013**, *5*(19), 2229–2232.

77. Nagavallemma, K. P.; Wani, S. P.; Padmaja, V. V.; Vineela, C.; Babu-Rao, M.; Sahrawat, K. L. *Vermicomposting: Recycling Wastes into Valuable Organic Fertilizer*; International Crops Research Institute For The Semi-Arid Tropics: Andhra Pradesh, India, **2004**, p 1–23.

78. Nair, R.; Varghese, S. H.; Nair, B. G.; Malkewa, Y.; Yoshida, D.; Kaur, S. Nanoparticulate Material Delivery to Plants. *Plant Sci.* **2010**, *179*, 154–163.

79. Neethirajan, S.; Jayas, D. S. Nanotechnology for the Food and Bioprocessing Industries. *Food Bioprocess Technol.* **2011**, *4*(1), 39–47.

80. Norman, S.; Hongda, C. Special Section on Nanobiotechnology, Part 2. *Ind. Biotechnol.* **2013**, *9*, 17–18.

81. Noubactep, C.; Care, S.; Crane, R. Nanoscale Metallic iron for Environmental Remediation: Prospects and Limitations. *Water, Air, Soil Pollut.* **2012**, *223*(3), 1363–1382.

82. Num, S. M.; Useh, N. M. Nanotechnology Applications in Veterinary Diagnostics and Therapeutics. *Sokoto J. Vet. Sci.* **2013**, *11*(2), 10–14.

83. Ocsoy, I.; Paret, M. L.; Ocsoy, M. A.; Kunwar, S.; Tao Chen, T.; Tan, W. Nanotechnology in Plant Disease Management: DNA-directed Silver Nanoparticles on Graphene Oxide as an Antibacterial Against Xanthomonas Perforans. *ACS Nano*, **2013**, *7*(10), 8972–8980.

84. Parisi, C.; Vigani, M.; Rodrigue-Cerezo, E. Agricultural Nanotechnologies: What are the Current Possibilities? *Nano Today*, **2015**, *10*(20), 124–127.

85. Parsad, R.; Kumar, V.; Parsad, K. S. Nanotechnology in Sustainable Agriculture: Present Concerns and Future Aspects. *Afr. J. Biotechnol.* **2014**, *13*(6), 705–713.

86. Patel, N.; Desai, P.; Patel, N.; Jha, A.; Gautam, H. Agro Nanotechnology for Plant Fungal Disease Management: A Review. *Int. J. Curr. Microbiol. Appl. Sci.* **2014**, *3*(10), 71–84.

87. Powell, J.; Harvey, R.; Sapwood, P.; Wolstencroft, R.; Gershwin, M.; Thompson, R. Immune Potentiation of Ultra-Fine Dietary Particles in Normal Subjects and Patients with Inflammatory Bowel Disease. *J. Autoimmun.* **2000**, *14*, 99–105.

88. Rai, M.; Ingle, A. Role of Nanotechnology in Agriculture with Special Reference to Management of Insect Pests. *App. Microbiol. Biotechnol.* **2012**, *94*, 287–293.

89. Rai, V.; Acharya, S.; Dey, N. Implications of Nanobiosensors in Agriculture. *J. Biomater. Nanobiotechnol.* **2012**, *3*, 315–324.

90. Raliya, R.; Tarafdar, J. C. ZnO Nanoparticle Biosynthesis and its Effect on Phosphorous-Mobilizing Enzyme Secretion and Gum Contents in Cluster Bean (*Cyamopsistetragonoloba* L). *Agri. Res.* **2013**, *2*(1), 48–57.

91. Raliya, R.; Tarafdar, J. C.; Gulecha, K.; Choudhary, K.; Ram, R.; Mal, P.; Saran, R. P. Scope of Nanoscience and Nanotechnology in Agriculture. *J. Appl. Biol. Biotechnol.* **2013**, *1*(3), 41–44.

92. Rameshaiah, G. N.; Pallavi, J.; Shabnam, S. Nano Fertilizers and Nano Sensors—an Attempt for Developing Smart Agriculture. *Int. J. Eng. Res. Gen. Sci.* **2015**, *3*(1), 314–320.

93. Ravichandran, R. Nanotechnology Applications in Food and Food Processing: Innovative Green Approaches, Opportunities and Uncertainties for Global Market. *Int. J. Green Nanotechnol. Phys. Chem.* **2010**, *1*, 72–96.

94. Razzaq, A.; Ammara, R.; Jhanazab, H.M.; Mahmood, T.; Hafeez, A.; Hussain, S. A Novel Nanomaterial to Enhance Growth and Yield of Wheat. *J. Nanosci. Technol.* **2016**, *2*(1), 55–58.

95. Rigi, K.; Sheikhpour, S.; Keshtehgar, A. Use of Biosensors in Agriculture. *Int. J. Far. Allied Sci.* **2013**, *2*(23), 1121–1123.

96. Rosen, J. E.; Yoffe, S.; Meerasa, A.; Verma, M. Nanotechnology and Diagnostic Imaging: New Advances in Contrastagent Technology. *J. Nanomed. Nanotechnol.* **2011**, *2*(5), 1–12.

97. Saxena, R.; Tomar, R. S.; Kumar, M. J. Exploring Nanobiotechnology to Mitigate Abiotic Stress in Crop Plants. *J. Pharm. Sci. Res.* **2016**, *8*(9), 974–980.

98. Scrinis, G.; Lyons, K. The Emerging Nano-Corporate Paradigm: Nanotechnology and the Transformation of Nature, Food and Agri-Food Systems. *Int. J. Socio. Agric. Food,* **2007,** *15*(2), 22–44.

99. Sekhon, B. S. Food Nanotechnology—an Overview. *Nanotechnol. Sci. Appl.* **2010,** *3*(1), 1–15.

100. Sekhon, B. S. Nanotechnology in Agri-Food Production: an Overview. *Nanotechnol. App. Sci.* **2014,** *7*(1), 31–53.

101. Shah, V.; Belozerova, I. Influence of Metal Nanoparticles on the Soil Microbial Community and Germination of Lettuce seeds. *Water, Air, Soil Pollut.* **2009,** *197*(1), 143–148.

102. Shiwen, H.; Ling, W.; Lianmeng, L.; Yuxuan, H.; Lu, L. Nanotechnology in Agriculture, Livestock, and Aquaculture in China. A Review. *Agron Sustain. Dev.* **2015,** *35,* 369–400.

103. Singh, A.; Singh, S.; Prasad, S. M. Scope of Nanotechnology in Crop Science: Profit or Loss. *Res. Rev. J. Bol. Sci.* **2016,** *5*(1), 1–4.

104. Singha, A.; Singha, N. B.; Hussaina, I.; Singha, H.; Singh, S. C. Plant-Nanoparticle Interaction: An Approach to Improve Agricultural Practices and Plant Productivity. *Int. J. Pharm. Sci. Invent.* **2015,** *4*(8), 25–40.

105. Sinha, R.; Herat, S.; Valani, D.; Chauhan, K. Environmental Economics of Crop Production by Vermiculture: Economically Viable & Environmentally Sustainable Over Chemical Agriculture. *Am-Eurasian J. Agric. Environ. Sci.* **2009,** *5*(S), 1–55.

106. Sonkaria, S.; Ahn, S. H.; Khare, V. Nanotechnology and its Impact on Food and Nutrition: a Review. *Recent Pat. Food, Nutr. Agric.* **2012,** *4*(1), 8–18.

107. Srilatha, B. Nanotechnology in Agriculture. *Nanomed. Nanotechnol.* **2011,** *2*(7), 1–5.

108. Srivastava, P.; Pandey, S.; Singh, P.; Singh, K. P. Nanotechnology and its Role in Pathogen Detection: A Short Review. *Int. J. Curr. Sci.* **2014,** *13,* 9–15.

109. Storrs, H. J. *Nano future: what's next for nanotechnology?* Prometheus Books: Amherst, NY, USA, **2006;** pp 551–554.

110. Tarafdar, J. C.; Sharma, S.; Raliya, R. Nanotechnology: Interdisciplinary Science of Applications. *Afr. J. Biotechnol.* **2013,** *12*(3), 219–226.

111. Thangavel, G.; Thiruvengadam, S. Nanotechnology in Food Industry—A Review. *Int. J. ChemTech Res.* **2014,** *6*(9), 4096–4101.

112. Thornton, P. K. Livestock Production: Recent Trends, Future Prospects. *Philos. Trans. R. Soc., B: Biol. Sci.* **2010,** *365,* 2853–2867.

113. Tillman, D.; Cassman, K. G.; Matson, P. A.; Naylor, R.; Polasky, S. Agricultural Sustainability and Intensive Production Practices. *Nature.* **2002,** *418,* 671–677.

114. Valsamma, K. M. Nanotechnology: The Flipside of Transition Technology. *Int. J. Appl. Phys. Math.* **2012,** *2*(3), 162–164.

115. Veronica, N.; Guru, T.; Thatikunta, R.; Reddy, N. Role of Nano Fertilizers in Agricultural Farming. *Int. J. Sci. Technol.* **2015,** *1*(1), 1–3.

116. Wang, Q.; Zhang, W.; Pei, H.; Chen, Y. The Impact of Cerium Oxide Nanoparticles on Tomato (*Solanum lycopersicum L.*) and its Implications for Food Safety. *Metallomics.* **2012,** *4,* 1105–1112.

117. Wang, P.; Lombi, E.; Zhao, F. J.; Kopittke, P. M. Nanotechnology: A New Opportunity in Plant Sciences. *Trends Plant Sci.* **2016**, *21*(8), 699–712.
118. Weiss, J.; Takhistov, P.; Julian, D. M. Functional Materials in Food Nanotechnology. *J. Food Sci.* **2006**, *71*(9), 107–116.
119. Zhang, J.; Chiodini, R.; Badr, A.; Zhang, G. F. The Impact of Next-Generation Sequencing on Genomics. *J. Genet.* Genomics, **2011**, *38*(3), 95–109.

CHAPTER 3

POTENTIAL OF NANOTECHNOLOGY IN DAIRY PROCESSING: A REVIEW

LOHITH KUMAR[1,*] AND PREETAM SARKAR[1,2]

[1]*Department of Food Process Engineering, National Institute of Technology at Rourkela, Rourkela, Odisha 769008, India*

[2]*Department of Food Process Engineering, National Institute of Technology at Rourkela, Rourkela, Odisha 769008, India*
E-mail: sarkarpreetam@nitrkl.ac.in; preetamdt@gmail.com

*Corresponding author. E-mail: lohithhanum8@gmail.com

CONTENTS

3.1   INTRODUCTION

Dairy technology is the study of milk biosynthesis and its transformation to dairy products. It encompasses various fields of study from biochemistry, molecular biology, engineering, microbiology, and organic chemistry. Milk remains as the most complicated food system and, therefore, extensive studies have been performed on a range of topics including synthesis, milk chemistry, and bioprocessing. In the last decade, there has been a growing interest in the study of foods as nanosystems. Milk has been considered as a rich source of nanomaterials because of the molecular structure of milk components. Therefore, the focus of this chapter is to study the nanoscience aspects of milk from biosynthesis to dairy food packaging.

The word "nano" means "dwarf" in Greek language. Nanotechnology is considered as the science of fabrication and characterization of nanomaterials that has at least one dimension approximately in the 1–100 nm range. Nanomaterials exhibit different chemical and physical properties compared to macro materials of the same substance. The word "nanotechnology" was coined by Norio Taniguchi.[81] Since then, nanotechnology developed into a multidisciplinary subject. Nanotechnology has application in many areas like cosmetics, electronics and devices, pharmaceuticals, agriculture, and foods. However, in recent years it has been making remarkable contribution in food and nutritional sciences, which holds great promise to provide benefits in enhancing food safety and quality with improved functional properties. Nanotechnology principles have been utilized in food matrix for encapsulation and protection of bioactive compounds, detection of foodborne pathogens, monitoring food quality, food packaging material development, fortification, and nano-additives fabrication. But the success of these principles will be dependent on the market demand and consumer acceptance.

Nanotechnology offers promising novel applications in almost all the disciplines. The doctrines of nanotechnology cleared the path to an unexplored science for studying individual nanoparticles and their application in the food industry. Nano foods provide a host of advantages such as improved shelf life, enhanced bioavailability of health promoting bioactive compounds, and enhanced safety of food against spoilage factors. Current research in the area of food nanotechnology is mainly focused on the designing of natural nanostructures for protecting nutrients in functional foods to improve their bioactivity and bioaccessibility at different environmental stresses. Naturally, foods contain nanocomponents such as proteins, carbohydrates, and lipids which determine their properties. Food processing conditions such as pH, temperature, pressure, and ionic strength changes affect the naturally occurring nanostructure in food and result in structural changes at nano and micro level. For instance, milk native protein β-lactoglobulin (β-Lg) has a characteristic length of 3.6 nm and it undergoes denaturation via heat, pH change, and pressure, resulting in the formation of reassembled larger structure.

This chapter explores potential of applications of nanotechnology in dairy engineering.

3.2 NANOTECHNOLOGY IN DAIRY PROCESSING: DAIRY FARM TO DAIRY FOODS PACKAGING

Dairy processing unit deals with both nano and micro level components. Nanotechnology can be applied in different phases of dairy processing (Fig. 3.1).

3.2.1 AGRICULTURE

Agriculture is having more demand for improving efficiency in the production of food. Application of nanotechnology in agriculture is comparatively underdeveloped. But, nanotechnology has the potential to provide solutions to these basic problems and opportunities to use nanofertilizers, nano-additives and pesticides that influence the crop health and yield. At nanoscale level, the fertilizers have increased surface area and lead to improved reactivity, faster dissolution, and uptake. Nano-additives include nutrients (Zn, pectin, rare earth oxides, Se, Fe, ammonium salts), pesticides, water retention materials, which are added to bulk products. Nanofertilizers are coated with polymers that help in controlled release of nutrients, for example, nanoclay and zeolites.[100] Nevertheless, it is crucial to evaluate the benefits and risks of nanotechnology in agriculture and nutrient management because agriculture is the main part in food chain and may influence the nanomaterial bioaccumulation in food- web and chain.

3.2.2 FEED

The concept and benefit behind the utilization of nanoparticles in feed as additives are due to low particle size, larger surface area, higher concentration per unit mass, and more particles at surface, which helps in faster

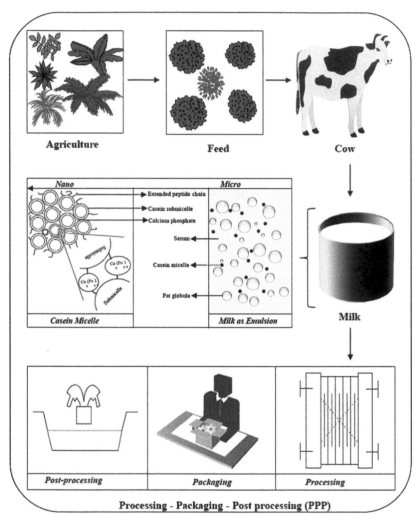

FIGURE 3.1 Schematic representation of nanotechnology interventions in dairy processing unit.

passive diffusion in body. Nanominerals namely, zinc, selenium, copper, silver have been used in feed formulations.[13,14,21,58,67] Chitosan, micelles, protein capsules, and liposomes were used to protect the potency of nanoadditive in feed. In recent studies, the use of nano zinc oxide improved the milk production and immunity of Holstein Friesian cross bred cows.[74]

3.2.3 COW

The udders of a cow can be considered as a natural nanodevice (i.e., device for producing nanosized food ingredients). A host of functions at the nano-level such as nanosized particle synthesis, dispensing proteins and fats into the milk in continuous phase, and assembling of particles are carried out in the cow's udder. Casein (300–400 nm) and fat globule membrane (100 nm–20 μm) and other milk proteins are nanosized materials ranging from nano to micrometer level which are produced in mammary epithelial cells.[84]

3.2.4 MILK

Milk is a naturally occurring nanostructure and it is also called a nano-colloid. Due to the presence of micro and nanocomponents in milk, it is considered as a complex food. The complexity of food, in general, helps in understanding the potential of using food components such as protein, carbohydrates, and fats as nanocarrier systems (Fig. 3.2). For example, casein micelles, milk fat globules, and other nanosized components of milk remain dispersed in the continuous phase and form a nanoemulsion. These milk nanostructures are natural carriers of essential micronutrients (phosphate and calcium), amino acids, and immune

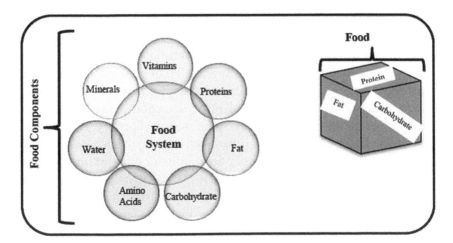

FIGURE 3.2 Importance of essential nutrients and food complexity.

modulators (lactoferrin and immunoglobulins). [85] In milk, casein is a proline rich protein with an open structure and adapt to any confirmations which is energetically favorable in the solution, hence it is known as a rheomorphic protein. Casein combines with calcium phosphate and forms nanoclusters of calcium phosphate which helps in protection against calcification in mammary gland caused by pathogens during lactation.[33] Casein demonstrates unique amphiphilic nature where α_{s1}, α_{s2}, and β-casein have serine phosphate for calcium sequestration. κ-casein is a glycoprotein having disulfide bridges between two cysteine molecules.[37,85] Moreover, 95% of the casein exists as self-assembled casein micelles in colloidal spherical shape[56]. Self-assembling properties of milk nanostructure influence the properties of milk protein-based nanotubes.[27,28] The potential application of a nanotube is encapsulation, template for nanowires, and development of scaffolds. In a recent study, α-lactalbumin was used to prepare a nanotube of diameter 20 nm and cavity 8 nm. The studies indicated that the prepared α-lactalbumin withstands freeze drying and pasteurization temperatures without structural modification.[28] Milk fat globules were used as nanocarriers of lipids and other lipophilic compounds, which is discussed in details under application section.

3.2.5   PROCESSING, PACKAGING, AND POST-PROCESSING (PPP)

Advancement in nanotechnology has supported in improving the efficiency of dairy food processing plants and nutritional status of dairy products. In dairy processing plants, the application of nanotechnology begins in platform tests for milk and ends in effluent treatment.[6] Even though the application of nanosensors and rapid detection assay for milk and milk product quality are in infancy stage, nanotechnology is growing in dairy processing tremendously. Utilization of nanoparticle surface energy and its area for reducing the organic load in dairy effluent are the promising steps towards healthy environment.[9,41] Nanoparticle properties are also being used in coating of heat transfer equipment surfaces of dairy to decrease adsorption of milk components, thereby increasing the production rate of plant.[11,40]

3.3   POTENTIAL APPLICATIONS OF NANOTECHNOLOGY IN DAIRY PROCESSING

3.3.1   ENCAPSULATION

Carbohydrates, proteins, fats, food grade surfactants and waxes are the majority of encapsulating materials used in food applications. In general, an encapsulation model is shown in Figure 3.3. Nano and micro ingredients properties of dairy materials make them extremely appropriate to use them in encapsulation of bioactive components in foods.[56,63,85] Milk fat, proteins (whey protein, casein, and lactoglobulin), and milk fat fractions, lactose and milk fat globule membrane (MFGM) constituents were used as encapsulating substances and in fabrication of encapsulation system.[48,60,63,103]

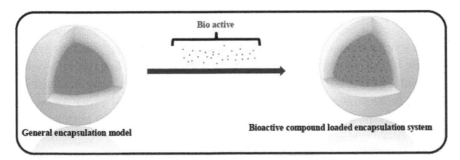

FIGURE 3.3   Schematic representation of general encapsulation model.

Designing a functional nano or micro encapsulation system for a bioactive compound requires knowledge about physicochemical properties, stability, and solubility of the encapsulating materials in food matrix. The encapsulation methods can be selected based on the processing methods and final structure of the food matrix. The controlled release of bioactive compounds from the encapsulated matrix can be manipulated and it can be triggered in response to pH, temperature, enzymatic action, or any other suitable environmental stress factors.

3.3.1.1   MILK COMPONENTS IN ENCAPSULATION

Milk ingredients have been used for the formulation of encapsulated systems for a multitude of bioactive components in the food matrix.

Properties like ability to gel, solubilize the bioactive compounds, stabilize an interfacial layer, and form different desired texture in the final food matrix make milk ingredients useful encapsulating agents. Different dairy ingredients and mode of encapsulation are tabulated in Table 3.1.

TABLE 3.1 Milk Ingredients and Mode of Encapsulation.

Milk ingredient	Mode of encapsulation	Reference
Casein	Particles, emulsions, hydrogel, coacervates	[56, 85]
Milk protein hydrolysates	Emulsion, nanotubes	[27, 28]
Milk fat	Emulsion	[46]
Milk fat globule membrane	Liposomes	[88, 95, 96]
Whey proteins (β-Lactoglobulin, lactoferrin)	Particles, emulsions, hydrogels, coacervates	[10, 29, 72]

β-Lg constitutes major portion of whey protein, and it is used as a nanocarrier system for hydrophobic molecules. Studies suggest that hydrophobic bioactive compounds (vitamin D, cholesterol) help in improving resistance to thermal degradation of proteins and this property makes β-Lg usable in food processing for the protection of thermal sensitive active compounds in the food matrix.[86] Whey proteins have the ability to gel under cold conditions and can be used for encapsulation of heat-labile molecules.[72] Lactoferrin is another protein component in whey proteins and has been used in the fabrication of delivery systems with chitosan and lecithin using electrostatic layer-by-layer deposition approach.[101]

Casein micelle is a natural nanoparticle present in milk which constitutes more than 80% of milk proteins and has a hydrodynamic radius of ~100 nm.[2] It is made up of calcium phosphate nanoclusters.[61] Casein micelle is a natural source of calcium and demonstrates the capacity to carry hydrophobic compounds.[85] Casein micelle demonstrates better encapsulation efficiency compared to that in the serum phase (~5.5 fold) and also improves the nutritional value of the food matrix.[19,30] β-casein is rich in lysine and is hydrophobic in nature.[19] Due to the presence of lysine, researchers derived nanoparticles from β-casein and reducing carbohydrate mediated through the Maillard browning reaction.[65]

Milk protein hydrolysates possess interfacial tension reducing property and, hence, it can be used in emulsions for encapsulation of functional materials.[56] Proteolysis of the milk proteins helps in improving the gelation property and surface active functions.[20,24] Limited and controlled hydro-lysis of casein proteins (2–10 %) and whey protein concentrate (10–27 %) increase the emulsifying activity. These properties of milk protein hydrolysates help in fabrication of novel delivery vehicles for effective controlled delivery of encapsulated functional materials.[56]

Triglycerides are the major components (98%) of milk fat.[69] Milk fat can be distinguished based on the functional and physical properties during fractionation process. The major functional properties considered in the milk fat are melting point and crystallization behavior, which gives different composition to the triglycerides. Hence, milk fat fractions could be used in encapsulation process with defined melting point.[68] Milk fat is also used for encapsulating lipophilic bioactive compounds in food matrices. Encapsulation of protease enzyme in milk fat improved cheese ripening rate compared to the control cheese.[38] Milk fat was used in encap-sulation of α-tocopherol in emulsion-based delivery vehicle in which it was used as an antioxidant against degradation.[90] Milk fat has found many applications in encapsulation of lipophilic compounds such as: flavor compounds, antimicrobials compounds, minerals, and vitamins.

Milk fat globule membrane is a blend of glycolipids, proteins, and phospholipids. Phospholipids in MFGM constitute more than 90% together with protein and 26–31% of total lipid percent.[69] Milk fat globule membrane has the ability to decrease the interfacial tension at the inter-face of two liquids, and it behaves differently compared to milk proteins and, hence, is used for stabilization in emulsion formation. The unique nature of MFGM is that it demonstrates resistance to interfacial displace-ment when low molecular weight emulsifiers are used compared to milk proteins in stabilized emulsion system. However, the phospholipid content also contributes to decrease the interfacial tension and attribute to strong interfacial interaction between MFGM and the dispersed oil.[68]

Liposomes are another type of encapsulation systems having lipid layer at the shell part, which can be used to assist the nutrient delivery in targeted food matrices. Liposomes can be prepared from MFGM and it possesses several benefits over other liposomes, including lower membrane permeability, thicker membrane, and high phase transition temperature.[95] Liposomes may be used as delivery system for various lipophobic and lipophilic functional molecules.[53] Potential application of MFGM as lipo-some includes enhancing efficacy of food bioactive substance, protection of sensitive molecules and masking undesirable flavors.[22]

3.3.1.2   *ENCAPSULATION OF MILK BIOACTIVE COMPOUNDS*

Milk is reflected as a near complete food. From nutrition point of view, bioactive peptides play a major role in human health and can be found in fermented dairy products, such as cheese, sour milk, and yoghurt.[82] Table 3.2 indicates bioactive peptides and other bioactive classes present in milk.

3.3.2   *NANO-FORTIFICATION AND ENRICHMENT*

Food security is the major concern in the world at present. Developed countries are utilizing improved technology in agriculture for enhancing crop yield and to meet food security. However, nutritional security is also a major concern where humans suffer from various malnutrition-based disorders. Hence, intervention of nanotechnology is important in improving the nutritional security of the world. In recent years, food processing industries are working to enhance the nutritional security through fortification and minimizing nutrient loss during processing.[8] Enrichment and fortification of food with nutraceuticals are essential in present conditions, but the effectiveness of its activity in body depends strongly on the bioaccessibility and bioavailability in different environmental conditions. Fortification of processed foods with nutrients encapsulated in nanosized delivery vehicles improve the appearance, nutritional profile, and boosts the textural properties of foods. For example, fortification of ice cream with nanoencapsulated bioactive curcumin enhanced ice cream characteristics such as stability at different process conditions (pH, heat).[51] Fortification of food at nano-level can be used for intensification of nutrition in the processed food. For example, the inclusion of "epigallocatechin gallate and green tea catechins" in nanocapsules, which are medically beneficial, will soon enable cheese to be marketed as health promoting dairy food.[76]

Enrichment of cheese matrices with omega-3 fatty acids, vitamins A, B_{12}, D_3, E, and CoQ_{10} was studied using nanoemulsion-based encapsulation system.[26] The inclusion of lipophilic bioactive compounds in cheese using emulsions improved the stability and functionality of bioactive components during storage with reduced lipid oxidation in the fortified cheese.[89,90] Recently, yogurt fortification was made with rice bran oil using nanoemulsion system. Rice bran oil is a rich source of tocopherol, gamma oryzanol, and tocotrienols that help in lowering serum cholesterol

TABLE 3.2 Different Bioactive Components in Milk and Their Bioactive Properties.

Class	Bioactive compound	Functional activity	Reference
Bioactive peptides and proteins	Lactoferrin	Iron binding	[4, 32, 35, 36, 44, 49, 59, 66, 83, 97]
	Lactoferricin B	Active against Gram's positive and Gram's negative bacteria	
	k-casein (casoxins)	Probiotic, opioid antagonists	
	Glycomacropeptide (GMP)	No aromatic amino acids, helps in inhibition of hemagglutination, prebiotic	
	Caseinophosphopeptides (CPPs)	Complexing and solubilization of calcium	
	Serorphin	Opioid agonist	
	β-Lactoglobulin	Vitamin carrier, fatty acid binding, potential antioxidant	
Bioactive lipids	Triglycerides	Helps in calcium absorption	[57, 64, 93]
	Conjugated linolenic acid (CLA)	Anticancer, immunomodulant, growth factor, antiatherogenic	
	Phospolipids	Antioxidative property, tumor suppressing properties (sphingomyelin, sphingosine)	
	Steroids (lanosterol, 7-dehydrocholesterol, dihydrolanosterol)	Provitamin D_3, precursor of steroid hormone synthesis	
	Alkyl-diacyl-glycerols and Alkyl-acyl-phospolipids	Antimetastatic activity, antineoplastic effect	
Bioactive oligosaccharides	Sialyl-oligosaccharide	Prebiotics, anti-pathogen	[5, 35, 104]
	Fucosylated-oligosaccharide		
Immunoglobulins	IgG, IgM, and IgA	Carriers of antibodies functions	[47]

in humans. However, fortification of cheese resulted in increased melting resistance and did not affect the native microflora, that is, *Lactobacillus* spp.[1] Besides using other nano-nutrient carriers for fortification of dairy products, we can use native nanostructures of milk such as casein micelles for improved functionality. Casein micelles successfully enhanced the bioavailability of encapsulated vitamin D_3 compared to its bioavailability in a synthetic emulsifier in fat free yogurt.[55]

In recent years, the use of biopolymer-based nanocarriers for entrapment of vitamins demonstrates high encapsulation efficiency.[78] High-amylose corn starch-based nanocarriers were used for carrying vitamin D_3, it improved controlled release of the vitamin compound during *in vitro* studies and the result of sensory evaluation showed improved homogeneity and taste of the fortified milk.[31] The size of a nutrient plays a major role in its bioavailability and was found that decreasing the size of calcium enhances the absorption and bioavailability in milk powder.[18] Beyond the importance of ensuring the safety of nano-additives in food, it is also useful to find the necessity of fortification of foods with nano nutrients from a public health perspective. Nevertheless, nano-fortification of food or beverage meets the individual dietary needs, since the nutrient is in nanoscale and it may lead to negative health impact.

3.3.3 NANO-FILTRATION

Nano-filtration is an alternative technique used for demineralization and concentration of whey. It has the additional advantages over combining conventional concentration and demineralization process (evaporation + electrodialysis).[34,45,98] Concurrent demineralization and concentration of whey can be achieved with nano-filtration with reduced cost, reduced energy, and effluent water[98] Membranes used in nano-filtration process have a very low permeability for organic compounds such as lactose, urea, and proteins. High permeability can be seen for monovalent salts such as potassium chloride and sodium chloride. The efficiency in the reduction of minerals in whey using nano-filtration is similar to that of electrodialysis, and it can be increased by the means of increased volume concentration ratio and batch processing.[91] Permeate obtained from the filtration process contains nitrogen salts and some amounts of lactose. The concentration of lactose and nitrogen in permeate is subjected to vary according to the characteristics of membrane, pretreatment conditions, and feed characteristics. Ion valency, ionic strength, salt composition, pH, and viscosity are considered in feed characteristics, while pore length, pore diameter, membrane charge, and membrane materials are important in membrane selection.[98]

Dairy liquids are considered as multicomponent solutions. The rejection of solution components/solutes can be seen in multicomponent solution due to salt concentration. This is also called negative rejections; which is due to the presence of potential gradient, initiated by the difference in ions mobility and partition properties.[98] Thus, it is difficult to forecast the separation properties of a filtration membrane used for these solutions. The rejection of salt could be ascribed to high flux, which indicates the significance of convective transport at high flux and diffusive transport at low flux.[102]

3.3.4 NANO-PACKAGING

Packaging is the essential unit operation in food industry, by which processed food is protected from the environment using packaging materials. The main aim of packaging is to improve the shelf life and keeping intact the quality of food system. Nanotechnology offers wide range of applications in food packaging where nanoedible coatings and nanoparticles are used to overcome the conventional packaging material problems such as high water vapor permeability (WVTR), biodegradability, and poor mechanical properties.[17,70]

3.3.4.1 NANOEDIBLE COATINGS

A nano coating consists of one or more multiple layers of nanomaterials. The complexation among layers is based on electric charge of the nanomaterials or complexation between the biopolymers at molecular level, where charged particles adsorb on oppositely charged layer. This type of nanoparticle laminated coatings can be used for encapsulation of various amphiphilic, hydrophilic, and bioactive functional compounds having properties such as anti-browning, antimicrobial, flavor retention, and antioxidants with enhanced strength of materials and barrier properties.[23] Recently, edible coating of *paneer* was studied using whey protein concentrate composite edible coating. The edible coating improved the shelf life of *paneer* up to 40 days; however uncoated paneer resulted in higher microbial load on 28th day.[3] Another study was conducted on the Coalho cheese shelf life, where alginate/lysozyme multilayer nano laminate was used as the coating system. The use of nano-laminate resulted in lesser cheese mass loss, lipid peroxidation, pH, microbial proliferation, and higher titratable acidity value in contrast with uncoated cheese.[62]

3.3.4.2 NANOPARTICLES-BASED PACKAGING

Nanomaterial utilization in food packaging is a novel concept in food packaging science. At nanoscale, these particles provide benefits in food processing. The advantage of using nanomaterials in food packaging includes the enhancement of physical–chemical properties of the packaging materials and plays a dynamic role in preservation of food. Silver, gold, zinc oxide, titanium dioxide, starch, and chitosan nanoparticles are extensively being used as nanoparticles in food packaging materials.[87] In a recent investigation, copper nanoparticles were embedded to polylactic acid to improve shelf life of cheese. The prepared polylactic acid-copper nanoparticles film showed antimicrobial effects and can be used as active packaging system for food contact applications to extend food quality.[12]

3.3.5 NANO-PRESERVATION

Dairy products are highly perishable due to high moisture content. Pasteurization is mainly used in liquid milk preservation; however, tetra packaging is gaining more importance in improving shelf life of milk and milk product preservation. Antimicrobials can also be used for the preservation of milk and milk products using encapsulation techniques. Encapsulation of antimicrobials such as nisin, silver nanoparticles, and eugenol may provide a potential alternative to pasteurization.[7,15] Nanoencapsulated antimicrobial such as nisin was studied in cheese matrix and the study showed that the nisin can improve the shelf life of cheese with inhibitory properties against pathogens, while protecting the cheese starter with improved stability.[7] Silver nanoparticles impregnated on aluminum surface were also used to preserve the raw milk, which are synthesized using Creighton's method. Good stability of raw milk was achieved up to 25 days using silver nanoparticles.[39]

3.3.6 DAIRY PRODUCTS QUALITY AND SAFETY ASSESSMENT AT NANO SCALE

Nanotechnology is an interdisciplinary science that builds the bridge between chemical, physical, engineering, and life sciences. The dairy industry needs the appropriate analytical procedures for assessment of the quality of raw or finished products in dairy processing lines. But due to cost, time, and complexity in the methods, concern arises for developing portable sensors for milk and milk product quality assessment. Detection of urea, hydrogen peroxide, and melamine can be done using sensors that uses nanotechnology interfaces. Nanoparticles are the major components of the sensors in this regard. Using Fe_3O_4 nanoparticle and carbon nano-tubes-based biosensor, hydrogen peroxide can be detected in the limit of 3.7 nm.[94] In a recent study, gold nanoparticles were used to detect the melamine content in milk using calorimetric detection. The detection was based on the color of the reaction, where the presence of melamine results in blue color and causes the nanoparticle aggregation, whereas wine red color indicates the absence of melamine.[50] Urea can be detected in milk using electrochemical biosensors where immobilized urease enzyme coupled with selective ammonium ion electrode is used, and the activity of substrate is measured using Nernst equation and the amount of urea is quantified in the milk sample.[99]

3.3.7 NANOPARTICLE SYNTHESIS

Synthesis of nanoparticles has developed into more mature scientific field. Though toxicity of nanoparticle is an issue in current age, their application is rapidly increasing in daily life. However, green synthesis of nanoparticles has an enormous influence on reducing environmental pollution, improved biological compatibility, and reduced physiological toxicity.[77] Naturally, nanoparticles exist in food and due to their complex shape, structure, morphologies, and bioactive functions, these particles play a significant role in controlling and understanding the structures, assemblies, or organizations, and functional activities of nanostructures in food.[79]

In recent years, green synthesis of silver nanoparticles has received substantial attention due to its antimicrobial activity in food and food-contact surfaces. Silver nitrite is used as substrate in green synthesis of silver nanoparticles, and reduction of silver nitrite from the native reducing compounds in the solution resulting in silver nanoparticle generation. Milk proteins have the ability to reduce the Ag^+ ions; hence, milk can be used as a bioresource material for synthesis. Silver nanoparticles synthesized using milk as a reduction medium resulted in yielding circular shaped nanoparticles with size range of 30–90 nm and also showed the antagonistic activity against *Monilinia* sp. and *Colletotrichum coccodes*.[54]

3.3.8   DAIRY EFFLUENT TREATMENT

Among the food industries, the dairy industry produces highest effluent with high load of organic materials. Each liter of milk needs water in the range of 2–5 L for cleaning and washing process. The volume of effluent produces also possesses high proportion of biodegradable organic materials and components with nutritional value.[75] Dairy effluent consists of milk components, essentially proteins and lactose. Due to increase in concern over economic problems and environmental impact of effluent, industries have adopted many processes to decrease organic materials' load in lactose-rich effluent from dairy and cheese industry.[16,25] These include activated sludge process and trickling filter, however, due to more space and operating conditions (pH, temperature) they are considered as inconvenient.[52,92] Hence, photocatalytic degradation of effluent can be employed; nanofiber membranes and composites can be used for exploiting photocatalysis. Nanoparticles such as zinc oxide, nickel oxide, titanium dioxide, silicone, exhibit photocatalytic properties and improve the stability in the high water flux treatment.[42,43] However, during preparation of nanofiber or nanocomposites using nanoparticles, researchers found some drawbacks, that is, when titanium alkoxide was used for the preparation of multilayered microspheres of titanium dioxide and silica dioxide, the titanium alkoxide hydrolyzes at a faster rate compared to silica alkoxide, resulting in reduced loading rate of titanium dioxide and formation of photoactive titanium dioxide due to sintering process.[73]

3.3.9   NANOCOATINGS FOR DAIRY EQUIPMENTS

Dairy industry always faces problem in fouling of heat exchangers due to adsorption of thermally denatured milk proteins on to the surface. Nevertheless, fouling at the surface can be reduced by efficient clean-in-place procedure in which water, electrical energy, and chemicals are required. If proper control over fouling is neglected, then it may cause the deterioration of milk quality. Even deposition causes the increase in pressure drop and reduces the heat transfer rate and, hence, disturbs the production rate of dairy plants.[40] It is known that milk is rich in water and, therefore, using hydrophobic nanocoatings on the surface can effectively reduce the fouling problem in heat exchangers.[40] Generally, hydrophobic materials refer those having water contact angle greater than 90°, and super-hydrophobic ones refer surfaces having more than 140° water contact angle.[71] Due to high contact angle, the wettability and surface energy of super-hydrophobic materials is very low. The application of these hydrophobic coatings in heat exchangers for milk and other liquid foods reduces the fouling rate. Food grade biopolymers, silica, diamond-like carbon, xylan, ion implantation (MoS_2, SiF^+), and Ni-P-PTFE were used for nano coating purposes.[40] Nano coating reduces the fouling rate along with reduced time for clean-in-place procedures. In a recent study, fouling caused by milk protein deposition in heat exchanger surface was decreased using carbon nanotubes and polytetrafluorethylene (PTFE) nanocomposite by reducing surface energy.[80]

3.4   CONCLUSIONS

Nanotechnology is considered as promising area in dairy science where it offers many benefits in storage, traceability, processing, and safety of dairy foods. The foremost application of nanotechnology principles in food and dairy industry is encapsulation of bioactive compounds for improving nutritional status. Dairy-based products are highly nutritious and exhibit lower shelf life. Currently, packaging sector shows largest market share in food and nanotechnology, spreading its application in dairy packaging. Nevertheless, the public awareness of nanotechnology applications in dairy industry is significant because of their alleged negative effects on health and environment. The nanomaterials used for fortification, encapsulation, or packaging must be environment friendly. Besides, the current nutritional requirements suggest the improvement and enhancement of the food processing methods to reach the required nutritional status. However, it seems to take more time without intervention of nanotechnology. Therefore to reach the nutritional security in the world, nanotechnology is a viable option and further studies are vital to examine the threats of nano-materials on human health.

3.5 SUMMARY

This chapter discusses the natural nanostructures in dairy ingredients with its properties and applications. Dairy industry involves multidisciplinary science concepts and, consequently, the whole dairy supply chain starting from raw materials production to final product processing is also discussed in the chapter. The potential applications of nanotechnology principles at various dairy processing unit operations are briefly discussed. In summary, authors have deliberated various key points in expanding the potentiality of nanotechnology in dairy processing.

KEYWORDS

- Casein
- emulsions
- immunoglobulins
- lactoferrin
- nanocoating
- nano-filtration
- nano-fortification

- nano-preservation
- nanoclay
- nanodevice
- nanoedible coating
- nanoencapsulation
- silver nanoparticle
- β-lacto globulins

REFERENCES

1. Alfaro, L.; Hayes, D.; Boeneke, C.; Xu, Z.; Bankston, D.; Bechtel, P. J.; Sathivel, S. Physical Properties of a Frozen Yogurt Fortified with a Nano-Emulsion Containing Purple Rice Bran Oil. *LWT-Food Sci. Technol.* **2015**, *62*(2), 1184–1191.

2. Anema, S. G.; Lowe, E. K.; Stockmann, R. Particle Size Changes and Casein Solubilisation in High-Pressure-Treated Skim Milk. *Food Hydrocolloids* **2005**, *19*(2), 257–267.

3. Archana, G. L.; Shyam, R. G. Impact of Edible Coating and Different Packaging Treatments on Microbial Quality of Paneer. *J. Food Process. Technol.* **2012**, *3*(6), 159.

4. Balcão, V. M.; Costa, C. I.; Matos, C. M.; Moutinho, C. G.; Amorim, M.; Pintado, M. E.; Gomes, A. P.; Vila, M. M.; Teixeira, J. A. Nanoencapsulation of Bovine Lactoferrin for Food and Biopharmaceutical Applications. *Food Hydrocolloids* **2013**, *32*(2), 425–431.

5. Barile, D.; Meyrand, M.; Lebrilla, C. B.; German, J. B. Examining Bioactive Components of Milk. Sources of Complex Oligosaccharides (Part 2). *Agro Food Ind. Hi-Tech* **2011**, *22*(4), 37.

6. Baruah, S.; Dutta, J. Nanotechnology Applications in Pollution Sensing and Degradation in Agriculture: A Review. *Environ. Chem. Lett.* **2009**, *7*(3), 191–204.

7. Benech, R. O.; Kheadr, E.; Laridi, R.; Lacroix, C.; Fliss, I. Inhibition of Listeria Innocua in Cheddar Cheese by Addition of Nisin Z in Liposomes or by in Situ Production in Mixed Culture. *App. Environ. Microbiol.* **2002**, *68*(8), 3683–3690.

8. Bieberstein, A.; Roosen, J.; Marette, S.; Blanchemanche, S.; Vandermoere, F. Consumer Choices for Nano-Food and Nano-Packaging in France and Germany. *Eur. Rev. Agric. Econ.* **2013**, *40*(1), 73–94.

9. Bora, T.; Dutta, J. Applications of Nanotechnology in Wastewater Treatment—A Review. *J. Nanosci. Nanotechnol.* **2014**, *14*(1), 613–626.

10. Chen, L. Y.; Remondetto, G. E.; Subirade, M. Food Protein-Based Materials as Nutraceutical Delivery Systems. *Trends Food Sci. Technol.* **2006**, *17*(5), 272–283.

11. Cho, S. I.; Kim, Y. R.; Lee, J. W.; So, D. S.; Cho, Y. J.; Suh, H. K.; San Park, T.; Oh, S. I.; Im, J. E. A Review on the Application of Nanotechnology in Food Processing and Packaging. *Food Eng. Prog.* **2010**, *14*(4), 283–291.

12. Conte, A.; Longano, D.; Costa, C.; Ditaranto, N.; Ancona, A.; Cioffi, N.; Scrocco, C. Sabbatini, L.; Conto, F.; Del Nobile, M. A. A Novel Preservation Technique Applied to Fiordilatte Cheese. *Innovative Food Sci. Emerging Technol.* **2013**, *19*, 158–165.

13. Cromwell, G.; Stahly, T.; Monegue, H. Effects of Source and Level of Copper on Performance and Liver Copper Stores in Weanling Pigs. *J. Anim. Sci.* **1989**, *67*(11), 2996–3002.

14. Cunningham, I. J. Some biochemical and physiological aspects of copper in animal nutrition. *Biochem. J.* **1931**, *25*(4), 1267.

15. Da Silva Malheiros, P.; Daroit, D. J.; Da Silveira, N. P.; Brandelli, A. Effect of Nanovesicle-Encapsulated Nisin on Growth of Listeria Monocytogenes in Milk. *Food Microbiol.* **2010**, *27*(1), 175–178.

16. Demirel, B.; Yenigun, O.; Onay, T. T. Anaerobic Treatment of Dairy Wastewaters: A Review. *Process Biochem.* **2005**, *40*(8), 2583–2595.

17. Dhineshkumar, V.; Ramasamy, D.; Sudha, K. Nanotechnology Application in Food and Dairy Processing. *Int. J. Farm Sci.* **2015**, *5*(3), 274–288.

18. Erfanian, A.; Mirhosseini, H.; Manap, M. Y. A.; Rasti, B.; Bejo, M. H. Influence of Nano-Size Reduction on Absorption and Bioavailability of Calcium from Fortified Milk Powder in Rats. *Food Res. Int.* **2014**, *66*, 1–11.

19. Esmaili, M.; Ghaffari, S. M.; Moosavi-Movahedi, Z.; Atri, M. S.; Sharifizadeh, A.; Farhadi, M.; Yousefi, R.; Chobert, J. M.; Haertlé, T.; Moosavi-Movahedi, A. A. Beta Casein-Micelle as a Nano Vehicle for Solubility Enhancement of Curcumin: Food Industry Application. *LWT-Food Sci. Technol.* **2011**, *44*(10), 2166–2172.

20. Foegeding, E. A.; Davis, J. P.; Doucet, D.; McGuffey, M. K. Advances in Modifying and Understanding Whey Protein Functionality. *Trends Food Sci. Technol.* **2002**, *13*(5), 151–159.

21. Fondevila, M.; Herrer, R.; Casallas, M.; Abecia, L.; Ducha, J. Silver Nanoparticles as a Potential Antimicrobial Additive for Weaned Pigs. *Anim. Feed Sci. Technol.* **2009**, *150*(3), 259–269.

22. Frenzel, M.; Krolak, E.; Wagner, A.; Steffen-Heins, A. Physicochemical Properties of Wpi Coated Liposomes Serving as Stable Transporters in a Real Food Matrix. *LWT-Food Sci. Technol.* **2015**, *63*(1), 527–534.

23. Gammariello, D.; Conte, A.; Buonocore, G. G.; Del Nobile, M. A. Bio-Based Nano-composite Coating to Preserve Quality of Fior di Latte Cheese. *J. Dairy Sci.* **2011**, *94*(11), 5298–5304.

24. Gauthier, S.; Paquin, P.; Pouliot, Y.; Turgeon, S. Surface Activity and Related Functional Properties of Peptides Obtained from Whey Proteins. *J. Dairy Sci.* **1993**, *76*(1), 321–328.

25. Gerardo, M.; Zacharof, M.; Lovitt, R. Strategies for the Recovery of Nutrients and Metals from Anaerobically Digested Dairy Farm Sludge Using Cross-Flow Microfiltration. *Water Res.* **2013**, *47*(14), 4833–4842.

26. Giroux, H. J.; Constantineau, S.; Fustier, P.; Champagne, C. P.; St-Gelais, D.; Lacroix, M.; Britten, M. Cheese Fortification Using Water-in-oil-in-Water Double Emulsions as Carrier for Water Soluble Nutrients. *Int. Dairy J.* **2013**, *29*(2), 107–114.

27. Graveland-Bikker, J.; De Kruif, C. Unique Milk Protein Based Nanotubes: Food and Nanotechnology Meet. *Trends Food Sci. Technol.* **2006**, *17*(5), 196–203.

28. Graveland-Bikker, J. F.; Koning, R. I.; Koerten, H. K.; Geels, R. B.; Heeren, R. M.; De Kruif, C. G. Structural Characterization of α-lactalbumin Nanotubes. *Soft Matter* **2009**, *5*(10), 2020–2026.

29. Gunasekaran, S.; Ko, S.; Xiao, L. Use of Whey Proteins for Encapsulation and Controlled Delivery Applications. *J. Food Eng.* **2007**, *83*(1), 31–40.

30. Haham, M.; Ish-Shalom, S.; Nodelman, M.; Duek, I.; Segal, E.; Kustanovich, M.; Livney, Y. D. Stability and Bioavailability of Vitamin D Nanoencapsulated in Casein Micelles. *Food Func.* **2012**, *3*(7), 737–744.

31. Hasanvand, E.; Fathi, M.; Bassiri, A.; Javanmard, M.; Abbaszadeh, R. Novel Starch Based Nanocarrier for Vitamin D Fortification of Milk: Production and Characterization. *Food Bioprod. Process.* **2015**, *96*, 264–277.

32. Hernández-Ledesma, B.; Dávalos, A.; Bartolomé, B.; Amigo, L. Preparation of Antioxidant Enzymatic Hydrolysates From α-lactalbumin and β-lactoglobulin. Identification of Active Peptides by HPLC-MS/MS. *J. Agric. Food Chem.* **2005**, *53*(3), 588–593.

33. Holt, C.; Wahlgren, N. M.; Drakenberg, T. Ability of a Beta-Casein Phosphopeptide to Modulate the Precipitation of Calcium Phosphate by Forming Amorphous Dicalcium Phosphate Nanoclusters. *Biochem. J.* **1996**, *314*, 1035–9.

34. Horton, B. Anaerobic Fermentation and Ultra-Osmosis. *Bull. Int. Dairy Federat.* **1987**, *212*, 77–83.

35. Hsieh, C. C.; Hernández-Ledesma, B.; Fernández-Tomé, S.; Weinborn, V.; Barile, D.; De Moura Bell, J. M. L. N. Milk Proteins, Peptides, and Oligosaccharides: Effects Against the 21st Century Disorders. *Bio. Med. Res. Int.* 2015.

36. Hu, B.; Ting, Y.; Zeng, X.; Huang, Q. Cellular Uptake and Cytotoxicity of Chitosan–Caseinophosphopeptides Nanocomplexes Loaded with Epigallocatechin Gallate. *Carbohydr. Polym.* **2012**, *89*(2), 362–370.

37. Jang, H. D.; Swaisgood, H. E. Disulfide Bond Formation Between Thermally Denatured β-Lactoglobulin and κ-Casein in Casein Micelles. *J. Dairy Sci.* **1990**, *73*(4), 900–904.

38. Kailasapathy, K.; Lam, S. Application of Encapsulated Enzymes to Accelerate Cheese Ripening. *Int. Dairy J.* **2005**, *15*(6), 929–939.

39. Kalaiselvi, A.; Gantz, S.; Ramalingam, C. Effects of Silver Nanoparticles on Storage Stability of Raw Milk. *Int. J. Pharm. Pharm. Sci.* **2013**, *5*(3), 274–277.

40. Kananeh, A. B.; Scharnbeck, E.; Kück, U.; Räbiger, N. Reduction of Milk Fouling Inside Gasketed Plate Heat Exchanger Using Nano-Coatings. *Food Bioprod. Process.* **2010**, *88*(4), 349–356.

41. Kanjwal, M. A.; Barakat, N. A.; Chronakis. I. S. Photocatalytic Degradation of Dairy Effluent Using AgTiO$_2$ Nanostructures/Polyurethane Nanofiber Membrane. *Ceram. Int.* **2015**, *41*(8), 9615–9621.

42. Kanjwal, M. A.; Alm, M.; Thomsen, P.; Barakat, N. A.; Chronakis, I. S. Hybrid Matrices of TiO$_2$ and TiO$_2$–Ag Nanofibers with Silicone for High Water Flux Photocatalytic Degradation of Dairy Effluent. *J. Ind. Eng. Chem.* **2016**, *33*, 142–149.

43. Kanjwal, M. A.; Chronakis, I. S.; Barakat, N. A. Electrospun nio, zno and Composite NiO–ZnO Nanofibers/Photocatalytic Degradation of Dairy Effluent. *Ceram. Int.* **2015**, *41*(9), 12229–12236.

44. Kawasaki, Y.; Isoda, H.; Shinmoto, H.; Tanimoto, M.; Dosako, S. I.; Idota, T.; Nakajima, I. Inhibition by κ-casein Glycomacropeptide and Lactoferrin of Influenza Virus Hemagglutination. *Biosci. Biotechnol. Biochem.* **1993**, *57*(7), 1214–1215.

45. Kelly, P.; Horton, B.; Burling, H.; Boer, R. D.; Jelen, P.; Puhan, Z. Partial Demineralization of Whey by Nanofiltration. *New Appl. Membr. Process.(International Dairy Federation-Special Issue)* **1992**, 130–140.

46. Keogh, M. K.; O'Kennedy, B. T. Milk Fat Microencapsulation Using Whey Proteins. *Int. Dairy J.* **1999**, *9*(9), 657–663.

47. Korhonen, H.; Marnila, P.; Gill, H. Milk Immunoglobulins and Complement Factors. *Br. J. Nutr.* **2000**, *84*(S1), 75–80.

48. Krasaekoopt, W.; Bhandari, B.; Deeth, H. Evaluation of Encapsulation Techniques of Probiotics for Yoghurt. *Int. Dairy J.* **2003**, *13*(1), 3–13.

49. Krissansen, G. W. Emerging Health Properties of Whey Proteins and Their Clinical Implications. *J. Am. Coll. Nutr.* **2007**, *26*(6), 713S–723S.

50. Kumar, N.; Seth, R.; Kumar, H. Colorimetric Detection of Melamine in Milk by Citrate-Stabilized Gold Nanoparticles. *Anal. Biochem.* **2014**, *456*, 43–49.

51. Kumar, D. D.; Mann, B.; Pothuraju, R.; Sharma, R.; BajajMinaxi, R. Formulation and Characterization of Nanoencapsulated Curcumin Using Sodium Caseinate and its Incorporation in Ice Cream. *Food Funct.* **2016**, *7*(1), 417–424.

52. Kushwaha, J. P.; Srivastava, V. C.; Mall, I. D. An Overview of Various Technologies for the Treatment of Dairy Wastewaters. *Crit. Rev. Food Sci. Nutr.* **2011**, *51*(5), 442–452.

53. Laridi, R.; Kheadr, E.; Benech, R. O.; Vuillemard, J.; Lacroix, C.; Fliss, I. Liposome Encapsulated Nisin Z: Optimization, Stability and Release During Milk Fermentation. *Int. Dairy J.* **2003**, *13*(4), 325–336.

54. Lee, K. J.; Park, S. H.; Govarthanan, M.; Hwang, P. H.; Seo, Y. S.; Cho, M.; Lee, W. H.; Lee, J. Y.; Kamala-Kannan, S.; Oh, B. T. Synthesis of Silver Nanoparticles Using Cow Milk and Their Antifungal Activity Against Phytopathogens. *Mater. Lett.* **2013**, *105*, 128–131.

55. Livney, Y. D. Final Report: Project 838-0576-12: Bioavailability of Vitamin D Encapsulated in Casein Micelles, Compared to its Bioavailability in a Synthetic Emulsifier Currently Used for Supplementation and Enrichment. Livney Sophia ish-shalom, 2012, 1–11.

56. Livney, Y. D. Milk Proteins as Vehicles for Bioactives. *Curr. Opin. Colloid Interface Sci.* **2010**, *15*(1), 73–83.

57. Lucas, A.; Quinlan, P.; Abrams, S.; Ryan, S.; Meah, S.; Lucas, P. Randomised Controlled Trial of a Synthetic Triglyceride Milk Formula for Preterm Infants. *Arch. Dis. Child. Fetal Neonat. Ed.* **1997**, *77*(3), F178–F184.

58. Ajith, M. K. S.; Anuraj, S. K. S. Synthesis of Zinc Oxide (ZnO) Nanoparticles From Zinc Sulphate, for Inclusion in Animal Feeds. *Global J. Res. Anal.* **2014**, *3*(7), 2.

59. Martınez-Gomis, J.; Fernández-Solanas, A.; Vinas, M.; Gonzalez, P.; Planas, M.; Sanchez, S. Effects of Topical Application of Free and Liposome-Encapsulated Lactoferrin and Lactoperoxidase on Oral Microbiota and Dental Caries in Rats. *Arc. Oral Biol.* **1999**, *44*(11), 901–906.

60. Matalanis, A.; Jones, O. G.; McClements, D. J. Structured Biopolymer-Based Delivery Systems for Encapsulation, Protection, and Release of Lipophilic Compounds. *Food Hydrocolloids* **2011**, *25*(8), 1865–1880.

61. McGann, T. C.; Donnelly, W. J.; Kearney, R. D.; Buchhemm, W. Composition and Size Distribution of Bovine Casein Micelles. *Biochimica et Biophysica Acta (BBA)-General Subjects* **1980**, *630*(2), 261–270.

62. Medeiros, B. G. D. S.; Souza, M. P.; Pinheiro, A. C.; Bourbon, A. I.; Cerqueira, M. A.; Vicente, A. A.; Carneiro-da-Cunha, M. G. Physical Characterisation of an Alginate/Lysozyme Nano-Laminate Coating and its Evaluation on "coalho" Cheese Shelf Life. *Food Bioprocess Technol.* **2014**, *7*(4), 1088–1098.

63. Mohamed, H. A., El -Salam.; El-Shibiny, S. Formation and Potential Uses of Milk Proteins as Nano Delivery Vehicles for Nutraceuticals: A Review. *Int. J. Dairy Technol.* **2012**, *65*(1), 13–21.

64. Molkentin, J. Occurrence and Biochemical Characteristics of Natural Bioactive Substances in Bovine Milk Lipids. *Br. J. Nutr.* **2000**, *84*(S1), 47–53.

65. Mu, M.; Pan, X.; Yao, P.; Jiang, M. Acidic Solution Properties of β-Casein-Graft-Dextran Copolymer Prepared Through Maillard Reaction. *J. Colloid Interface Sci.* **2006**, *301*(1), 98–106.

66. Nagpal, R.; Behare, P.; Rana, R.; Kumar, A.; Kumar, M.; Arora, S.; Morotta, F.; Jain, S.; Yadav, H. Bioactive Peptides Derived from Milk Proteins and Their Health Beneficial Potentials: An Update. *Food Funct.* **2011**, *2*(1), 18–27.

67. Oldfield, J.; Schubert, J.; Muth, O. Feed Additives, Implications of Selenium in Large Animal Nutrition. *J. Agric. Food Chem.* **1963**, *11*(5), 388–390.

68. Oliver, C. M.; Augustin, M. A. Using Dairy Ingredients for Encapsulation. Chapter 22, In *Dairy-Derived Ingredients: Food and Nutraceutical Uses*; Corredig, M. Eds.; Woodhead Publishers: Cambridge, **2009**, p. 565–588.

69. Park, Y. W. *Bioactive Components in Milk and Dairy Products*, John Wiley & Sons: New York, 2009; p. 440.

70. Patel, R.; Prajapati, J.; Balakrishnan, S. Recent Trends in Packaging of Dairy and Food Products. *National Seminar on Indian Dairy Industry—Opportunities and Challenges;* AAU, Anand: Gujarat, January 8–9, 2015, pp 118–124.

71. Phan, H. T.; Caney, N.; Marty, P.; Colasson, S.; Gavillet, J. Surface Wettability Control by Nanocoating: The Effects on Pool Boiling Heat Transfer and Nucleation Mechanism. *Int. J. Heat Mass Transfer* **2009**, *52*(23), 5459–5471.

72. Picot, A.; Lacroix, C. Encapsulation of Bifidobacteria in Whey Protein-based Micro-capsules and Survival in Simulated Gastrointestinal Conditions and in Yoghurt. *Int. Dairy J.* **2004**, *14*(6), 505–515.

73. Pinho, L. M.; Mosquera, J. Photocatalytic Activity of TiO_2–SiO_2 Nanocomposites Applied to Buildings: Influence of Particle Size and Loading. *Appl. Catalysis B-Environ.* **2013**, *134*, 205–221.

74. Ramakrishnan, D. R. G.; Thomas, K. S. K. S. Enhancing the Milk Production and Immunity in Holstein Friesian Crossbred Cow by Supplementing Novel Nano Zinc Oxide. *Res. J. Biotechnol.* **2013**, *8*(5), 11–17.

75. Ramasamy, E.; Abbasi, V. S. A. Energy Recovery from Dairy Waste-Waters: Impacts of Biofilm Support Systems on Anaerobic cst Reactors. *App. Energy* **2000**, *65*(1–4), 91–98.

76. Rashidinejad, A.; Birch, E. J.; Sun-Waterhouse, D.; Everett, D. W. Delivery of Green tea Catechin and Epigallocatechin Gallate in Liposomes Incorporated into Low-Fat Hard Cheese. *Food Chem.* **2014**, *156*, 176–83.

77. Raveendran, P.; Fu, J.; Wallen, S. L. Completely "green" Synthesis and Stabilization of Metal Nanoparticles. *J. Am. Chem. Soci.* **2003**, *125*(46), 13940–13941.

78. Renard, D.; Robert, P.; Lavenant, L.; Melcion, D.; Popineau, Y.; Gueguen, J.; Duclairoir, C.; Nakache, E.; Sanchez, C.; Schmitt, C. Biopolymeric Colloidal Carriers for Encapsulation or Controlled Release Applications. *Int. J. Pharm.* **2002**, *242*(1), 163–166.

79. Rogers, M. A. Naturally Occurring Nanoparticles in Food. *Curr. Opin. Food Sci.* **2016**, *7*, 14–19.

80. Rungraeng, N.; Cho, Y. C.; Yoon, S. H.; Jun, S. Carbon Nanotube-Polytetrafluoroethylene Nanocomposite Coating for Milk Fouling Reduction in Plate Heat Exchanger. *J. Food Eng.* **2012**, *111*(2), 218–224.

81. Sandhu, A. Who Invented Nano? *Nat. Nanotechnol.* **2006**, *1*(2), 87–87.

82. Sarkar, P.; Lohith Kumar, D. H.; Dhumal, C.; Panigrahi, S. S.; Choudhary, R. Traditional and Ayurvedic Foods of Indian Origin. *J. Ethnic Foods* **2015**, *2*(3), 97–109.

83. Schlimme, E.; Meisel, H. Bioactive Peptides Derived From Milk Proteins. Structural, Physiological and Analytical Aspects. *Food Nahrung* **1995**, *39*(1), 1–20.

84. Sekhon, B. S. Food Nanotechnology—An Overview. *Nanotechnol. Sci. Appl.* **2010**, *3*, 1–15.

85. Semo, E.; Kesselman, E.; Danino, D.; Livney, Y. D. Casein Micelle as a Natural Nano-Capsular Vehicle for Nutraceuticals. *Food Hydrocolloids* **2007**, *21*(5), 936–942.

86. Shpigelman, A.; Israeli, G.; Livney, Y. D. Thermally-Induced Protein–Polyphenol co-assemblies: Beta Lactoglobulin-Based Nanocomplexes as Protective Nanovehicles for EGCG. *Food Hydrocolloids* **2010**, *24*(8), 735–743.

87. Silvestre, C.; Duraccio, D.; Cimmino, S. Food Packaging Based on Polymer Nano-materials. *Prog. Polym. Sci.* **2011**, *36*(12), 1766–1782.

88. Singh, H. The Milk Fat Globule Membrane—A Biophysical System for Food Applications. *Curr. Opin. Colloid Interface Sci.* **2006**, *11*(2), 154–163.

89. Stratulat, I.; Britten, M.; Salmieri, S.; Fustier, P.; St-Gelais, D.; Champagne, C. P.; Lacroix, M. Enrichment of Cheese with Bioactive Lipophilic Compounds. *J. Funct. Foods* **2014**, *6*, 48–59.

90. Stratulat, I.; Britten, M.; Salmieri, S.; Fustier, P.; St-Gelais, D.; Champagne, C. P.; Lacroix, M. Enrichment of Cheese with Vitamin D3 and Vegetable Omega-3. *J. Funct. Foods* **2015**, *1*, 300–307.

91. Suárez, E.; Lobo, A.; Alvarez, S.; Riera, F.A.; Álvarez, R. Demineralization of Whey and Milk Ultrafiltration Permeate by Means of Nanofiltration. *Desalination* **2009**, *241*(1–3), 272–280.

92. Tawfik, A.; Sobhey, M.; Badawy, M. Treatment of a Combined Dairy and Domestic Wastewater in an Up-Flow Anaerobic Sludge Blanket (UASB) Reactor Followed by Activated Sludge (as System). *Desalination* **2008**, *227*(1), 167–177.

93. Tetens, I. *European Food Safety Authority; Response to Comments on the Scientific Opinion of the Efsa Panel on Dietetic Products, Nutrition and Allergies (NDA) on the Scientific Substantiation of a Health Claim Related to Beta-Palmitate and Increased Calcium Absorption Pursuant to Article 14 of Regulation (EC) No 1924/2006*; European Food Safety Authority: Parma, Italy, 2011, p. 233.

94. Thandavan, K.; Gandhi, S.; Nesakumar, N.; Sethuraman, S.; Rayappan, J. B. B.; Krishnan, U. M. Hydrogen Peroxide Biosensor Utilizing a Hybrid Nano-Interface of Iron Oxide Nanoparticles and Carbon Nanotubes to Assess the Quality of Milk. *Sens. Actuators B: Chem.* **2015**, *215*, 166–173.

95. Thompson, A.; Singh, H. Preparation of Liposomes from Milk Fat Globule Membrane Phospholipids Using a Microfluidizer. *J. Dairy Sci.* **2006**, *89*(2), 410–419.

96. Thompson, A. K.; Haisman, D.; Singh, H. Physical Stability of Liposomes Prepared from Milk Fat Globule Membrane and Soya Phospholipids. *J. Agr. Food Chem.* **2006**, *54*(17), 6390–6397.

97. Tomita, M.; Bellamy, W.; Takase, M.; Yamauchi, K.; Wakabayashi, H.; Kawase, K. Potent Antibacterial Peptides Generated by Pepsin Digestion of Bovine Lactoferrin. *J. Dairy Sci.* **1991**, *74*(12), 4137–4142.

98. Van der Horst, H. C.; Timmer, J. M. K.; Robbertsen, T.; Leenders, J. Use of Nano-filtration for Concentration and Demineralization in the Dairy Industry: Model for Mass Transport. *J. Membr. Sci.* **1995**, *104*(3), 205–218.

99. Verma, N.; Singh, M. A Disposable Microbial Based Biosensor for Quality Control in Milk. *Biosens. Bioelectron.* **2003**, *18*(10), 1219–1224.

100. Vijayalakshmi, C.; Chellaram, C.; Kumar, S. L. Modern Approaches of Nanotechnology in Agriculture-A Review. *Biosci. Biotechnol. Res. Asia* **2015**, *12*(1), 327–331.

101. Ye, A.; Singh, H. Formation of Multilayers at the Interface of Oil-in-Water Emulsion Via Interactions Between Lactoferrin and β-Lactoglobulin. *Food Biophys.* **2007**, *2*(4), 125–132.

102. Yorgun, M. S.; Balcioglu, I. A.; Saygin, O. Performance Comparison of Ultrafiltration, Nanofiltration and Reverse Osmosis on Whey Treatment. *Desalination* **2008**, *229*(1–3), 204–216.

103. Young, S.; Sarda, X.; Rosenberg, M. Microencapsulating Properties of Whey Proteins, II: Combination of Whey Proteins with Carbohydrates. *J. Dairy Sci.* **1993**, *76*(10), 2878–2885.

104. Zivkovic, A.; Barile, M. D. Bovine Milk as a Source of Functional Oligosaccharides for Improving Human Health. *Adv. Nutr.: Int. Rev. J.* **2011**, *2*(3), 284–289.

PART II

Emerging Issues, Challenges and Specific Examples of Nanotechnology for Sustainable Biological Systems

CHAPTER 4

PLANT-NANOPARTICLES (NP) INTERACTIONS—A REVIEW: INSIGHTS INTO DEVELOPMENTAL, PHYSIOLOGICAL, AND MOLECULAR ASPECTS OF NP PHYTOTOXICITY

SHWETA JHA

Department of Botany (UGC-Centre of Advance Study), Jai Narain Vyas University, Jodhpur, Rajasthan 342001, India

Corresponding author. E-mail: jha.shweta80@gmail.com,
sj.bo@jnvu.edu.in

CONTENTS

4.1   INTRODUCTION

Nanotechnology is an emerging science which deals with the synthesis and manipulation of nanoscale size particles. They have one or more external dimensions in the range of 1–100 nm and possess a high surface/volume ratio.[95] According to the European Union (EU), the definition of nano-materials is: "Nanomaterial means a natural, incidental, or manufactured material containing particles, in an unbound state or as an agglomerate and where, for 50% or more of the particles in the number size distribution, one or more external dimensions are in the size range 1–100 nm. In specific cases and where warranted by concerns for the environment, health, safety or competitiveness the number size distribution threshold of 50% may be replaced by a threshold between 1 and 50%."[35] Similarly, engineered nanomaterials were defined as "…any intentionally produced material that has one or more dimensions of the order of 100 nm or less or that is composed of discrete functional parts, either internally or at the surface, many of which have one or more dimensions of the order of 100 nm or less, including structures, agglomerates or aggregates, which may have size above the order of 100 nm but retain properties that are characteristics of the nanoscale."[36] The extremely small size and greater relative surface area of nanoparticles (NPs) result in unique physicochemical, electrical and optical properties, quantum effect, altered biological properties, higher solubility, and catalytic reactivity different from their corresponding larger counterparts.[50]

Nanoparticles can be classified on the basis of their dimensionality, morphology, and composition. They can be spherical, tubular, irregularly shaped and can also exist in aggregated or agglomerated forms.[109] Based on the dimensions, NPs can be classified into three categories:

- Materials having one dimension in the nanoscale, for example layers, graphene, thin films, and surface coatings
- Materials having two dimensions in the nanoscale, for example nanowire and nanotube
- Materials having all three dimensions in nanoscale, for example NPs, dendrimers, fullerenes, colloids and quantum dots (QDs)

According to the United States Environmental Protection Agency,[138] NPs can be classified on the basis of their composition as follows:

- Carbon-based materials, for example, carbon nanotubes (CNTs) and fullerene
- Metal-based substances, for example, metal oxides and QDs
- Nanosized polymers with branched units, for example, dendrimers
- Composites integrating NPs/complexes with other bulk-scale materials, for example, titanium with attached deoxyribonucleic acid (DNA) strands

Nanoparticles can be produced by natural processes such as volcanic eruptions, photochemical reactions, fire, erosion, and biocompatible by-products of plants and animals, for example, soil organic matter, nanoclay, carotenoid, lycopene, lipoproteins, ferritin, and so forth.[19] Nanomaterials can also be manufactured by anthropogenic processes and called as manufactured/engineered NPs (MNP/ENP), for example, carbon NPs, metal oxide NPs, zero-valence metal NPs, QDs, and dendrimers. Engineered NPs can enter the environment through unintentional accidental release (e.g., during combustion, atmospheric emissions, domestic wastewater, agriculture, chemical manufacturing/transport), or intentional release (e.g., during environmental remediation efforts); therefore they may cause toxicity to the local flora and fauna as well as human health by entering in the food chain.[109]

This chapter reviews the available literature on the phytotoxic effects of nanomaterials on plant development, physiological behavior and molecular basis behind this phytotoxicity and NP-plant interaction.

4.2 APPLICATIONS OF NANOMATERIALS IN AGRICULTURE

Due to their wider composition and nature, NPs present multiple functions and almost infinite applications in different fields such as biotechnology, biomedical, pharmaceuticals, cosmetics, electronics, optics, material science, textiles, energy sectors, water treatment technology, environmental remediation, and food and agriculture industry. Nanotechnology is one of the most important tools in modern agriculture and opens up novel applications for crop protection and improvement.[111] Agri-nanotechnology promises to improve current agriculture practices and is anticipated to become a driving economic force in the near future (Fig. 4.1). The key applications of nanotechnology in agriculture are described in the following sections:

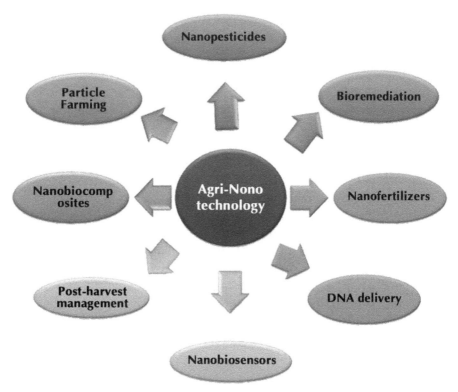

FIGURE 4.1 Schematic of applications of nanotechnology in agriculture.

4.2.1 NANOPESTICIDES

Currently, nanopesticides are the most promising strategies being used for protection of host plants against pests and address the problems of commonly used traditional pesticides.[6,61] Nano-agrochemicals are mostly reformulations of existing pesticides and fungicides. Nanopesticides cover a wide variety of products including organic polymers, surfactants, and metal NPs. Nanomaterials including polymeric NPs, iron oxide NPs, gold NPs, and silver ions have been exploited as pesticides. Researchers have reported various aspects of NP formulation, their characterization, and potential use for management of plant diseases.[6,115] The lack of water solubility is one of the limiting factors in the development of nanopesticides. Nanoencapsulation has been used as a versatile tool for hydrophobic pesticides, enhancing their apparent solubility and protection against premature degradation, and allows a controlled slow and prolonged release of the active ingredients in a targeted manner, thereby reducing the application dosage and improving efficiency.[61]

Silver NPs have received significant attention as a pesticide for agricultural applications.[3] Researchers have reported applications of silver and titanium dioxide NP in plant-disease management exhibiting significant larvicidal activity.[130] Similarly, amorphous nanosilica, nanocopper particles, nanostructured alumina, alumina-silicate NPs and DNA-tagged gold NPs may also provide a cheap and reliable alternative for controlling insect pests compared to commercially available insecticides.[115] Likewise, efficient antifungal and antimicrobial activity of sulfur NPs, polymer-based copper nanocomposites and silica–silver NPs were also shown.[24] Pheromones immobilized in a nanogel have potential as eco-friendly biological control agents, and nano-silicon carrier has been used for delivery of pesticides and herbicides in plants.[13]

4.2.2 NANOFERTILIZERS

The application of nanofertilizers substituted for traditional fertilizers is an innovative way to release nutrients gradually into the soil in a controlled way, thus preventing negative effects caused by the excessive consumption of commercially available fertilizers.[103] In nanofertilizers, nutrients can be encapsulated by nanomaterials, coated with a thin protective film, or delivered as emulsions. Nanofertilizers have unique features such as ultrahigh light absorption and photo energy transmission, causing significant expansion in the leaf surface area and rise in photosynthesis.[25] Thus, the use of nanofertilizer may lead to increase in production along with reduction in the frequency of fertilizer application and soil toxicity.

Nanocomposites consisting of nitrogen, phosphorus, potassium, and other micronutrients have shown to enhance the uptake and use of nutrients by crops. A compound of silicon dioxide (SiO_2) and titanium-dioxide (TiO_2) NPs increased the activity of nitrate reductase in soybean and intensified plant absorption capacity, making the use of water and fertilizer more efficient.[90] Zinc–aluminum layered double-hydroxide nanocomposites have been employed for the controlled release of plant growth regulators.[56] Similarly, environmentally sustainable nano-organic iron-chelated fertilizers, SiO_2, TiO_2, and zinc oxide (ZnO NPs), urea-modified hydroxyapatite NP, and CNTs and nanoporous zeolite can be used as alternate strategy to improve the efficiency of nutrient use in crop production systems.[63,67,93] Enhancement in plant growth and development as well as fruit yield and nutritional quality has been observed in tomato by foliar application of TiO_2 and ZnO NPs as fertilizer.[117]

4.2.3   NANOBIOSENSORS

Nanobiosensors can be effectively utilized to detect the presence of a wide variety of diseases and pathogens inside the stored grain bulk, such as insects, fungus, viruses, contaminants, herbicide, pesticide, insecticide, as well as moisture, pH and nutrient level in soil.[116] Potential nanosensors may indicate temperature, freshness, ripeness, and contaminant/pathogen status on the food package.[110] They can be used as indicators to monitor the environmental stresses and crop conditions. Nanobarcodes and nanoprocessing could also be used to monitor the quality of agricultural products.[81,152] Thus, nanobiosensors could help in the efficient use of agricultural natural resources such as water, nutrients, and chemicals, which can further support sustainable agriculture for enhancing crop productivity.

4.2.4   NANO-BIOREMEDIATION

The use of nanomaterials can help in improvement of quality of the environment and detection of polluted sites and its remediation. For example, the levels of environmental pollution can be evaluated quickly by nano-smart dust (tiny wireless sensors) and gas sensors, and can be used to remediate contaminated soil and groundwater.[100]

4.2.5   NANOCOMPOSITES

Nanocomposites are eco-friendly and biodegradable polymers reinforced with small quantities of nano-sized particles (e.g., nano-zinc oxide and nano-magnesium oxide), which have high aspect ratios and improve the properties and performance of a polymer. Nanocomposites have a wide range of applications in various fields including agriculture and food packaging.[21,135] The use of nanocomposites with new thermal and gas barrier properties can prolong the post-harvest life of food, and facilitate the transportation and storage of food.[131] For agricultural applications, nanocomposites can be used to retain soil moisture and control weeds.[120]

4.2.6 NANOBIOFUELS

Nanotechnology is one of the appropriate technologies for production of future biofuels. The use of nanotechnology has been described in transesterification, pyrolysis, and hydrogenation.[118] Mostly, liquid biofuels are produced by conversion of cellulose to ethanol and biodiesel. Cellulose nanocrystals have been shown to reinforce the biocomposites for industrial applications.[70] Nanomaterials such as calcium oxide and magnesium oxide NPs can be used as biocatalysts in oil transesterification to biodiesel. Biodiesel production from heterotrophic microalgae through transesterification using nanotechnology has been also reported.[153]

4.2.7 NANODEVICES FOR GENETIC ENGINEERING OF PLANTS

Nanotechnology provides novel tools and techniques for delivery of genes and drug molecules to specific sites at cellular levels, and possesses great potential to augment crop productivity through genetic improvement of plants. This enables NPs, nanofibers, and nanocapsules to carry foreign DNA and gene-modifying chemicals.[68] Nanoparticle-mediated gene transfer in plants has been reported for the development of insect-resistant varieties and to increase product shelf life.[42] This technique has been successfully applied to introduce DNA into isolated cells of tobacco and corn plants using chemically coated mesoporous silica NPs.[136] Similarly, cellular "injection" with carbon nanofibers containing foreign DNA has been used to genetically modify golden rice.[42] In addition, novel plant varieties may be developed using these approaches.[123]

Other major applications of agri-nanotechnology include nanoherbicides, nanofoods, postharvest management, soil improvement and water purification.[111]

4.3   NANOTECHNOLOGY: ASSOCIATED RISKS AND THEIR ASSESSMENT FOR DEVELOPMENTAL AND PHYSIOLOGICAL PHYTOTOXICITY

Despite potential applications and benefits of nanotechnology in the agri-food sector, the extensive and indiscriminate use of NPs could lead to greater risk of their release into the environment,[47] posing a serious threat to the ecological system and living beings including humans.[14]

To date, little is known about environmental implications of nanotechnology. Plants being primary producers serve as an important potential route for NP transport, and the release of engineered NPs may cause adverse effects on edible plants. Nanoparticles entering wastewater streams may predominantly be incorporated into sewage sludge and applied to agricultural fields. It may affect the whole food chain through bioaccumulation at the first trophic level through soil or water.[156] Metal oxide NPs are finding increasing application in various commercial products, leading to concerns for their environmental fate and potential toxicity. The use of nanosilver as an antibacterial agent in consumer and health care products has also raised public concerns about safety.[15] Tomato plants exposed to cerium oxide NPs have shown to contain elevated cerium content in shoots and edible fruits, demonstrating long-term impact of NPs on plant health and its implications for our food security.[143]

Due to having considerable impact on human health and the environment, ENPs should be treated as a new group of contaminants, and it is necessary to develop new risk-assessment methods for the estimation of their toxicological properties. There is an urgent need for regulatory systems capable of managing any risks associated with the use of nanotechnology in the agriculture.

Concern over the potentially harmful effects of NPs has stimulated research on nanotoxicology with a focus on few model organisms.[121] The majority of the nanotoxicological studies have focused on mammalian cytotoxicity or impacts on animals and bacteria. The impact of ENPs has recently been investigated on a range of species, including algae, crustaceans, fish, bacteria, yeast, nematodes, protozoa and mammalian cell lines,[16] pathogenic/fresh water bacteria,[27,32,57,71] soil microbial community,[51,58,69] algae,[48,97,107,141] invertebrates such as nematodes *Caenorhabditis elegans*[122,142] and crustaceans,[41,52,154] and vertebrates such as fish[37,124,146] and rats.[33,74] However, the impact of NPs on terrestrial plants is still not well understood.[12] Plants constitute a significant link in ecotoxicological studies; therefore, assessment of the potential phytotoxicity of ENPs is particularly crucial. Plant nanotoxicology is an emerging and less-explored area of research for the plant scientists. The accumulation of NPs at high levels in the plant could not only affect their growth and metabolism, but also could pose a route for contamination in the food chain. Thus, the fate of the NPs in the environment needs to be addressed by adequate toxicological studies on ecologically important plants.[84]

Phytotoxicity profile of NPs[137] is mainly measured in terms of developmental and physiological parameters, as shown in Figure 4.2. A limited number of studies have been performed to assess the impact of various types of ENPs on higher plants, which varies greatly in different plant species and showed both positive and the negative effects on plant's growth and development. A comprehensive review of the available literature describing the effects on plants exposed to different types of nanomaterials is presented below (Table 4.1). These findings may provide a link between the plant phenotypes and the mechanism behind the associated toxic effects of NPs on plants.

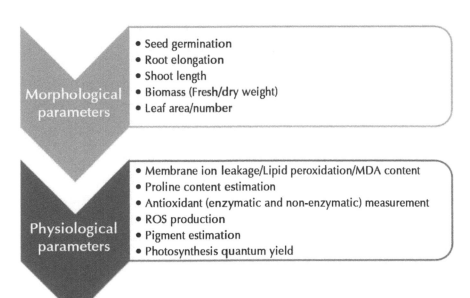

FIGURE 4.2 Phenological, physiological, and biochemical parameters to measure toxicological effects of nanoparticles on plants.

TABLE 4.1 Risk Assessment Studies for Nanoparticle-Mediated Phytotoxicity: Effect of Different NPs on Plants.

S. No.	Type of NP	Size of NP	Dose of NP	Exposure time	Model plant	Toxicity assay	Parameters studied	Reference
1	Ag NPs	20 nm, 100 nm	0, 5, 10, 20, 40, 80, and 160 mg l^{-1}	7 d, 14 d	Lemna minor	Developmental	Relative growth rate, frond number, dry weight	[49]
2	Ag NPs	25 nm	50 µg ml^{-1}, 500 µg ml^{-1}, and 1000 µg ml^{-1}	12 d	Oryza sativa	Morphological	Microscopy, TEM	[96]
3	Ag NPs	5–25 nm (average 10 nm)	Agar 0, 5, 10, 20, and 40 mg l^{-1}; soil: 0, 100, 300, 500, 1000, and 2000 mg kg$^-$ dry soil	2–5 d	Phaseolus radiatus, Sorghum bicolor	Developmental	Seedling growth in both agar and soil media, shoot and root lengths	[77]
4	Ag NPs, Pt NPs	Pt NP 5–35 nm, Ag NP 3–50 nm	Not specified	5 d	Solanum lycopersicum, Raphanus sativus	Developmental	Relative root length; relative seed germination rate; germination index	[128]
5	Ag NPs, ZnO NPs	Ag NP 40 nm, ZnO NP 20 nm	0, 25, 50, and 100 mg l^{-1}	1 d, 5 d, 15 d, 20 d	Lemna gibba	Developmental, physiological	Growth parameters, root elongation and frond area, water content, total chlorophylls, accumulation of GA$_3$ and production of IAA, activity of antioxidant enzymes: catalase (CAT), glutathione peroxidase (GPX) and glutathione reductase (GR)	[38]

TABLE 4.1 (Continued)

S. No.	Type of NP	Size of NP	Dose of NP	Exposure time	Model plant	Toxicity assay	Parameters studied	Reference
6	MWCNTs, Ag, Cu, ZnO NPs	Not specified	500, 100 and 1000 mg l^{-0}	15 d, 17	Cucurbita pepo	Developmen	Seed germination, root elongation, biomass and transpiration volume	[132]
7	ZnO NPs	30 nm	10, 50, and 100 mg l^{-1}	72 h	Pseudokirchneriella subcapitata	Flow cytometer	Algal cell counts	[39]
8	ZnO NPs	10 nm	500–4000 mg l^{-1}	15 d	Prosopis juliflora-velutina	Physiological, micro X-ray fluorescence (µXRF)	Bioaccumulation, specific activity of catalase (CAT) and ascorbate peroxidase (APX)	[53]
9	ZnO NPs, MPs	ZnO NPs < 50 nm, ZnO MPs < 5 µm	10–2000 mg l^{-1}	5 d	Fagopyrum esculentum	Developmental, physiological	Plant growth, bioaccumulation, and antioxidative enzyme activity, reactive oxygen species (ROS) generation by ZnO NPs was estimated as the reduced glutathione level and catalase activity	[78]
10	ZnO NPs	< 100 nm	200, 500, 1000, 1500 mg l^{-1}	96 h	Brassica juncea	Developmental, physiological	Plant growth, biomass and bioaccumulation. Proline content and lipid peroxidation, estimation of the antioxidant enzyme CAT, APX, GR, SOD, generation of ROS and photosynthetic pigments	[119]

TABLE 4.1 (*Continued*)

S. No.	Type of NP	Size of NP	Dose of NP	Exposure time	Model plant	Toxicity assay	Parameters studied	Reference
11	ZnO, TiO$_2$ and Ni NPs	Nano-ZnO <100 nm, nano-TiO$_2$ <21 nm, nano-Ni <100 nm	Aq. solution: 10, 100, and 1,000 mg l^{-1}, bulk counterpart: 100 mg l^{-1}. Soil: 10, 100, and 1,000 mg kg^{-1}. Bulk counterpart: 100 mg kg^{-1}	3 d	*Lepidium sativum*	Developmental	Germination inhibition and root elongation in aqueous solutions and in soil	[59]
12	Co and ZnO NPs	50 nm	5, 10, 20 µg ml^{-1}	3 d	*Allium cepa*	Developmental	Root elongation, root morphology and cell morphology	[45]
13	CuO NPs and ZnO NPs	CuO <50 nm, ZnO <100 nm, bulk CuO 8000–9000 nm, bulk ZnO <1000 nm	500 mg Cu and Zn/kg	14 d	*Triticum aestivum*	Developmental, physiological	Root/shoot growth, lipid peroxidation, oxidized glutathione content, chlorophyll content, peroxidases and catalase activities, and ROS production	[28]
14	CuO NPs and ZnO NPs	50 nm	10, 50, 100, 500, 1000 mg l^{-1}	5 d	*Cucumis sativus*	Developmental, physiological	Bioaccumulation in plants, biomass, ROS production, and enzyme activities such as SOD, CAT, and POD	[66]

TABLE 4.1 (Continued)

S. No.	Type of NP	Size of NP	Dose of NP	Exposure time	Model plant	Toxicity assay	Parameters studied	Reference
15	CuO NPs	50 ± 20 nm	2000 mg l^{-1}	5 d	Glycine max, Cicer arietinum	Developmental	Seed germination, shoot and root growth, dry matter yield	[2]
16	CuO NPs	<50 nm	0.5, 1, 1.5 mM	7 d, 14 d	O. sativa	Developmental, physiological	Seed germination, root length, root weight, shoot length and shoot weight; root cells viability; membrane damage, in vivo ROS production, foliar H_2O_2 and proline accumulation, activity of APX, SOD, GR, DHAR, MDAR; carotenoid, ascorbate and glutathione contents	[127]
17	CuO NPs	6–8 nm	0.05, 0.1, 0.5, 1 mg l^{-1}	15 d	Vigna radiata	Developmental, physiological	Root length, shoot length, fresh weight, dry weight; pigment contents, anti-oxidative activities and photosynthetic activity	[114]
18	CuO, NiO, TiO$_2$, Fe$_2$O$_3$, Co$_3$O$_4$ NPs	TiO$_2$ NPs 30–50 nm, Fe$_2$O$_3$ NPs 20–40 nm, CuO NPs 30–50 nm, NiO NPs 30 nm, Co$_3$O$_4$ NPs 10–30 nm	CuO:20–100 or 100–500 mgl^{-1}, NiO; 20–100 or 200–1000 mg l^{-1}, TiO$_2$, Fe$_2$O$_3$, Co$_3$O$_4$: 1000–5000 mg l^{-1}	3 d	Lactuca sativa, R. sativus, C. sativus	Developmental	Seed germination and root elongation assay. Small seeds (i.e., lettuce) were the most sensitive to CuO and NiO NPs	[147]

TABLE 4.1 (Continued)

S. No.	Type of NP	Size of NP	Dose of NP	Exposure time	Model plant	Toxicity assay	Parameters studied	Reference
19	MWCNTs	Not specified	10, 20, 40 µg ml^{-1}	20–27 d	S. lycopersicum	Developmental	Germination and growth rates	[65]
20	Al$_2$O$_3$, SiO$_2$, Fe$_3$O$_4$, and ZnO NPs	nAl$_2$O$_3$ 150 nm, nSiO$_2$ 42.8 ± 3.9 nm, nFe$_3$O$_4$ < 50 nm, nZnO 44.4 ± 6.7 nm, Larger ZnO particles-820±8 nm	400, 2000, and 4000 mg l^{-1}	18 d	Arabidopsis thaliana	Developmental	Seed germination, root elongation and number of leaves	[76]
21	γ-Fe$_2$O$_3$ (IONPs)	6 nm	500, 1000, and 2000 mg l^{-1}	5 d, 14 weeks	O. sativa L. var. Koshihikari	Developmental, Physiological	Seed germination, root/shoot growth, dry weight of roots, shoots, and grains. Chlorophyll content, SPAD indices, stomatal conductance, intercellular CO$_2$ concentration, and transpiration rate	[5]
22	Si NPs	14, 50, 200 nm	250 and 1000 mg l^{-1}	3, 6 week	A. thaliana	Developmental	Reduced rosette diameter, biomass, and stem length. Root length, chlorosis	[129]

TABLE 4.1 (Continued)

S. No.	Type of NP	Size of NP	Dose of NP	Exposure time	Model plant	Toxicity assay	Parameters studied	Reference
23	ZnO, CeO$_2$ NPs	7 nm	0, 500, 1000, 2000, and 4000 mg l^{-1}	6–9 d	Medicago sativa, Zea mays, C. sativus, S. lycopersicum	Developmental	Germination rate, biomass production, shoot and root elongation	[87]

TEM: transmission electron microscopy, Ag NPs: silver nanoparticles, SPAD: soil plant analysis development, NPs: nanoparticles, APX: ascorbate peroxidase, GR: glutathione reductase, SOD: superoxide dismutase, Ni NPs: nickel nanoparticles, NiO NPs: nickel oxide nanoparticles, Co NPs: cobalt nanoparticles, CuO NPs: copper oxide nanoparticles, DHAR: dehydroascorbate reductase, MDAR: monodehydroascorbate reductase, POD: peroxidase, TiO$_2$: titanium dioxide, Fe$_2$O$_3$: iron(III) oxide, Co$_3$O$_4$: cobalt(II, III) oxide, Si NPs: silicon nanoparticles, CeO$_2$ NPs: cerium oxide nanoparticles, GA$_3$: gibberellic acid, IAA: Indole-acetic acid, ZnO NPs: zinc oxide nanoparticles, TiO$_2$ NPs: titanium dioxide nanoparticles

4.3.1   SILVER NANOPARTICLES (Ag NPs)

Among the various metal oxide NPs, silver nanoparticles (Ag NPs) is by far the most well studied NP whose toxicity has been tested in different crop systems. Silver NPs may readily transform into the environment through various pathways,[80] thus posing a significant potential risk to the environment. Ag NPs of size 29 nm exerted visible reduction effects on the germination of cucumber and lettuce seeds.[12] Phytotoxicity of Ag NPs on *Oryza sativa* was studied by directly exposing 1000 µg ml^{-1} concentrations that leads to adverse effects on plant species. Penetration of large particles, breakage in cell wall and vacuole and intracellular depositions was also observed in plant cells.[96] The toxic effects of Ag NPs were also described for *Lolium multiflorum*.[151] Similarly, phytotoxicity effect of Ag NPs was studied on aquatic plant *Lemna minor* showing significant inhibition of plant growth after exposure to small (20 nm) and larger (100 nm) Ag NPs at lower concentrations (5 mg l^{-1}), and this effect became more acute with a longer exposure time.[49] In addition, El-Temsah and Joner[34] have reported reduced seed germination of ryegrass barley exposed to Ag NPs. The toxicity of the silver and platinum nanoparticles (Pt NPs) was assessed on *Solanum lycopersicum* and *Raphanus sativus*. Interestingly, Pt NPs exhibit excellent germination index and do not cause harmful effects whereby Ag NPs significantly reduced root elongation and germination index in both plants.[128]

The application of NPs in soil is a matter of great importance to elucidate the terrestrial toxicity of NPs as phytotoxicity, bioaccumulation, and dissolution of NPs influence by exposure medium. The toxicity and effect of exposure media of Ag NPs in the crop plant species *Phaseolus radiatus* and *Sorghum bicolor* were evaluated in both agar and soil. The seedling growth of test species was adversely affected by exposure to Ag NPs.[77] In the agar tests, both species showed a concentration-dependent growth inhibition effect while the properties of NPs have been shown to change in soil. The lower bioavailability of Ag NPs corresponds to less toxicity of Ag NPs in soil than that in agar. On the other hand, the growth responses of *Lemna gibba* exposed to different concentrations of Ag NPs and ZnO NPs have shown to be contrasting.[38] Ag NPs showed acute toxicity and caused significant inhibition of growth rate, root elongation, and frond area in a dose-dependent manner, while ZnO NPs showed stimulating effect on both root growth and frond area. Total chlorophyll and gibberellic acid (GA_3) concentration were significantly decreased progressively with increasing concentration of Ag NPs; while Indole-acetic acid (IAA) accumulation and activity of antioxidant enzymes catalase (CAT), glutathione peroxidase (GP) and glutathione reductase (GR) were stimulated.[38]

The toxicity effect of Ag NPs could not be simply attributed to the release of metal ions from the nanomaterial. The toxic effect of Ag NPs was more pronounced than bulk silver solutions in agricultural plant *Cucurbita pepo*.[101] Similarly, the exposure to five NPs (multi-walled carbon nanotubes (MWCNTs), Ag, Cu, ZnO, and Si) significantly reduced root elongation, biomass, and transpiration in the *C. pepo* as compared to their corresponding bulk counterparts, while seed germination remained unaffected.[132] Additionally, the cytotoxic effects of Ag NPs have been reported in *Allium cepa* showing chromatin bridge, stickiness, disturbed metaphase, and other such effects.[72]

4.3.2 TITANIUM-OXIDE NANOPARTICLES (TIO_2 NPs)

Both growth-promoting and inhibitory effects were demonstrated for TiO_2 NPs on plants. An overall toxic effect of TiO_2 NPs was found in the algal species *Desmodesmus subspicatus*.[55] TiO_2 NPs can inhibit leaf growth and transpiration via impairing root water transport.[9] Application of TiO_2 NPs caused various inhibitory effects on the transpiration rate, seed germination, and root elongation of willow cuttings.[125] TiO_2 and ZnO NPs negatively affect wheat growth and soil enzyme activities.[31] A concentration-dependent alteration in seed germination, seedling development, mitotic activity, and chromosomal aberrations was observed as an effect of nano-TiO_2 in *Vicia narbonensis* and *Zea mays*, indicating genotoxic effects of NPs.[22] In addition, the biological effects of various metal oxide NPs (copper oxide (CuO), nickel oxide (NiO), TiO_2, iron(III) oxide (Fe_2O_3), and cobalt(II, III) oxide (Co_3O_4) were assessed on *Lactuca sativa*, *R. sativus* and *Cucumis sativus* based on the seed germination and root elongation.[147] Only CuO and NiO showed deleterious impacts. Phytotoxicity of TiO_2, Fe_2O_3, and Co_3O_4 to the tested seeds was not significant, and Co_3O_4 NP solution was proven to improve root elongation of radish seedling. It was discovered that smaller sized seeds such as lettuce seeds are more sensitive to toxic NPs.

On the other hand, nano-sized TiO_2 was reported to enhance nitrogen photo reduction/metabolism and photosynthesis by altering ribulose-1,5-bisphosphate carboxylase activase activity, and thus improving spinach growth.[43,79,149,155] The positive impact of nano-SiO_2 and nano-TiO_2 mixture was demonstrated in *Glycine max*, and enhancement in seed germination and seedling growth along with increased nitrate reductase activity, absorption potential and antioxidant system were observed.[90] Recently, the impact of TiO_2 and ZnO NPs of similar size (25 ± 3.5 nm) over a range of concentrations (0–1000 mg kg^{-1}) was evaluated mechanistically for uptake, translocation, and accumulation of NPs in *Solanum lycopersicum*.[117] The results indicated improvement in plant's growth and development up to a critical concentration of TiO_2 and ZnO NPs, along with enhanced fruit yield, nutritional quality, and lycopene content. Aerosol-mediated application was found to be more effective than the soil-mediated application on the uptake of the NPs in plants.

4.3.3   ZINC OXIDE NANOPARTICLES (ZnO NPs)

To date, several reports have described the toxic effect of ZnO NPs on plants. Toxicity experiments using the freshwater alga *Pseudokirchneriella subcapitata* revealed comparable toxicity for nanoparticulate ZnO, bulk ZnO, and $ZnCl_2$, attributable solely to dissolved zinc.[39] The developmental phytotoxicity is exerted by five types of NPs: MWCNTs, nAl, nAl_2O_3, nZn, and nZnO and were investigated on six different plant species, i.e., radish, rape, ryegrass, lettuce, corn, and cucumber to address the effect of elemental composition.[84] The study reported significant inhibition of ryegrass germination by nZn and corn germination by nZnO or nAl_2O_3, whereas no inhibition was observed for MWCNTs. Interestingly, nAl caused both positive and negative effects on root elongation, depending on the plant species.[84] Another study on rye-grass indicated that both the seed germination and root growth were inhibited by the Zn and ZnO, and NPs phytotoxicity cannot be explained only on the basis of dissolution of ZnO NPs from bulk materials alone.[85] Similar research was undertaken on the toxicology of ZnO NPs on *Arabidopsis thaliana*, and the results showed that nano-ZnO at 400 mg l^{-1} could inhibit germination.[76] Phytotoxicity of Co and ZnO NPs was investigated using the roots of *A. cepa* as an indicator organism.[46] With increasing concentrations of both types of NPs, the elongation of the roots was severely inhibited. Zinc oxide NPs caused damage due to severe accumulation in both the cellular and the chromosomal modules, thus signifying their highly hazardous phytotoxic nature. Similarly, Boonyanitipong et al.[17] investigated the toxicity of ZnO and TiO$_2$ NPs on germinating rice seedlings revealing more toxicity of nano-ZnO on root growth and development as compared to TiO$_2$. In another study, phytotoxicity of CuO and ZnO NPs was investigated in *Cucumis sativus*, which showed significant reduction in seedling biomass and increase in superoxide dismutase (SOD), CAT, and peroxidase (POD) activities by both CuO and ZnO NPs.[66] The effects of ZnO NPs and micropar-ticles (MPs) on plant growth, biomass, bioaccumulation, and antioxidative enzyme activity in pseudocereal food crops and medicinal herb *Fagopyrum esculentum* (buckwheat) were estimated under hydroponic culture revealing significant biomass reduction and reactive oxygen species (ROS) genera-tion along with reduced glutathione level and catalase activity.[78] Similarly, a significant reduction in plant growth and biomass was recorded with gradual increase in proline content, antioxidant enzyme activities and lipid peroxidation upon exposure to ZnO NPs in *Brassica juncea*.[119] In contrast, Hernandez-Viezcas et al.[53] recently reported the effect of ZnO NPs in a wild desert velvet mesquite plant (*Prosopis juliflora-velutina*). The results of this study indicate that mesquite plants absorbed Zn from the NPs treatments. Zinc oxide increased the specific activity of stress enzymes CAT in root, stem, and leaves; and APX only in stem and leaves. However, mesquite plants showed no chlorosis, necrosis, stunting or wilting, even after 30 days of treatment, suggesting that this desert plant may display some tolerance to ZnO NPs and to the zinc ion released from these particles.

4.3.4 COPPER OXIDE NANOPARTICLES (CuO NPs)

Response to copper stress has been studied extensively over the past decade at physiological, molecular, and proteomic levels.[4,133,134] Nevertheless, the impact of CuO NPs on plants is not at all well explored. Copper oxide NPs mediate DNA damage in terrestrial plants.[10] The impact of commercial CuO NPs (<50 nm) and ZnO NPs (<100 nm) was demonstrated on sand-grown wheat.[28] Amendment of the sand with CuO and ZnO NPs signif-icantly reduced root growth, but only CuO NPs impaired shoot growth. Oxidative stress in the NP-treated plants was evidenced by increased lipid peroxidation, oxidized glutathione, increased production of ROS, higher POX and CAT activities and decreased chlorophyll content. The effect of CuO NPs (< 50 nm) was also described for germination and growth of seeds of soybean and chickpea.[2] Results showed that germination was not prevented up to 2000 ppm CuO, but the root elongation was severely inhibited with increasing concentration of CuO NPs and exhibited necrosis due to massive adsorption on root surface. It was found that the accumulation and uptake of NPs were dependent on the exposure concentrations. Phytotoxicity of nano-CuO was assessed on a monocot-rice (*O. sativa* cv. Swarna) seedling in terms of growth performance and modulation of physiological parameters.[127] Results showed significant reduction in seed germination under nano-copper stress along with the loss of root cells viability, enhanced proline accumulation and membrane damage, reduction in DHAR activity and carotenoids level and severe oxidative damage, as revealed by increase in ROS, antioxidant enzymes activities, and contents of antioxidant metabolites of ascorbate–glutathione cycle. On the other hand, CuO NPs treatment did not impart any toxicity to the plant system including morphological and/or physiological alterations in mung bean, and enhanced photosynthetic activity and carbon assimilatory pathway.[114]

4.3.5   ALUMINA NANOPARTICLES (AL$_2$O$_3$ NPs)

The use of alumina nanoparticles (Al$_2$O$_3$ NPs) may give rise to environmental toxic response. In this context, Doshi et al.[30] have evaluated ecotoxicological assessment of two types of nano-alumina injected into a sand column. Alumina NPs showed inhibition of the root elongation and growth of five crop species: corn, cucumber, soybean, cabbage, and carrot in hydroponic culture medium.[149] Conversely, phytotoxicity evaluation of four different metal oxide NPs, aluminum oxide (nAl$_2$O$_3$), silicon dioxide (nSiO$_2$), magnetite (nFe$_3$O$_4$) and zinc oxide (nZnO) on the development of *A. thaliana* showed maximum toxic effects of ZnO on seed germinations and root elongations, followed by Fe$_3$O$_4$ and SiO$_2$, whereas Al$_2$O$_3$ NPs exhibited no toxicity even at the concentration of 4000 mg l^{-1}, in respect to root elongation and the development.[76]

4.3.6   SILICA NANOPARTICLES (SiO$_2$ NPs OR SiNPs)

As for other NPs, silica nanoparticles (SiO$_2$ NPs or SiNPs) also showed both positive and negative effects on plants. Silica NPs enhanced the growth of Changbai larch (*Larix olgensis*) with increasing NPs concentration.[83] Likewise, an increase in shoot/root ratio has been reported in *L. sativa* plants after exposure to SiNPs.[126] Gold-capped mesoporous silica nanoparticles (MSNs) were found to be able penetrate cell wall and deliver DNA into plant cell.[136] Nair et al.[104] confirmed uptake of the fluorescein isothiocyanate (FITC)-labeled SiNPs into rice seedlings and determined that the particles had no effect on germination up to 50 mg l^{-1}. Phytotoxicity assays with *C. pepo* showed no significant difference in germination percent, root elongation, or biomass after exposure to SiNPs.[132] On the other hand, the phytotoxicity of SiNPs was evaluated as a function of particle size (14, 50, and 200 nm), concentration (250 and 1000 mg l^{-1}), and surface composition in *A. thaliana* plants.[129] Reduced development and chlorosis were observed for plants exposed to highly negative SiNPs (−20.3 and −31.9 mV for the 50 and 200 nm SiNPs, respectively) regardless of particle concentration. The phytotoxic effects observed for the negatively charged 50 and 200 nm SiNPs were attributed to pH effects and the adsorption of macro- and micro-nutrients to the silica surface. In another study, the toxicity of SiNPs to green alga *Pseudokirchneriella subcapitata* and *Scenedesmus obliquus* was shown to reduce the growth by 20% and decrease chlorophyll content.[54,145]

4.3.7　CARBON NANOTUBES (CNTs)

There is an extensive interest to investigate the ability of NPs to penetrate plant cell walls and work as smart delivery systems in plants. Several research groups reported that different types of NPs are able to penetrate plant cell walls. Carbon nanotubes were found to penetrate tomato seeds and enhance their germination and growth rates.[63] Similarly, single-walled carbon nanotubes (SWCNTs) can also penetrate the cell wall and cell membrane of tobacco cells.[86] Carbon nanotubes toxicity and risk assessment was reviewed by Lam et al.[74] The effects of functionalized and non-functionalized SWCNTs were investigated on root elongation of six crop species—cabbage, carrot, cucumber, lettuce, onion, and tomato.[20] Both CNTs and functionalized carbon nanotubes (fCNTs) affected root elongation of crop species, but phytotoxicity varied between CNTs and fCNTs. On the other hand, MWCNTs exerted positive effects through germination and seedling growth of *B. juncea* and *Phaseolus mungo*.[45]

4.3.8 RARE EARTH OXIDE NANOPARTICLES

Some rare earth oxide NPs severely inhibited the root elongation of higher plant species.[91] Phytotoxicity of nanoceria (CeO_2 NPs) (0–4000 mg l^{-1}) was determined on *Medicago sativa, Z. mays, C. sativus, S. lycopersicum* and *G. max* showing significantly reduced seed germination (about 20–30% at 2000 mg l^{-1}) and genotoxic effects.[88,87] The root growth was significantly promoted by nanoceria in cucumber and corn but reduced in alfalfa and tomato. At almost all concentrations, nanoceria promoted shoot elongation in the four plant species. These results clearly demonstrated species-specific toxic effects of CeO_2 NPs.

4.3.9 OTHER NANOPARTICLES

Influence of soil type and environmental conditions were determined for the Ni NPs phytotoxicity.[59] Various methods for assessment of toxicity were applied to compare the phytotoxicity of three ENPs: nano-Ni, nano-ZnO, and nano-TiO_2 on *Lepidium sativum* in both aqueous medium and soil.[60] Depending on the kind of ENPs, concentration, and the matrix, the phytotoxicity (measured in terms of root elongation and seed germination) differed between water and soil. Conversely, nickel(II) hydroxide ($Ni(OH)_2$) NPs showed no adverse effect on growth of mesquite plant,[112] whereas manganese nanoparticles (Mn NPs) have been used to enhance the photosynthetic efficacy in mung bean system.[113] The phytotoxicity potential of 6 nm γ-Fe_2O_3 nanoparticles (IONPs) was determined in terms of root elongation and the physiological performance, mainly photosynthetic parameters of rice plants (*O. sativa* L. var. Koshihikari), demonstrating the lower toxicity of nanosized iron oxide compared to a microsized preparation under reductive conditions.[5]

4.4 MOLECULAR MECHANISM OF PLANT-NANOPARTICLE INTERACTIONS

The modes of action of NPs on cellular structures are multiple, complex, and still poorly understood. The most commonly described mechanisms of NP toxicity in plants include cell surface coating causing mechanical damage or clogging of pores,[9,26] increased production of ROS causing oxidative stress,[98] and release of toxic metal ions.[26,98] However, little progress has been made in studying ENP-plant interaction at molecular level.

The ability of the NPs to cross barriers and their interaction with subcellular structures due to their small size and high surface reactivity contribute to potential cellular toxicity. It is also critical to note that NPs upon dissolution act as metal ions, able to interact with the sulfhydryl and carboxyl groups of proteins and alter their activity. They can transfer electrons to molecular oxygen and cause cytotoxicity via the generation of most destructive ROS.[89] Plants have special mechanisms to remove or inactivate ROS such as H_2O_2, OH^-, and O^{2-} radicals; however, excess ROS can cause damage to cellular components such as protein breakdown, lipid peroxidation in membranes, and DNA injury resulting in necrosis and apoptosis. Thus, the oxidative stress potential in plants might be a better indicator of phytotoxicity in plants exposed to NPs. Up till now, very little data exist regarding the effects of ROS in relation to NPs on the plant cells.[26,108]

Analysis of the changes in gene expression through high-throughput methods such as complementary DNA microarrays or quantitative real-time PCR (qRT-PCR) constitutes a powerful approach for understanding the mechanisms of toxicity and molecular responses in cells exposed to nanomaterials.[148] Gene expression analyses are typically conducted to complement morphological and/or physiological investigations and provide unique information on specific toxicity pathways and modes of action that cannot be obtained directly from other approaches.[7]

Although transcriptional analyses have been widely used to study the molecular basis of NP toxicity in a variety of organisms including microbes, humans, mammalian cell lines, and other model organisms,[8] only limited investigations have been conducted to assess the molecular mechanism of the ENP-plant interactions and NP phytotoxicity. For example, gene expression analyses of the model plant *A. thaliana* by RT-PCR have provided new insights into the molecular mechanisms of plant response to Ag NPs. Dimkpa et al.[29] investigated that exposure of commercial Ag NPs to wheat in a sand growth matrix induced plant defense response in *Arabidopsis* plants, as revealed by significant upregulation of pathogenesis-related (*PR1, PR2, and PR5*) genes involved in systemic acquired resistance (SAR).[23] Similarly, the transcriptional response of *A. thaliana* exposed to Ag NPs were analyzed using whole genome cDNA expression microarrays, which resulted in upregulation of 286 genes and downregulation of 81 genes as compared to the control.[62] Real-time PCR analysis showed significant transcriptional modulation of genes involved in sulfur assimilation and glutathione biosynthesis ,i.e., adenosine triposphate sulfurylase (*ATPS*), 3'-phosphoadenosine 5'-phosphosulfate reductase (*APR*), sulfite reductase (*SiR*), cysteine synthase (*CS*), glutamate-cysteine ligase (*GCL*), glutathione synthetase (*GS2*) with upregulation of glutathione S-transferase (*GSTU12*), glutathione reductase (*GR*), and phytochelatin synthase (*PCS1*) genes.[105] In another study by the same group, the expression of cell cycle genes proliferating cell nuclear antigen (*PCNA*) and DNA mismatch repair (*MMR*) were found to be modulated as the results of oxidative stress caused by Ag NPs exposure in *A. thaliana*.[106] MicroRNAs (miRNAs) are a newly discovered post-transcriptional gene regulators, which belong to small endogenous class of noncoding RNAs (□ 20–22 nt). Interestingly, miRNAs have also been shown to play an important role in plant response to NPs by regulating gene expression. In a study by Frazier et al.,[40] nano-TiO_2 exposure to tobacco plants significantly affected the expression profiles of miRNAs, with *miR395* and *miR399* exhibiting the greatest fold changes of 285-fold and 143-fold, respectively.

Recently, García-Sánchez et al.[44] evaluated the transcriptome changes of *A. thaliana* in response to the different types of NPs (metallic TiO_2 and Ag; carbonaceous MWCNTs). A set of 16 comparable transcriptome profiles were produced to monitor early changes in gene expression upon NP and stress exposure, resulting in downregulation of a significant number of genes involved in the responses to microbial pathogens and activation of phosphate-starvation, and root hair development genes. Similarly, Landa et al.[75] reported gene expression changes upon long-term exposure to TiO_2 NPs, ZnO NPs and fullerene soot (FS) using oligonucleotide microarrays. Nano-ZnO was found to be the most toxic and resulted in 660 up- and 826 downregulated genes, whereas FS caused differential gene expression with 232 up- and 189 downregulated genes. Only mild changes in gene expression were observed upon TiO_2 NP exposure, which resulted in only 80 upregulated and 74 downregulated genes, mainly involved in responses to biotic and abiotic stimuli. Khodakovskaya et al.[65] have demonstrated that the growth of tobacco cell culture (callus) can be highly enhanced by the introduction of MWCNTs in the growth medium and is directly correlated to the overexpression of marker genes for cell division (*CycB*), cell wall extension (*NtLRX1*), and water transport (aquaporin gene *NtPIP1*) in tobacco cells exposed to MWCNTs. Similarly, the observed physiological responses of tomato plants exposed to MWCNTs were linked with the complex sets of information provided by microarray analysis showing upregulation of tomato aquaporin gene *LeAqp2* and a number of other genes related to plant responses to environmental stress.[64] On the other hand, Lahiani et al.[73] demonstrated that MWCNTs can activate seed germination and growth of seedlings in barley, corn, and soybean by stimulating expression of different gene families of aquaporins such as plasma membrane intrinsic proteins (PIP), tonoplast intrinsic proteins (TIP), and small and basic intrinsic proteins (SIP). Recently the functions and the regulatory networks of the genes involved in cadmium sulfide (CdS) QDs uptake, translocation, detoxification, and accumulation were determined using both a genome-wide top-down and a bottom-up approach.[94] Similar molecular studies were performed to understand the toxic effects of exposure to other NPs in plants. The graphene oxide (GO) exposure combined with drought/salt stress induced severe alterations in the expression patterns of genes required for root development (*PIN2, PIN7, SHR,* and *SCR*) and abiotic stress (*ABI4, ABI5, AREB1, HKT1, SOS1, RD29A*) in *Arabidopsis* seedlings.[144]

Expression profile of certain miRNAs (*miR395, miR397, miR398, and miR399*) was significantly upregulated upon exposure of tobacco plants to Al_2O_3 NPs,[18] providing the ability of plants to withstand stress to Al_2O_3 NPs in the environment. In addition, Ma et al.[92] evaluated molecular responses triggered by exposure of cerium oxide (CeO_2) and indium oxide (In_2O_3) NPs on *Arabidopsis* showing alteration of gene expression related to sulfur metabolism, glutathione (GSH) biosynthesis pathway and detoxification of metal toxicity.

Proteomics is emerging as a powerful technique to be applied to the field of crop abiotic stress tolerance research.[11] Studies of cellular reactions at protein level upon NPs exposure can significantly contribute to our understanding of physiological mechanisms underlying plant-NP interaction. Nanotechnology-based proteomic biomarker development is still in its infancy, and as yet only a few proteomic studies have examined the effects of NPs on different organisms. It has been used to determine cytotoxicity of different NPs for bacteria, fungi, rats, mice, *Daphnia*, and human cell lines.[1] However, in plants, the reports describing proteomic studies in response to NP exposure are very rare, and mostly deal with investigation of Ag NPs phytotoxicity.

Proteomic responses of *Eruca sativa* roots exposed to Ag NPs revealed accumulation of proteins related to sulfur metabolism, stress/defense response, cell cycle, protein folding, transport and activation of enzymatic and non-enzymatic pathways of ROS detoxification machinery and down-regulation of binding protein 1 (BiP1), heat shock protein 70–2, and two vacuolar-type proton ATPase (VATPase) subunits.[139] Similarly, gel-based proteomics analysis of *O. sativa* identified a total of 28 responsive proteins that change in abundance upon exposure to different concentrations of Ag NPs colloidal suspension.[99] The identified proteins were involved in normal cell metabolic processes such as transcription, protein synthesis/degradation, cell division, and apoptosis, along with proteins related to Ca^{2+} regulation and signaling, oxidative stress tolerance and direct damage to cell walls and DNA/RNA/proteins. In another study, genomic and proteomic changes induced by Ag NPs were analyzed in wheat seedlings using amplified fragment length polymorphism (AFLP) technique and two-dimensional electrophoresis 2-DE coupled with liquid chromatography LC–electrospray ionization ESI–tandem mass spectroscopy MS/MS[140] revealing altered expression of proteins involved in primary metabolism, protein synthesis/folding, cell defense and stress responses in multiple cellular compartments. Recently, Ag NPs-induced changes in the proteome profiles of roots and cotyledons of soybean exposed to flooding stress were evaluated using a gel-free proteomic technique to elucidate the underlying mechanism of NP-mediated growth promotion of soybean under flooding stress.[102] Silver NPs primarily affected the abundances of 107 root proteins predominantly associated with stress, signaling, and cell metabolism.

4.5   CONCLUSIONS AND FUTURE PROSPECTS

Concerns over the potential hazardous effects of NPs has stimulated increasing amount of research on ecotoxicity. The present review summarizes the limited conflicting data regarding the NPs toxicity and accumulation in various plants. The phytotoxicity profile of NPs has been investigated mostly via seed germination and root elongation tests as suggested by U.S. Environmental Protection Agency, which evaluated the acute effects of NPs on plant metabolism. However, these standard phytotoxicity tests may not be sensitive enough when evaluating chronic toxicity to plant species. On the other hand, impacts of NPs on plant development and metabolism has been found to be highly dependent on both plant type and NP properties such as nature/composition of particles, size, concentration and exposure time.[26] The morphology, surface characteristics such as area and charge, coating, purity and particle solubility also play important roles in the fate and toxicity of the NPs in plant systems.[12] In addition, the degree of aggregation/agglomeration of ENPs strongly influences the availability of NPs for uptake into cells.[82]

The reports from few recent studies have advanced our knowledge of toxicological impact of several types of nanomaterials. There are still many unresolved issues and challenges concerning the biological effects of NPs. For a full estimation of the risk related with the presence of NPs in the environment, the use of the current ecotoxicological methods may prove insufficient. The studies indicate sometimes extremely different effects of the ENPs on the plants in relation to the method applied. This may lead to erroneous estimations, either underestimating or overestimating the potential threat related with ENPs. This necessities further research on the potential impacts of manufactured NPs on agricultural and environmental systems. The development of universal methods for the estimation of toxicity of ENPs would contribute to the assessment of the scale of the problem in various regions of the world and for the development of legislative regulations. Clearly, more work is needed to be done to clarify the ecotoxicological effects of NP exposure in soils and under field conditions, as well as to characterize potential risk associated with food chain contamination through agricultural species. Further studies need to be focused on the effects of concentrations and the mode of action of different types of NPs on various plant species, the uptake mechanism and translocation of NPs.

High-throughput "omics" techniques such as transcriptomics, proteomics, and metabolomics may provide new insights into the biochemical and molecular responses of an organism and play a key role in understanding the mechanisms of cellular toxicity of ENPs, as well as other environmental contaminants. These profiling techniques could be used to support aspects of regulatory decision making in ecotoxicology. However, these techniques have many parallel challenges with regard to data collection, integration, and interpretation, and mostly rely on advanced expertise and expensive resources. Although tools exist to detect alterations in transcriptome or proteome profile in various organisms, the genomes of most plant species have not been fully sequenced and annotated till date. Ongoing genome sequencing projects, in long term, would obviate issues related to the global identification of gene products in key test species used for ecotoxicological risk assessments.

Nevertheless, future perspectives on NP-plant interaction will depend on a thorough understanding of the molecular mechanisms responsible for the specific response triggered by engineered nanomaterials. Along with a boost of new methodological approaches, it could be expected that these approaches would contribute to a detailed characterization towards better understanding of NP phytotoxicity.

4.6   SUMMARY

The rapid development and widespread utilization of NPs (NPs) in various fields have raised considerable environmental concerns because they are often released into the environment and pose toxicity risk. As plants are essential components of the food chain, the potential uptake of NPs by agricultural plants have important implications for the human health. Therefore it is necessary to address environmental fate, uptake and potential phytotoxicity of NPs. In addition, an important aspect of the risk assessment of nanomaterials is to understand the NP-plant interactions. In this chapter, a brief summary of the current available literature have been presented for NP-mediated developmental and physiological phytotoxicity and the possible molecular mechanism of NP-plant interactions. Both positive and inhibitory effects of NPs on plant growth and development have been documented. Future perspectives have also been discussed to refine our knowledge on this topic.

4.7   ACKNOWLEDGMENTS

The author gratefully acknowledges Science and Engineering Research Board, Department of Science and Technology, Government of India for DST-SERB Young Scientist grant (SB/YS/LS-39/2014), University Grants Commission, India for UGC start-up grant (F.30-50/2014-BSR), UGC-Special Assistance Program (UGC-SAP-CAS) and DST-FIST in Department of Botany, J. N. V. University, Jodhpur. *The author declares no financial or commercial conflict of interest.*

KEYWORDS

- Agri-nanotechnology
- developmental phytotoxicity
- engineered nanoparticles
- ENP-plant interactions
- food packaging
- gene expression
- genetic engineering
- mass spectrometry
- metabolomics
- microarray
- nano ecotoxicology
- nanobiocomposite
- nanobioremediation
- nanodevices for DNA delivery
- nanofertilizer
- nanoparticle farming
- nanoparticle phytotoxicity
- nanopesticide
- nano-toxicogenomics
- nano-toxicoproteomics
- next-generation sequencing
- physiological phytotoxicity
- post-harvest management
- quantitative RT-PCR
- risk assessment
- targeted drug delivery
- transcriptomics
- two-dimensional gel electrophoresis

REFERENCES

1. Abdelhamid, H. N.; Wu, H. F. Proteomics Analysis of the Mode of Antibacterial Action of Nanoparticles and Their Interactions with Proteins. *Trends Anal. Chem.* **2015**, *65*, 30–46.
2. Adhikari, T.; Kundu, S.; Biswas, A. K.; Tarafdar, J. C.; Rao, A. S. Effect of Copper Oxide Nano Particle on Seed Germination *of Selected Crops. J. Agric. Sci. Technol.* **2012**, *2*, 815–823.
3. Afrasiabi, Z.; Eivazi, F.; Popham, H.; Stanley, D.; Upendran, A.; Kannan, R. *Silver Nanoparticles as Pesticides.* The Capacity Building Grants Program Project Director's Meeting, National Institute of Food and Agriculture Huntsville: AL., Sept 16–19, 2012.
4. Ahsan, N.; Lee, D. G.; Lee, S. H.; Kang, K. Y.; Lee, J. J.; Kim, P. J.; Yoon, H. S.; Kim, J. S.; Lee, B. H. Excess Copper Induced Physiological and Proteomic Changes in Germinating Rice Seeds. *Chemosphere* **2007**, *67*(6), 1182–1193.
5. Alidoust, D.; Isoda, A. Phytotoxicity Assessment of γ-Fe_2O_3 Nanoparticles on Root Elongation and Growth of Rice Plant. *Environ. Earth Sci.* **2014**, *71*, 5173–5182.
6. Al-Samarrai, A. M. Nanoparticles as Alternative to Pesticides in Management Plant Diseases—A Review. *Int. J. Sci. Res. Publ.* **2012**, *2*(4), 1–4.

7. Ankley, G. T.; Daston, G. P.; Degitz, S. J.; Denslow, N. D.; Hoke, R. A.; Kennedy, S. W.; Miracle, A. L.; Perkins, E. J.; Snape, J.; Tillitt, D. E. et al. Toxicogenomics in Regulatory Ecotoxicology. *Environ. Sci. Technol.* **2006**, *40*, 4055–4065.

8. Asharani, P. V.; Mun, G. L. K.; Hande, M. P.; Valiyaveettil, S. Cytotoxicity and Genotoxicity of Silver Nanoparticles in Human Cells. *ACS Nano* **2009**, *3*, 279–290.

9. Asli, S.; Neumann, P. M. Colloidal Suspensions of Clay or Titanium Dioxide Nanoparticles can Inhibit Leaf Growth and Transpiration via Physical Effects on Root Water Transport. *Plant Cell Environ.* **2009**, *32*, 577–584.

10. Atha, D. H.; Wang, H.; Petersen, E. J.; Cleveland, D.; Holbrook, R. D.; Jaruga, P.; Dizdaroglu, M.; Xing, B.; Nelson, B. C. Copper Oxide Nanoparticle Mediated DNA Damage in Terrestrial Plant Models. *Environ. Sci. Technol.* **2012**, *46*, 1819–1827.

11. Barkla, B. J.; Vera-Estrella, R.; Pantoja, O. Progress and Challenges for Abiotic Stress Proteomics of Crop Plants. *Proteomics* **2013**, *13*, 1801–1815.

12. Barrena, R.; Casals, E.; Colón, J.; Font, X.; Sánchez, A.; Puntes, V. Evaluation of the Ecotoxicity of Model Nanoparticles. *Chemosphere* **2009**, *75*, 850–857.

13. Bhagat, D.; Samanta, S. K.; Bhattacharya, S. Efficient Management of Fruit Pests by Pheromone Nanogels. *Scientific Rep.* **2013**, *3*(1294), 1–8.

14. Bhatt, I.; Tripathi, B. N. Interaction of Engineered Nanoparticles with Various Components of the Environment and Possible Strategies for Their Risk Assessment. *Chemosphere* **2011**, *82*, 308–317.

15. Boholm, M.; Arvidsson, R. Controversy over Antibacterial Silver: Implications for Environmental and Sustainability Assessments. *J. Clean Prod.* **2014**, *68*, 135–143.

16. Bondarenko, O.; Juganson, K.; Ivask, A.; Kasemets, K.; Mortimer, M.; Kahru, A. Toxicity of Ag, CuO and ZnO Nanoparticles to Selected Environmentally Relevant Test Organisms and Mammalian Cells in vitro: A Critical Review. *Arch. Toxicol.* **2013**, *87*, 1181–1200.

17. Boonyanitipong, P.; Kositsup, B.; Kumar, P.; Baruah, S.; Dutta, J. Toxicity of ZnO and TiO_2 Nanoparticles on Germinating Rice Seed *Oryza sativa* L. *Int. J. Biosci. Biochem. Bioinf.* **2011**, *1*(4), 282–285.

18. Burklew, C. E.; Ashlock, J.; Winfrey, W. B.; Zhang, B. H. Effects of Aluminum Oxide Nanoparticles on the Growth, Development, and MicroRNA Expression of Tobacco (*Nicotiana tabacum*). *PLoS ONE* **2012**, *7*(5), e34783.

19. Buzea, C.; Pacheco, I. I.; Robbie K. Nanomaterials and Nanoparticles: Sources and Toxicity. *Biointerphases* **2007**, *2*(4), MR17–71.

20. Canas, J.; Long, M.; Nations, S.; Vadan, R.; Dai, L.; Luo, M. et al. Effects of Functionalized and Non-Functionalized Single-Walled Carbon Nanotubes on Root Elongation of Select Crop Species. *Environ. Toxicol. Chem.* **2008**, *27*, 1992–1931.

21. Cao, X.; Habibi, Y.; Magalhães, W. L. E.; Rojas, O. J.; Lucia, L. A. Cellulose Nanocrystals-Based Nanocomposites: Fruits of a Novel Biomass Research and Teaching Platform. *Curr Sci.* **2011**, *100*(8), 1172–1176.

22. Castiglione, M. R.; Giorgetti, L.; Geri, C.; Cremonini, R. The Effects of Nano-TiO_2 on Seed Germination, Development and Mitosis of Root Tip Cells of *Vicia narbonensis* L. and *Zea mays* L. *J. Nanopart. Res.* **2011**, *13*, 2443–2449.

23. Chu, H.; Kim, H. J.; Kim, J. S.; Kim, M. S.; Yoon, B. D.; Park, H. J.; Kim, C. Y. A Nanosized Ag-silica Hybrid Complex Prepared by Gamma-Irradiation Activates the Defense Response in Arabidopsis. *Radiat. Phys. Chem.* **2012**, *81*, 180–184.

24. Cioffi, N.; Torsi, L.; Ditaranto, N. et al. Antifungal Activity of Polymer-Based Copper Nanocomposite Coatings. *Appl. Phys. Lett.* **2004**, *85*(12), 2417–2419.
25. DeRosa, M. C.; Monreal, C.; Schnitzer, M.; Walsh, R.; Sultan, Y. Nanotechnology in Fertilizers. *Nat. Nanotechnol.* **2010**, *5*(2), 91.
26. Dietz, K. J.; Herth, S. Plant Nanotoxicology. *Trends Plant Sci.* **2011**, *16*(11), 582–589.
27. Dimkpa, C. O.; McLean, J. E.; Britt, D. W.; Johnson, W. P.; Arey, B.; Lea, A. S.; Anderson, A. J. Nanospecific Inhibition of Pyoverdine Siderophore Production in *Pseudomonas chlororaphis* O6 by CuO Nanoparticles. *Chem. Res. Toxicol.* **2012a**, *25*, 1066–1074.
28. Dimkpa, C. O.; McLean, J. E.; Latta, D. E.; Manango´n, E.; Britt, D. W.; Johnson, W. P.; Boyanov, M. I.; Anderson, A. J. CuO and ZnO Nanoparticles: Phytotoxicity, Metal Speciation, and Induction of Oxidative Stress in Sand-Grown Wheat. *J. Nanopart. Res.* **2012b**, *14*, 1125.
29. Dimkpa, C. O.; McLean, J. E.; Martineau, N.; Britt, D. W.; Haverkamp, R.; Anderson, A. J. Silver Nanoparticles Disrupt Wheat (*Triticum aestivum* L.) Growth in a Sand Matrix. *Environ. Sci. Technol.* **2013**, *47*, 1082–1090.
30. Doshi, R.; Braida, W.; Christodoulatos, C.; Wazne, M.; O'Connor, G. Nano-aluminum: Transport Through Sand Columns and Environmental Effects on Plants and Soil Communities. *Environ. Res.* **2008**, *106*, 296–303.
31. Du, W.; Sun, Y.; Ji, R.; Zhu, J.; Wu, J.; Guo, H. TiO_2 and ZnO Nanoparticles Negatively Affect Wheat Growth and Soil Enzyme Activities in Agricultural Soil. *J. Environ. Monit.* **2011**, *13*, 822–828.
32. Durán, N.; Marcato, P. D.; Conti, R. D.; Alves, O. L.; Costa, F. T. M.; Brocchi, M. Potential Use of Silver Nanoparticles on Pathogenic Bacteria, Their Toxicity and Possible Mechanisms of Action. *J. Braz. Chem. Soc.* **2010**, *21*(6), 949–959.
33. Elgrabli, D.; Floriani, M.; Abella-Gallart, S.; Meunier, L.; Gamez, C.; Delalain, P.; Rogerieux, F.; Boczkowski, J.; Lacroix, G. Biodistribution and Clearance of Instilled Nanotubes in Rat Lung. *Part. Fibre Toxicol.* **2008**, *5*, 20.
34. El-Temsah, Y. S.; Joner, E. J. Impact of Fe and Ag Nanoparticles on Seed Germination and Differences in Bioavailability During Exposure in Aqueous Suspension and Soil. *Environ. Toxicol.* **2012**, *27*, 42–49.
35. European Commission (EU). Commission Recommendation of 18 October 2011 on the Definition of Nanomaterial (2011/696/EU). *Official J. Euro. Union L.* 275/38. http://eurlex.europa.eu/LexUriServ/LexUriServ.do?uri=OJ:L:2011:275:0038:0040:EN:PDF.
36. European Commission. Regulation (EU) No 1169/2011 of the European Parliament and of the Council of 25 October 2011 on the Provision of Food Information to Consumers. *Official J. Euro. Union L.* 304/26. http://eurlex. europa.eu/LexUriServ/LexUriServ.do?uri=OJ:L:2011:304:0018:0063:EN:PDF.
37. Farkas, J.; Christian, P.; Urrea, J. A. G.; Roos, N.; Hassellöv, M.; Tollefsen, K. E.; Thomas, K. V. Effects of Silver and Gold Nanoparticles on Rainbow Trout (*Oncorhynchus mykiss*) Hepatocytes. *Aquat. Toxicol.* **2010**, *96*, 44–52.
38. Farrag H. F. Evaluation of the Growth Responses of *Lemna gibba* l. (duckweed) Exposed to Silver and Zinc Oxide Nanoparticles. *World Appl. Sci. J.* **2015**, *33*(2), 190–202.
39. Franklin, N. M.; Rogers, N. J.; Apte, S. C.; Batley, G. E.; Gadd, G. E.; Casey, P. S. Comparative Toxicity of Nanoparticulate ZnO, bulk ZnO, and $ZnCl_2$ to a Freshwater Microalga (*Pseudokirchneriella subcapitata*): The Importance of Particle Solubility. *Environ. Sci. Technol.* **2007**, *41*, 8484–8490.

40. Frazier, T. P.; Burklew, C. E.; Zhang, B. H. Titanium Dioxide Nanoparticles Affect the Growth and MicroRNA Expression of Tobacco (*Nicotiana tabacum*). *Funct. Integr. Genomics* **2014**, *14*, 75–83.

41. Gaiser, B. K.; Biswas, A.; Rosenkranz, P.; Jepson, M. A.; Lead, J. R.; Stone, V.; Tyler, C. R.; Fernandes, T. F. Effects of Silver and Cerium Dioxide Micro- and Nano-Sized Particles on *Daphnia magna*. *J. Environ. Monit.* **2011**, *13*, 1227–1235.

42. Galbraith, D. W. Nanobiotechnology: Silica Breaks Through in Plants. *Nat. Nanotechnol.* **2007**, *2*(5), 272–273.

43. Gao, F.; Liu, C.; Qu, C.; Zheng, L.; Yang, F.; Su, M. et al. Was Improvement of Spinach Growth by Nano-TiO$_2$ Treatment Related to the Changes of Rubisco Activase?. *BioMetals* **2008**, *21*, 211–217.

44. García-Sánchez, S.; Bernales, I.; Cristobal, S. Early Response to Nanoparticles in the *Arabidopsis* Transcriptome Compromises Plant Defence and Root-Hair Development Through Salicylic Acid Signalling. *BMC Genomics* **2015**, *16*, 341.

45. Ghodake, G.; Seo, Y. D.; Park, D. H.; Lee, D. S. Phytotoxicity of Carbon Nanotubes Assessed by *Brassica juncea* and *Phaseolus mungo*. *J. Nanoelectron. Optoelectron.* **2010**, *5*, 157–160.

46. Ghodake, G.; Seo, Y. D.; Lee, D. S. Hazardous Phytotoxic Nature of Cobalt and Zinc Oxide Nanoparticles Assessed Using *Allium cepa*. *J. Hazard. Mater.* **2011**, *186*, 952–955.

47. Gottschalk, F.; Nowack, B. The Release of Engineered Nanomaterials to the Environment. *J. Environ. Monit.* **2011**, *13*, 1145–1155.

48. Griffitt, R. J.; Luo, J.; Gao, J.; Bonzongo, J. C.; Barber, D. S. Effects of Particle Composition and Species on Toxicity of Metallic Nanomaterials in Aquatic Organisms. *Environ. Toxicol. Chem.* **2008**, *27*, 1972–1978.

49. Gubbins, E. J.; Batty, L. C.; Lead, J. R. Phytotoxicity of Silver Nanoparticles to *Lemna minor* L. *Environ. Pollut.* **2011**, *159*, 1551–1559.

50. Handy, R. D.; Kamme, F.; Lead, J. R.; Hasselov, M.; Owen, R.; Crane, M. The Ecotoxicology and Chemistry of Manufactured Nanoparticles. *Ecotoxicology* **2008**, *17*, 287–314.

51. Hänsch, M.; Emmerling, C. Effects of Silver Nanoparticles on the Microbiota and Enzyme Activity in Soil. *J. Plant Nutr. Soil Sci.* **2010**, *173*, 554–558.

52. Heinlaan, M.; Ivask, A.; Blinova, I.; Dubourguier, H.-C.; Kahru, A. Toxicity of Nanosized and Bulk ZnO, CuO and TiO$_2$ to Bacteria *Vibrio fischeri* and Crustaceans *Daphnia magna* and *Thamnocephalus platyurus*. *Chemosphere* **2008**, *71*, 1308–1316.

53. Hernandez-Viezcas, J. A.; Castillo-Michel, H.; Servin, A. D.; Peralta-Videa, J. R.; Gardea-Torresdey, J. L. Spectroscopic Verification of Zinc Absorption and Distribution in the Desert Plant *Prosopis juliflora-velutina* (velvet mesquite) Treated with ZnO Nanoparticles. *Chem. Eng. J.* **2011**, *170*(1–3), 346–352.

54. Hoecke, K. V.; de Schamphelaere, K. A. C.; van der Meeren, P.; Lucas, S.; Janssen, C. R. Ecotoxicity of Silica Nanoparticles to the Green Alga *Psuedokirchneriella subcapitata*: Importance of Surface Area. *Environ. Toxicol. Chem.* **2008**, *27*, 1948–1957.

55. Hunde-Rinke, K.; Simon, M. Ecotoxic Effect of Photocatalytic Active Nanoparticles TiO$_2$ on Algae and Daphnids. *Environ. Sci. Pollut. Res.* **2006**, *13*, 225–232.

56. Hussein, M. Z.; Sarijo, S. H.; Yahaya, A. H.; Zainal, Z. Synthesis of 4-chlorophenoxyacetate-zinc-aluminium-layered Double Hydroxide Nanocomposite:

Physico-Chemical and Controlled Release Properties. *J. Nanosci. Nanotechnol.* **2007**, *7*(8), 2852–2862.

57. Jiang, W.; Mashayekhi, H.; Xing, B. Bacterial Toxicity Comparison Between Nano- and Micro-Scaled Oxide Particles. *Environ. Pollut.* **2009**, *157,* 1619–1625.

58. Johansen, A.; Pedersen, L. A.; Jensen, A. K.; Karlson, U.; Hansen, M. B.; Scott-Fordsmand, J. J.; Winding, A. Effects of C60 Fullerene Nanoparticles on Soil Bacteria and Protozoans. *Environ. Toxicol. Chem.* **2008**, *27,* 1895–1903.

59. Joško, I.; Oleszczuk, P. Influence of Soil Type and Environmental Conditions on the ZnO, TiO$_2$ and Ni Nanoparticles Phytotoxicity. *Chemosphere* **2013**, *92,* 91–99.

60. Joško, I.; Oleszczuk, P. Phytotoxicity of Nanoparticles-Problems with Bioassay Choosing and Sample Preparation. *Environ. Sci. Pollut. Res.* **2014**, *21,* 10215–10224.

61. Kah, M.; Beulke, S.; Tiede, K.; Hofmann T. Nanopesticides: State of Knowledge, Environmental Fate, and Exposure Modeling. *Crit. Rev. Environ. Sci. Technol.* **2013**, *43*(16), 1823–1867.

62. Kaveh, R.; Li, Y. S.; Ranjbar, S.; Tehrani, R.; Brueck, C. L.; Van Aken, B. Changes in *Arabidopsis thaliana* Gene Expression in Response to Silver Nanoparticles and Silver Ions. *Environ. Sci. Technol.* **2013**, *47,* 10637–10644.

63. Khodakovskaya, M.; Dervishi, E.; Mahmood, M. et al. Carbon Nanotubes are Able to Penetrate Plant Seed Coat and Dramatically Affect Seed Germination and Plant Growth. *ACS Nano* **2009**, *3*(10), 3221–3227.

64. Khodakovskaya, M. V.; de Silva, K.; Nedosekin, D. A.; Dervishi, E.; Birisa, A. S.; Shashkov, E. V.; Galanzha, E. I.; Zharov, V. P. Complex Genetic, Photothermal, and Photoacoustic Analysis of Nanoparticle-Plant Interactions. *Proc. Natl. Acad. Sci. U.S.A.* **2011**, *108*(3), 1028–1033.

65. Khodakovskaya, M. V.; de Silva, K.; Biris, A. S.; Dervishi, E.; Villagarcia, H. Carbon Nanotubes Induce Growth Enhancement of Tobacco Cells. *ACS Nano* **2012**, *6,* 2128–2135.

66. Kim, S.; Lee, S.; Lee, I. Alteration of Phytotoxicity and Oxidant Stress Potential by Metal Oxide Nanoparticles in *Cucumis sativus*. *Water Air Soil Pollut.* **2012**, *223,* 2799–2806.

67. Kottegoda, N.; Munaweera, I.; Madusanka, N.; Karunaratne, V. A Green Slow-Release Fertilizer Composition Based on Urea-Modified Hydroxyapatite Nanoparticles Encapsulated Wood. *Curr. Sci.* **2011**, *101*(1), 73–78.

68. Kovalchuk, I.; Ziemienowicz, A.; Eudes, F. Inventors; Plantbiosis Ltd, Assignee. T-DNA/Protein Nano-Complexes for Plant Transformation. U. S. patent 20,120,070,900 A1, March 22, 2012.

69. Kumar, N.; Shah, V.; Walker, V. K. Perturbation of an Arctic Soil Microbial Community by Metal Nanoparticles. *J. Hazard. Mater.* **2011**, *190,* 816–822.

70. Kumar, A.; Negi, Y. S.; Choudhary, V.; Bhardwaj, N. K. Characterization of Cellulose Nanocrystals Produced by Acid-Hydrolysis from Sugarcane Bagasse as Agro-Waste. *J. Mater. Phys. Chem.* **2014a**, *2*(1), 1–8.

71. Kumar, D.; Kumari, J.; Pakrashi, S.; Dalai, S.; Raichur, A. M.; Sastry, T. P.; Mandal, A. B.; Chandrasekaran, N.; Mukherjee, A. Qualitative Toxicity Assessment of Silver Nanoparticles on the Fresh Water Bacterial Isolates and Consortium at Low Level of Exposure Concentration. *Ecotoxicol. Environ. Saf.* **2014b**, *108,* 152–160.

72. Kumari, M.; Mukherjee, A.; Chandrasekaran, N. Genotoxicity of Silver Nanoparticles in *Allium cepa*. *Sci. Total Environ.* **2009**, *407*, 5243–5246.

73. Lahiani, M. H.; Dervishi, E.; Chen, J.; Nima, Z.; Gaume, A.; Biris, A. S.; Khodakovskaya, M. V. Impact of Carbon Nanotube Exposure to Seeds of Valuable Crops. *ACS Appl. Mater. Interfaces* **2013**, *5*, 7965–7973.

74. Lam, C.-W.; James, J. T.; McCluskey, R.; Arepalli, S.; Hunter, R. L. A Review of Carbon Nanotube Toxicity and Assessment of Potential Occupational and Environmental Health Risks. *Crit. Rev. Toxicol.* **2006**, *36*, 189–217.

75. Landa, P.; Vankova, R.; Andrlova, J.; Hodek, J.; Marsik, P.; Storchova, H.; White, J. C.; Vanek, T. Nanoparticle-Specific Changes in *Arabidopsis thaliana* Gene Expression After Exposure to ZnO, TiO_2, and Fullerene Soot. *J. Hazard. Mater.* 2012, *241*, 55–62.

76. Lee, C. W.; Mahendra, S.; Zodrow, K.; Li, D.; Tsai, Y. C.; Braam, J. et al. Developmental Phytotoxicity of Metal Oxide Nanoparticles to *Arabidopsis thaliana*. *Environ. Toxicol. Chem.* **2010**, *29*, 669–675.

77. Lee, W. M.; Kwak, J. I.; An, Y. J. Effect of Silver Nanoparticles in Crop Plants *Phaseolus radiatus* and *Sorghum bicolor*: Media effect on Phytotoxicity. *Chemosphere* **2012**, *86*, 491–499.

78. Lee, S.; Kim, S.; Kim, S.; Lee, I. Assessment of Phytotoxicity of ZnO NPs on a Medicinal Plant, *Fagopyrum esculentum*. *Environ. Sci. Pollut. Res.* **2013**, *20*, 848–854.

79. Lei, Z.; Fashui, H.; Shipeng, L.; Liu, C. Effect of Nano-TiO_2 on Strength of Naturally Aged Seeds and Growth of Spinach. *Biol. Trace Elem. Res.* **2005**, *104*, 83–91.

80. Levard, C.; Hotze, E. M.; Lowry, G. V.; Brown, G. E. Jr. Environmental Transformations of Silver Nanoparticles: Impact on Stability and Toxicity. *Environ. Sci. Technol.* **2012**, *46*, 6900–6914.

81. Li, Y.; Cu, Y. T.; Luo, D. Multiplexed Detection of Pathogen DNA with DNA-based Fluorescence Nanobarcodes. *Nat. Biotechnol.* **2005**, *23*(7), 885–889.

82. 82. Limbach, L. K.; Yuchun, L.; Grass, R. N.; Brunner, T. J.; Hintermann, M. A.; Muller, M.; Gunther, D.; Stark, W. J. Oxide Nanoparticle Uptake in Human Lung Fibroblasts: Effects of Particle Size, Agglomeration, and Diffusion at Low Concentrations. *Environ. Sci. Technol.* **2005**, *39*, 9370–9376.

83. Lin, B. S.; Diao, S. Q.; Li, C. H.; Fang, L. J.; Qiao, S. C.; Yu, M. Effect of TMS (nanostructured silicon dioxide) on Growth of Changbai Larch Seedlings. *J. For. Res. –CHN* **2004**, *15*, 138–140.

84. Lin, D.; Xing, B. Phytotoxicity of Nanoparticles: Inhibition of Seed Germination and Root Growth. *Environ. Pollut.* **2007**, *150*, 243–250.

85. Lin, D.; Xing, B. Root Uptake and Phytotoxicity of ZnO Nanoparticles. *Environ. Sci. Technol.* **2008**, *42*, 5580–5585.

86. Liu, Q.; Chen, B.; Wang, Q.; Shi, X.; Xiao, Z.; Lin, J.; Fang, X. Carbon Nanotubes as Molecular Transporters for Walled Plant Cells. *Nano Lett.* **2009**, *9*, 1007–1010.

87. Lo´pez-Moreno, M.; de la Rosa, G.; Hernandez-Viezcas, J.; Peralta-Videa, J.; Gardea-Torresdey, J. X-ray Absorption Spectroscopy (XAS) Corroboration of the Uptake and Storage of CeO_2 Nanoparticles and Assessment of Their Differential Toxicity in Four Edible Plant Species. *J. Agric. Food Chem.* **2010a**, *58*, 3689–3693.

88. Lo´pez-Moreno, M.; de la Rosa, G.; Hernandez-Viezcas, J.; Castillo- Michel, H.; Botez, C.; Peralta-Videa, J.; Gardea-Torresdey, J. Evidence of the Differential

Biotransformation and Genotoxicity of ZnO and CeO$_2$ Nanoparticles on Soybean (*Glycine max*) Plants. *Environ. Sci. Technol.* **2010b**, *44*, 7315–7320.

89. Long, T. C.; Saleh, N.; Tilton, R. D.; Lowry, G. V.; Veronesi, B. Titanium Dioxide (P25) Produces Reactive Oxygen Species in Immortalized Brain Microglia (BV2): Implications for Nanoparticles Neurotoxicity. *Environ. Sci. Technol.* **2006**, *40*, 4346–4352.

90. Lu, C. M.; Zhang, C. Y.; Wen, J. Q.; Wu, G. R.; Tao, M.X. Research of the Effect of Nanometer Materials on Germination and Growth Enhancement of *glycine max* and its Mechanism. *Soybean Sci.* **2002**, *21*(3), 168–171.

91. Ma, Y.; Kuang, L.; He, Z.; Ding, W. B. Y.; Zhang, Z.; Zhao, Y.; Chai, Z. Effects of Rare Earth Oxide Nanoparticles on Root Elongation of Plant. *Chemosphere*, **2010**, *78*, 273–279.

92. Ma, C. X.; Chhikara, S.; Xing, B. S.; Musante, C.; White, J. C.; Dhankher, O. P. Physiological and Molecular Response of *Arabidopsis thaliana* (L.) to Nanoparticle Cerium and Indium Oxide Exposure. *ACS Sustainable Chem. Eng.* **2013**, *1*, 768–778.

93. Manikandan, A.; Subramanian, K. S. Fabrication and Characterization of Nanoporous Zeolite Based N Fertilizer. *Afr. J. Agric. Res.* **2014**, *9*(2), 276–284.

94. Marmiroli, M.; Pagano, L.; Sardaro, M. L. S.; Villani, M.; Marmiroli, N. Genome-Wide Approach in *Arabidopsis thaliana* to Assess the Toxicity of Cadmium Sulfide Quantum Dots. *Environ. Sci. Technol.* **2014**, *48*, 5902–5909.

95. Maynard, A. D.; Aitken, R. J.; Butz, T. et al. Safe Handling of Nanotechnology. *Nature* **2006**, *444*(7117), 267–269.

96. Mazumdar, H.; Ahmed, G. U. Phytotoxicity Effect of Silver Nanoparticles on *Oryza sativa*. *Int. J. Chem. Technol. Res.* **2011**, *3*(3), 1494–1500.

97. Miao, A. J.; Luo, Z.; Chen, C. S.; Chin, W. C.; Santschi, P. H.; Quigg, A. Intracellular Uptake: a Possible Mechanism for Silver Engineered Nanoparticles Toxicity to a Freshwater Alga *Ochromonas danica*. *PLoS ONE* 2010, *5*, e15196.

98. Miralles, P.; Church, T. L.; Harris, A. T. Toxicity, Uptake, and Translocation of Engineered Nanomaterials in Vascular Plants. *Environ. Sci. Technol.* **2012**, *46*, 9224–9239.

99. Mirzajani, F.; Askari, H.; Hamzelou, S.; Schober, Y.; Römpp, A.; Ghassempour, A.; Spengler, B. Proteomics Study of Silver Nano Particles Toxicity on *Oryza sativa* L. *Ecotoxicol. Environ. Saf.* **2014**, *108*, 335–339.

100. Mura, S.; Seddaiu, G.; Bacchini. F.; Roggero. P. P.; Greppi, G. F. Advances of Nanotechnology in Agro-Environmental Studies. *Ital. J. Agron.* **2013**, *8*(3), e18.

101. Musante, C; White J. C. Toxicity of Silver and Copper to *Cucurbita pepo*: Differential Effects of Nano and Bulk-Size Particles. *Environ. Toxicol.* **2012**, *27*(9), 510–517.

102. Mustafa, G.; Sakata, K.; Hossain, Z.; Komatsu, S. Proteomic Study on the Effects of Silver Nanoparticles on Soybean Under Flooding Stress. *J. Proteomics* **2015**, *122*, 100–118.

103. Naderi, M. R.; Danesh-Shahraki, A. Nanofertilizers and Their Roles in Sustainable Agriculture. *Int. J. Agric. Crop Sci.* **2013**, *5*(19), 2229–2232.

104. Nair, R. P.; Nagaoka, Y.; Yoshida, Y.; Maekawa, T.; Kumar, D. S. Uptake of FITC-Labeled Silica Nanoparticles and Quantum Dots by Rice Seedlings: Effects on Seed Germination and Their Potential as Biolabels for Plants. *J. Fluoresc.* **2011**, *21*, 2057–2068.

105. Nair, P. M. G.; Chung, I. M. Assessment of Silver Nanoparticle-Induced Physiological and Molecular Changes in *Arabidopsis thaliana*. *Environ. Sci. Pollut. Res.* **2014a**, *21*, 8858–8869.

106. Nair, P. M. G.; Chung, I. M. Cell Cycle and Mismatch Repair Genes as Potential Biomarkers in *Arabidopsis thaliana* Seedlings Exposed to Silver Nanoparticles. *Bull. Environ. Contam. Toxicol.* **2014b**, *92*, 719–725.

107. Navarro, E.; Piccipetra, F.; Wagner, B.; Marconi, F.; Kaegi, R.; Odzak, N.; Sigg, L.; Behra, R. Toxicity of Silver Nanoparticles to *Chlamydomonas reinhardtii*. *Environ. Sci. Technol.* **2008**, *42*, 8959–8964.

108. Nel, A.; Xia, T.; Moedler, L.; Li, N. Toxic Potential of Materials at Nano Level. *Science* **2006**, *311*(5761), 622–627.

109. Nowack, B.; Bucheli, T. D. Occurrence, Behavior and Effects of Nanoparticles in the Environment. *Environ. Pollut.* **2007**, *150*(1), 5–22.

110. Otles, S.; Yalcin, B. Nano-Biosensors as New Tool for Detection of Food Quality and Safety. *LogForum* **2010**, *6*(4), 67–70.

111. Parisi, C.; Vigani, M.; Rodríguez-Cerezo, E. Agricultural Nanotechnologies: What are the Current Possibilities?. *Nano Today* **2014**, *407*, 1–4.

112. Parsons, J. G.; Lopez, M. L.; Gonzalez, C. M.; Peralta-Videa, J. R.; Gardea-Torresdey, J. L. Toxicity and Biotransformation of Uncoated and Coated Nickel Hydroxide Nanoparticles on Mesquite Plants. *Environ. Toxicol. Chem.* **2010**, *29*, 1146–1154.

113. Pradhan, S.; Patra, P.; Das, S.; Chandra, S.; Mitra, S.; Dey, K. K.; Akbar, S.; Palit, P.; Goswami, A. Photochemical Modulation of Biosafe Manganese Nanoparticles on *Vigna radiata*: A Detailed Molecular, Biochemical, and Biophysical Study. *Environ. Sci. Technol.* **2013**, *47*, 13122–13131.

114. Pradhan, S.; Patra, P.; Mitra, S.; Dey, K. K.; Basu, S.; Chandra, S.; Palit, P.; Goswami, A. Copper Nanoparticle (CuNP) Nanochain Arrays with a Reduced Toxicity Response: A Biophysical and Biochemical Outlook on *Vigna radiata*. *J. Agric. Food Chem.* **2015**, *63*, 2606–2617.

115. Rai, M.; Ingle, A. Role of Nanotechnology in Agriculture with Special Reference to Management of Insect Pests. *Appl. Microbiol. Biotechnol.* **2012a**, *94*(2), 287–293.

116. Rai, V.; Acharya, S.; Dey, N. Implications of Nanobiosensors in Agriculture. *Journal of Biomater. Nanobiotchnol.* **2012b**, *3*, 315–324.

117. Raliya, R.; Nair, R.; Chavalmane, S.; Wang, W. N.; Biswas P. Mechanistic Evaluation of Translocation and Physiological Impact of Titanium Dioxide and Zinc Oxide Nanoparticles on the Tomato (*Solanum lycopersicum* L.) Plant. *Metallomics* **2015**, *7*(12), 1584–1594.

118. Ramsurn, H.; Gupta, R. B. Nanotechnology in Solar and Biofuels. *ACS Sustainable Chem. Eng.* **2013**, *1*(7), 779–797.

119. Rao, S.; Shekhawat, G. S. Toxicity of ZnO Engineered Nanoparticles and Evaluation of Their Effect on Growth, Metabolism and Tissue Specific Accumulation in *Brassica juncea*. *J. Environ. Chem. Eng.* **2014**, *2*, 105–114.

120. Ray, S. S. *Environmentally Friendly Polymer Nanocomposites: Types, Processing and Properties*, 1st ed.; Woodhead Publishing: Cambridge, UK, 2013; p 512.

121. Roco, M. C. Environmentally Responsible Development of Nanotechnology. *Environ. Sci. Technol.* **2005**, *39*, 106–112A.

122. Roh, J.-Y.; Sim, S. J.; Yi, J.; Park, K.; Chung, K. H.; Ryu, D.-Y.; Choi, J. Ecotoxicity of Silver Nanoparticles on the Soil Nematode *Caenorhabditis elegans* Using Functional Ecotoxicogenomics. *Environ. Sci. Technol.* **2009**, *43*, 3933–3940.

123. Sastry, R. K.; Rao, N. H. Emerging Technologies for Enhancing Indian Agriculture-Case of Nanobiotechnology. *Asian Biotechnol. Dev. Rev.* **2013**, *15*(1), 1–9.

124. Scown, T. M.; Santos, E. M.; Johnston, B. D.; Gaiser, B.; Baalousha, M.; Mitov, S.; Lead, J. R.; Stone, V.; Fernandes, T. F.; Jepson, M.; van Aerle, R; Tyler, C. R. Effects of Aqueous Exposure to Silver Nanoparticles of Different Sizes in Rainbow Trout. *Toxicol. Sci.* **2010**, *115,* 521–534.

125. Seeger, E. M.; Baun, A.; Kästner, M.; Trapp, S. Insignificant Acute Toxicity of TiO$_2$ Nanoparticles to Willow Trees. *J. Soils Sediment* **2009**, *9,* 46–53.

126. Shah, V. B. I. Influence of Metal Nanoparticles on the Soil Microbial Community and Germination of Lettuce Seeds. *Water, Air Soil Pollut.* **2009**, *197,* 143–148.

127. Shaw, A. K.; Hossain, Z. Impact of Nano-CuO Stress on Rice (*Oryza sativa* L.) Seedlings. *Chemosphere* **2013**, *93,* 906–915.

128. Shiny, P. J.; Mukerjee, A.; Chandrasekaran, N. Comparative Assessment of the Phytotoxicity of Silver and Platinum Nanoparticles. *Proceedings of the International Conference on Advanced Nanomaterials and Emerging Engineering Technologies,* Sathyabama University: Chennai, India, 2013; 391–393.

129. Slomberg, D. L.; Schoenfisch, M. H. Silica Nanoparticle Phytotoxicity to *Arabidopsis thaliana. Environ. Sci. Technol.* **2012**, *46,* 10247–10254.

130. Soni, N.; Prakash, S. Efficacy of Fungus Mediated Silver and Gold Nanoparticles Against *Aedes aegypti* Larvae. *Parasitol. Res.* **2012**, *110*(1), 175–184.

131. Sozer, N.; Kokini, J. L. Nanotechnology and its Applications in the Food Sector. *Trends biotechnol.* **2009**, *27*(2), 82–89.

132. Stampoulis, D.; Sinha, S. K.; White, J. C. Assay-Dependent Phytotoxicity of Nanoparticles to Plants. *Environ. Sci. Technol.* **2009**, *43,* 9473–9479.

133. Sudo, E.; Itouga, M.; Yoshida-Hatanaka, K.; Ono, Y.; Sakakibara, H. Gene Expression and Sensitivity in Response to Copper Stress in Rice Leaves. *J. Exp. Bot.* **2008**, *59*(12), 3465–3474.

134. Thounaojam, T. C.; Panda, P.; Mazumdar, P.; Kumar, D.; Sharma, G. D.; Sahoo, L.; Panda, S. K. Excess Copper Induced Oxidative Stress and Response of Antioxidants in Rice. *Plant Physiol. Biochem.* **2012**, *53,* 33–39.

135. Tiwari, A. *Recent Developments in Bio-Nanocomposites for Biomedical Applications,* 1st ed.; Nova Science Publishers, Inc.: Hauppauge, NY, 2010.

136. Torney, F.; Trewyn, B. G.; Lin, V. S.; Wang, K. Mesoporous Silica Nanoparticles Deliver DNA and Chemicals into Plants. *Nat. Nanotechnol.* **2007**, *2*(5), 295–300.

137. United States Environmental Protection Agency (US EPA). *Ecological Test Guidelines (OPPTS 850, 4200): Seed Germination/Root Elongation Toxicity Test;* US EPA: Washington, DC, USA, 1996.

138. United States Environmental Protection Agency (US EPA). *Nanotechnology White Paper: Tech. Rep. EPA 100/B-07/001*; Science Policy Council, US EPA: Washington, DC, USA, 2007.

139. Vannini, C.; Domingo, G.; Onelli, E.; Prinsi, B.; Marsoni, M.; Espen, L.; Bracale, M. Morphological and Proteomic Responses of *Eruca sativa* Exposed to Silver Nanoparticles or Silver Nitrate. *PLoS ONE* **2013**, *8*(7), e68752.

140. Vannini, C.; Domingo, G.; Onelli, E.; De Mattia, F.; Bruni, I.; Marsoni, M.; Bracale, M. Phytotoxic and Genotoxic Effects of Silver Nanoparticles Exposure on Germinating Wheat Seedlings. *J. Plant Physiol.* **2014**, *171,* 1142–1148.

141. Wang, J.; Zhang, X.; Chen, Y.; Sommerfeld, M.; Hu, Q. Toxicity Assessment of Manufactured Nanomaterials Using the Unicellular Green Alga *Chlamydomonas reinhardtii*. *Chemosphere* **2008**, *73*, 1121–1128.

142. Wang, H.; Wick, L. R.; Xing, B. Toxicity of Nanoparticulate and Bulk ZnO, Al_2O_3 and TiO_2 to the Nematode *Caenorhabditis elegans*. *Environ. Pollut.* **2009**, *157*, 1171–1177.

143. Wang, Q.; Ma, X.; Zhang, W.; Pei, H.; Chen, Y. The Impact of Cerium Oxide Nanoparticles on Tomato (*Solanum lycopersicum* L.) and its Implications for Food Safety. *Metallomics* **2012**, *4*(10), 1105–1112.

144. Wang, Q.; Zhao, S.; Zhao, Y.; Rui, Q.; Wang, D. Toxicity and Translocation of Graphene Oxide in *Arabidopsis* Plants Under Stress Conditions. *R. Soc. Chem. Adv.* **2014**, *4*, 60891–60901.

145. Wei, C. Z. Y.; Guo, J.; Han, B.; Yang, X.; Yuan, J. Effects of Silica Nanoparticles on Growth and Photosynthetic Pigment Contents of *Scenedesmus obliquus*. *J. Environ. Sci.* **2010**, *22*, 155–160.

146. Wise, Sr. J. P.; Goodale, B. C.; Wise, S. S.; Craig, G. A.; Pongan, A. F.; Walter, R. B.; Thompson, W. D.; Ng, A. K.; Aboueissa, A. M.; Mitani, H.; Spalding, M. J.; Mason, M. D. Silver Nanospheres are Cytotoxic and Genotoxic to Fish Cells. *Aquat. Toxicol.* **2010**, *97*, 34–41.

147. Wu, S. G.; Huang, L.; Head, J.; Chen, D. R.; Kong, I. C.; Tang, Y. J. Phytotoxicity of Metal Oxide Nanoparticles is Related to both Dissolved Metals Ions and Adsorption of Particles on Seed Surfaces. *J. Pet. Environ. Biotechnol.* **2012**, *3*, 4.

148. Xu, L. M.; Takemura, T.; Xu, M. S.; Hanagata, N. Toxicity of Silver Nanoparticles as Assessed by Global Gene Expression Analysis. *Mater. Express*, **2011**, *1*, 74–79.

149. Yang, L.; Watts, D. J. Particle Surface Characteristics May Play an Important Role in Phytotoxicity of Alumina Nanoparticles. *Toxicol. Lett.* **2005**, *158*, 122.

150. Yang, F.; Liu, C.; Gao, F.; Su, M.; Wu, X.; Zheng, L. et al. The Improvement of Spinach Growth by Nano-Anatase TiO_2 Treatment is Related to Nitrogen Photo Reduction. *Biol. Trace Elem. Res.* **2007**, *119*, 77–88.

151. Yin, L. C. Y.; Espinasse, B.; Colman, B. P.; Auffan, M.; Wiesner, M.; Rose, J.; Liu, J.; Bernhardt, E. S. More than the Ions: The Effects of Silver Nanoparticles on *Lolium multiflorum*. *Environ. Sci. Technol.* **2011**, *45*, 2360–2367.

152. Yotova, L.; Yaneva. S.; Marinkova, D. Biomimetic Nanosensors for Determination of Toxic Compounds in Food and Agricultural Products. *J. Chem. Technol. Metall.* **2013**, *48*(3), 215–227.

153. Zhang, X. L.; Tyagi, Y. R. D.; Surampalli, R. Y. Biodiesel Production from Heterotrophic microalgae Through Transesterification and Nanotechnology Application in the Production. *Renewable Sustainable Energy Rev.* **2013**, *26*, 216–223.

154. Zhao, C.-M.; Wang, W.-X. Comparison of acute and chronic toxicity of silver nanoparticles and silver nitrate to *Daphnia magna*. *Environ. Toxicol. Chem.* **2011**, *30*, 885–892.

155. Zheng, L.; Hong, F.; Lu, S.; Liu, C. Effect of Nano-TiO_2 on Strength of Naturally Aged Seeds and Growth of Spinach. *Biol. Trace Elem. Res.* **2004**, *101*, 1–9.

156. Zhu, H.; Han, J.; Xiao, J. Q.; Jin, Y. Uptake, Translocation, and Accumulation of Manufactured Iron Oxide Nanoparticles by Pumpkin Plants. *J. Environ. Monit.* **2008**, *10*, 713–717.

CHAPTER 5

EXTRACELLULAR AND INTRACELLULAR SYNTHESIS OF SILVER NANOPARTICLES

PRASANNA SUBRAMANIAN[1,2] AND
KIRUBANANDAN SHANMUGAM[1,*]

[1]*Centre for Biotechnology, Anna University, Chennai, Tamil Nadu 600020, India*

[2]*Centre for Biotechnology, Anna University, Chennai, Tamil Nadu 600020, India*
E-mail: prasanarengarajan@gmail.com

[*]Corresponding author. E-mail: skirubanandan80@gmail.com

CONTENTS

This chapter is a part and edited version of, "Prasanna, S. Extracellular and intracellular synthesis of silver nanoparticles, M. Tech. thesis, Anna University, Madras, India, 2006, pages 5–30"; and it was published by the authors with the same title as an open access article in,: "http://innovareacademics.in/journals/index.php/ajpcr/article/download/13302/8140."

5.1   INTRODUCTION

The application of nanotechnology in life sciences is called nanobiotechnology, and it is already having an impact on diagnostics and drug delivery/targeting. Now, researchers are starting to use nanotechnology in the field of drug discovery. Some of these have already established in research through well-known technologies such as biosensors and biochips. Nanoparticles are still used extensively for developing diagnostics and some of the assays for drug discovery. Nanoscale assays can contribute significantly to cost saving in screening campaigns. In addition, some nanosubstances could be potential drugs in the future. Although, there might be some safety concerns with respect to the in vivo use of nanoparticles, studies are in place to determine the nature and extent of adverse effects. Future prospects for the application of nanotechnology in healthcare and for the development of personalized medicine appear to be excellent. For example, nanoparticulate technology could prove to be very useful in cancer therapy allowing effective and targeted drug delivery by overcoming the many biological, biophysical, and biomedical barriers that the body stages against a standard intervention such as the administration of drugs or contrast agents. Among nanoparticles, in the biological sciences, many applications for metal nanoparticles are being explored, including biosensors,[17] labels for cells and biomolecules,[28] and cancer therapeutics.[6]

Though a variety of chemical routes have been followed to synthesize nanoparticles, yet natural routes are always desired. It has been known for a long time that in nature a variety of nanomaterials are synthesized by biological processes. For example, the magnetotactic bacteria synthesize intracellular magnetite or granitite nanocrystallites.[22] Similarly, certain yeasts, when challenged with toxic metals such as cadmium, synthesize intracellular CdS nanocrystallites as a mechanism of detoxification.[3] Bacteria involved in silver leaching have been reported to accumulate silver sulfide within their membrane[6] and natural biofilms of sulfate reducing bacteria were shown to deposit nanocrystalline sphalerite (ZnS).[11] Owing to the reported particle size uniformity in such 'biosynthetic' processes and their inherent environment friendly nature, there is a renewed interest in biological routes of nanoparticle synthesis.[7,8,9] The interest also extends to the assembly of nanoparticles using biological templates such as DNA or proteins.[17]

It has been demonstrated that in the case of noble-metal nanocrystals, the electromagnetic, optical, and catalytic properties are highly influenced by shape and size.[2,13,16] Among noble-metal nanomaterials, silver nanoparticles have received considerable attention due to their attractive physicochemical properties. The surface plasmon resonance and large effective scattering cross section of individual silver nanoparticles make them ideal candidates for molecular labeling,[25] where phenomena such as surface-enhanced Raman scattering (SERS) can be exploited. In addition, the strong toxicity that silver exhibits in various chemical forms to a wide range of microorganisms is very well known[4,12,18]; and silver nanoparticles have recently been proved to be a promising antimicrobial material.[26] It has been found that silver nanoparticles inhabit the virus by binding to host cells, as demonstrated "in vitro."[15]

Different approaches of the synthesis of silver nanoparticles have been conducted previously.[23,24,28] Silver nanoparticles were synthesized biologically using fungus and other microorganisms.[1,10] Silver nanoparticles with different sizes were synthesized by electrochemical reduction.[21] Biologically synthesized silver nanoparticles could have many applications, such as spectrally selective coatings for solar energy absorption and intercalation material for electrical batteries,[9] as optical receptors,[25] catalysts in chemical reactions, bio-labelling,[5] and so forth.

In this chapter, authors have presented the research study on the synthesis of silver nanoparticle using human cells. The cancerous and noncancerous cell lines were chosen for the study. These synthesized silver nanoparticles have potential application in drug discovery and drug delivery/targeting.

5.2   METHODS AND MATERIALS

5.2.1   PREPARATION OF SILVER NITRATE SOLUTION

Silver ion was prepared in the concentration of 10^{-3} M. The solution was prepared in 4-(2-hydroxyethyl)-1-piperazineethanesulfonic acid (HEPES) (21 mm) buffer. Sodium nitrate (137 mm) and sodium hydroxide were added to adjust the pH to physiological pH.

5.2.2   PREPARATION OF CELL CULTURE

The cultures are viewed to assess the degree of confluency and the absence of bacterial and fungal contaminants is confirmed. The spend medium is removed. The cell monolayer is washed with HEPES using a volume equivalent to half the volume of culture medium. The wash step is repeated if the cells are known to adhere strongly. The trypsin/ethylenediaminetet-raacetate (EDTA) is pipetted on to the washed cell monolayer using 1 ml per 25 cm^2 of the surface area. The flask is rotated to cover the monolayer with trypsin. The excess trypsin is removed by decantation. The flask is returned to the incubator and left for 2–10 min. The cells are examined by an inverted microscope to ensure that all the cells are detached and floating. The side of the flasks may be gently tapped to release any remaining attached cells. The cells are suspended in a small volume of fresh medium containing medium (1 ml of 10% serum containing Dulbeco's Modified Eagles Medium—DMEM) to inactivate the trypsin. The solution is centrifuged at 3000 RPM for 5 min. The cells are resuspended in 1 ml of 10% serum containing DMEM. The cell count is performed (the wells are seeded with same initial value of 3 × 10^5 cells per well for each kind of cell line). The required number of cells is transferred to a new labeled flask containing pre-warmed medium. It is incubated at 37°C in 5% CO$_2$ environment for the duration of 48 h, before ionic solution can be added.

5.2.3 ADDITION OF SILVER ION SOLUTION TO CELL LINES

Three cell lines (cancerous as well as noncancerous) were grown to 70% confluency level after seeding in a 6 well plate (9 cm^2, TPP oxygen). The cell lines utilized in this study were: human embryonic kidney cell (HEK 293), HeLa cell line and SiHa cell line. During the growth period, culture medium containing DMEM is used. Just before the addition of silver ions to the well, the culture medium is removed. The cells were washed twice with HEPES. Then silver ion solution in HEPES is filter sterilized using MILLEX © GV Durapore ©PVDF membrane filter (pore size: 0.22 μm), and is added to the well. To each well, 2 ml of the solution is added. The cells are then kept in a 5% CO$_2$ incubator at 37°C. The biotransformation was observed by periodic sampling of aliquots (200 μl diluted to 1 ml of the aqueous component and measuring the UV-visible spectrum of the solution). The absorbance values are compared over a period of 4 days. On last day, the cells are scrapped off the well surface and sample preparation is done for transmission electron microscopy (TEM).

5.2.4   SAMPLE PREPARATION FOR TEM

The Ag^{2+} HEPES solution is removed from the wells containing cells. It is washed with HEPES and HEPES is discarded (repeat twice). Trypsin (2 ml) is added to the wells. The cells are incubated for 5 min at 37°C in 5% CO_2. The medium is added with serum (2 ml) to the cells. The liquid from well is transferred to a centrifuge tube and centrifugation for 5 min at 3000 rpm is carried out. The pellet is collected after centrifugation. The pellet is resuspended in HEPES. It is centrifuged again under the same conditions. The pellet is collected and washed with HEPES twice. This final pellet is collected and resuspended in 1% glutaraldehyde and is stored at 4°C.

5.2.5   PROTOCOL FOR CELL LYSIS

The solution is aspirated. The cells are scarped from the surface of the well using a cell scraper. The solution contained in the well is transferred into a centrifuge tube. The cells are sonicated at duty cycle value of 50% and in output mode and 3 pulses are given for 10 s, each pulse separated by 20 s of cooling in ice. It is carried out centrifugation at 5000 rpm for 5 min at 4°C.

5.2.6   ULTRA-MICROTOME AND TEM

- Fixation: Primary fixation was done with 1% glutaraldehyde (as described above). After primary fixation the cells were washed repeatedly with phosphate buffer and were subjected to secondary fixation with 1% osmium tetroxide for 1 h followed by the repeated washing with phosphate buffer.
- Dehydration: Before proceeding for dehydration the cells were suspended in molten agar (2%). Small blocks of solidified agar (1-2 mm) were cut and passed through the series of 30, 50, and 100% ethanol v/v for 15 min each. The lower grade ethanol was prepared in deuterium-depleted water (DD water).
- Embedding: The dehydrated agar blocks were suspended in propylene oxide for 20 min at 4°C (2 changes) followed by the treatment with 1:1 mixture of propylene oxide and Araldite A for 1 h at 60°C.
  Araldite A: Araldite (resin) Cy-212: 10 ml
  Dodecenyl succinic anhydride (hardener): 12 ml
  Dibutyl-phthalate (plasticizer): 1 ml

The agar blocks containing cells were then incubated in Araldite A for 1 h at 60°C followed by overnight incubation at room temperature. The next day Araldite B was prepared freshly as follows:
Araldite A: 23 ml
Trimethyl aminomethyl phenol: 0.4 ml (DMP- 30 accelerator)
Araldite A was poured out and freshly prepared Araldite B was added to the agar blocks. This was followed by incubation for 2 h at 60°C for infiltration. The blocks were then finally removed and placed in refined beam capsule containing Araldite B and incubated for polymerization for 48–72 h, at 60°C.

- Block trimming and sectioning: The resin blocks were carefully trimmed to expose the underlying agar blocks. Sections of various thickness (200, 300, 500, and 1000 nm) were cut using Leica Ultra cut UCT microtome and transferred to 300 mesh copper grids.
- Uranyl acetate staining: A 10% alcoholic solution of uranyl acetate was prepared and centrifuged to remove any precipitate in there. The sections on copper grids were stained for 1 h with uranyl acetate in dark at room temperature and then washed with distilled water thoroughly.
- TEM: The sections on grids were observed with a transmission electron microscope working at 120 keV.

5.2.7   INSTRUMENTS USED

The optical spectroscopic studies were done on a Perkin Elmer Lambda 25 UV-visible spectrophotometer, scanning was done from 1100 to 200 nm. The bright-field TEM images were taken using a JEOL 200 keV TEM. The samples for TEM were prepared by dropping a dispersion of the particles on copper grid supported Formvar films. Microtome was done on the selected samples as described. X-ray powder diffractograms were measured with a Shimadzu diffractometer with Cu Kα radiation.

5.3   RESULTS AND DISCUSSION

5.3.1   NANOPARTICLE CHARACTERIZATION

Figure 5.1 shows the UV-visible spectrum of silver nanoparticles synthesized by chemical process. The spectrum shows the characteristic peak of silver nanoparticles at 415 nm.

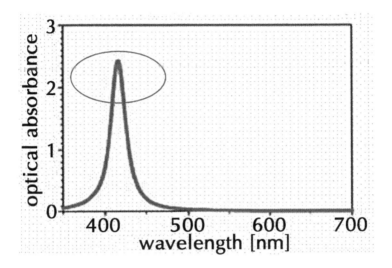

FIGURE 5.1   UV-visible spectrum of citrate capped silver nanoparticles.

5.3.2   TIME BOUND CHARACTERIZATION OF SILVER ION SOLUTION TAKEN FROM THE CELL WELL

The process of Ag^{2+} reduction on exposure to human cells can easily be tracked by monitoring UV-visible spectroscopy measurements. Authors used the silver ions in HEPES. This medium has been found to be stable to the cells. Sodium nitrate was used to maintain ionic balance. Subsequently, pH was adjusted to 7.2 using sodium hydroxide. The UV-visible spectrum of the Ag^{2+} in HEPES is taken for 24 and 96 h which show the absence of peak corresponding to silver nanoparticle (Fig. 5.3). The UV-visible spectrum of the Ag^{2+} sample after 24, 55, 72, 96 h of exposure to HEK 293, HeLa , SiHa (Figs. 5.2, 5.4–5.8) showed a well-defined surface plasmon band centered around 420–435 nm with characteristic of colloidal silver. Further, the steady increase in the peak absorbance value for all the experimental samples over a period of 96 h suggest that reduction process is slow and is a biological process. Solution at this stage becomes yellow in color. The cells due to over stressful condition, detaches itself from the well. The intracellular reduction of cell samples is also easily tracked by UV-visible spectroscopy measurements. The cells attached to the wells were detached by trypsinization and were subsequently lysed following experimental protocol for cell lysis. The UV-visible spectra of

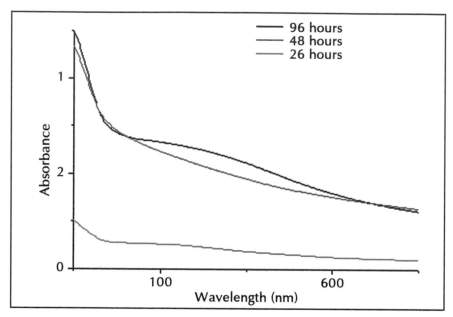

FIGURE 5.2   Time bound UV-visible measurements for HEK cells with 0.001 M silver ions in HEPES.

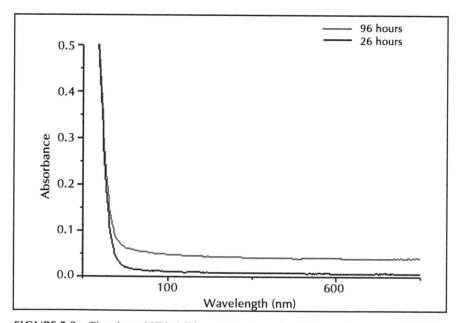

FIGURE 5.3   Time bound UV-visible measurements for 0.001 M silver ions in HEPES.

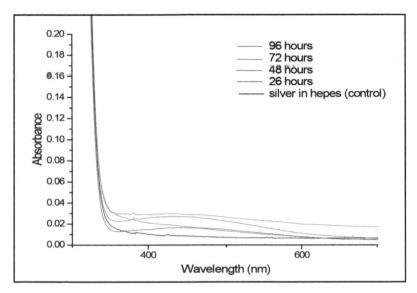

FIGURE 5.4 Time bound UV-visible measurements for HeLa cells with 0.001 M silver ions in HEPES.

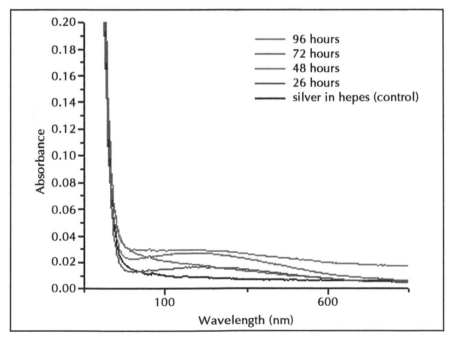

FIGURE 5.5 Time bound UV-visible measurements for SiHa cells with 0.001 M silver ions in HEPES.

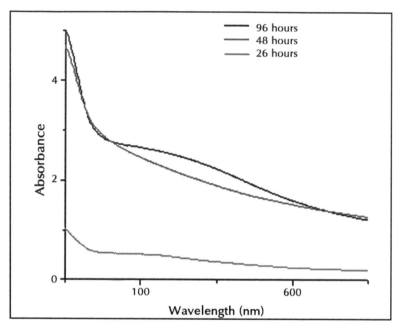

FIGURE 5.6   Time bound UV-visible measurements for lysed HEK cells treated with 0.001 M silver ions in HEPES.

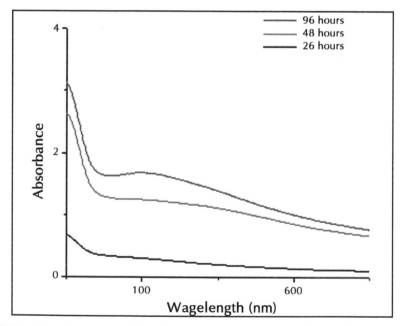

FIGURE 5.7   Time bound UV-visible measurements for lysed HeLa cells treated with 0.001 M silver ions in HEPES.

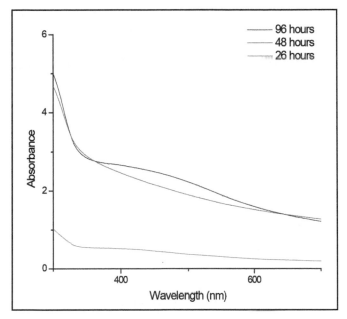

FIGURE 5.8   Time bound UV-visible measurements for lysed SiHa cells treated with 0.001 M silver ions in HEPES.

lysed samples were taken after 24, 55, 72, 96 h of exposure to Ag^{2+} solution. The peak value is subsequently higher than the corresponding spectra obtained by extracellular reduction at different time intervals.

The fact that presence of nanoparticles was showing up in the samples collected from the solution (external to the cell membrane) proves the presence of silver nanoparticles in the solution without any aggregation. Additionally, the detached cells from the well after exposure to Ag^{2+} solution were intensely yellowish brown in color at all-time intervals. Before detaching, the cells present in the well were found to be yellowish in color after 96 h (Figs. 5.9–5.14).

Figure 5.15 shows the XRD patterns for HEK samples obtained after lysing the cells and plating the lysate on a glass slide. It clearly shows three prominent peaks which are characteristic of crystal lattice of silver nanoparticles. Peak at 38 is due to rejection from (111), peak at 44 is due to reflection from (200), and peak at 65 is due to reflection from (220).

Electron diffraction patterns of ultramicrotome samples show the presence of silver crystals (Fig. 5.16). It is also well known that various kinds of proteins present inside the cytoplasm of cell have amine and/or thiol

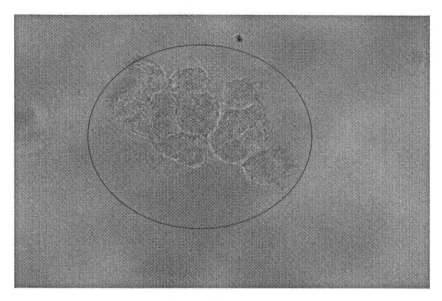

FIGURE 5.9   HEK 293 cells with silver ions (Day 1, 40×).

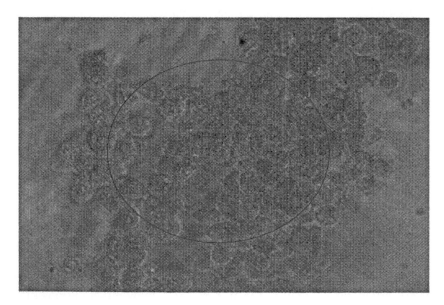

FIGURE 5.10   HEK 293 cells with silver ions (Day 4, 40×).

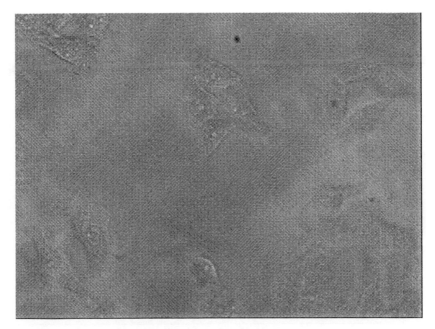

FIGURE 5.11 HeLa cells with silver ions (Day 1, 40×).

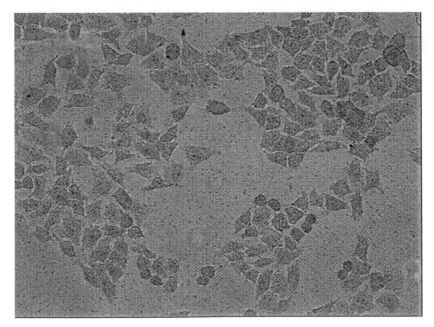

FIGURE 5.12 HeLa cells with silver ions (Day 4, 40×).

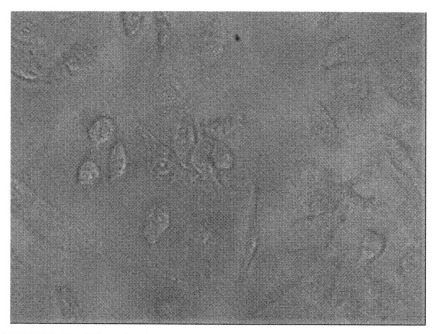

FIGURE 5.13   SiHa cells with silver ions (Day 1, 40×).

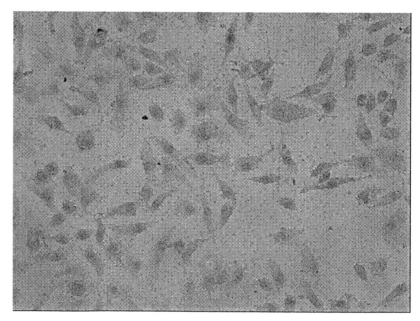

FIGURE 5.14   SiHa cells with silver ions (Day 4, 20×).

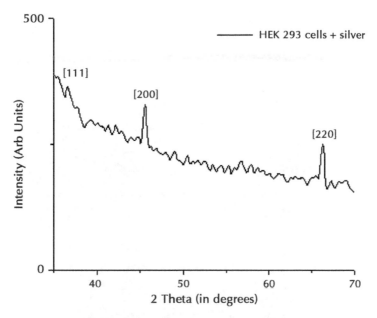

FIGURE 5.15   X-ray diffractogram of Hek 293 cells after treatment with silver ion solution (lysed).

groups present in them. The presence of these two groups crates favorable binding interactions between proteins and nanoparticles surfaces. The presence of capping agent, in addition to causing size changes, also contributes to stabilization of nanoparticles and shift in characteristic peak value position in UV-visible spectrometry measurements. The appearance of peak at above 420 nm suggests that silver nanoparticles are capped with certain molecular species containing amine group or cysteine residues in protein, and so forth

As observed from the images, the cells had become hugely stressed out (Figs. 5.9–5.14), due to the absence of growth medium and their involvement in reduction of silver ions. There was no further growth observed in the cells after day 0 as nutrient medium removed. The changes in cell morphology were observable. TEM images show the additional data of the presence of silver nanoparticles in the cell lines (Figs. 5.17–5.20).

It is reported that silver nanoparticles are useful in the fields of biomolecular detection and diagnostic and therapeutic agents and antimicrobials for drug delivery and targeting. Various strategies are involved in the preparation of silver nanoparticles by physical and chemical methods which are

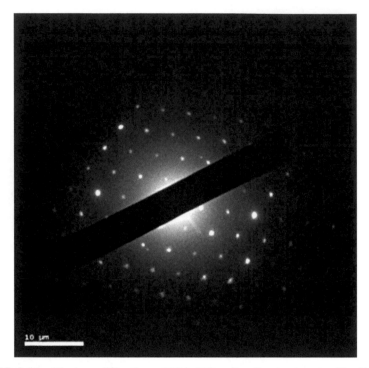

FIGURE 5.16   Electron diffraction of Hek 293 cells after treatment with silver ion solution (lysed).

useful to produce nanoparticles with low yield and environmental issues like usage of volatile solvents and high temperature process like pyrolysis and high attrition process. Biological process for the preparation of silver nanoparticle is an alternative to replace conventional method, and it will be expected that the biological process could produce clean and high yield of silver nanoparticles. Biosynthesis approach utilizing living cells for production of silver nanoparticles have emerged as a simple, clean, and viable alternative for conventional processes. It is testified that a number of biological system including microbial and plant cells is used for synthesis of nanoparticles.[20]

The utilization of mammalian cells used for biosynthesis of silver nanoparticles is a novel approach in this study. The cancer and noncancerous cell lines are used for preparation of silver nanoparticles. The biosynthesized silver nanoparticles were characterized by optical absorption spectroscopy. When metal nanoparticles are exposed to

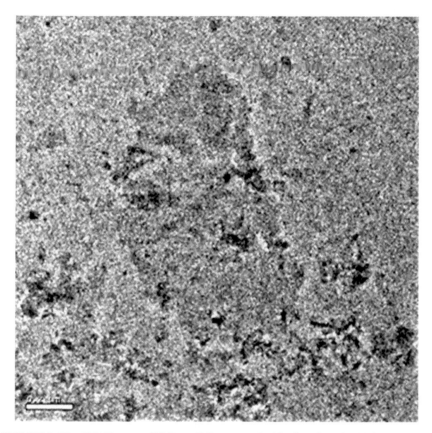

FIGURE 5.17   TEM image of HeLa cell line treated with silver ions for 96 h.

electromagnetic waves, the oscillating electric field causes electrons in the conduction band to oscillate. When the frequency of the oscillation of electrons becomes equal to the frequency of incident light, then resonance takes place which results in the plasmon absorption peak. This happens in the visible region of silver nanoparticles (Fig. 5.1).

The exact mechanism of silver ion reduction, transport into cells and its subsequent nucleation is yet to be studied thoroughly.[14,19,27] It is well known that sugars and enzymes present on the cell wall and within cytoplasm can accomplish the reduction process. The secreted proteins from the cells to the silver solution medium might also assist in the reduction process. Though the cells become dead, the subsequent increase in the plasmon peaks of lysed samples at increasing time intervals, suggest

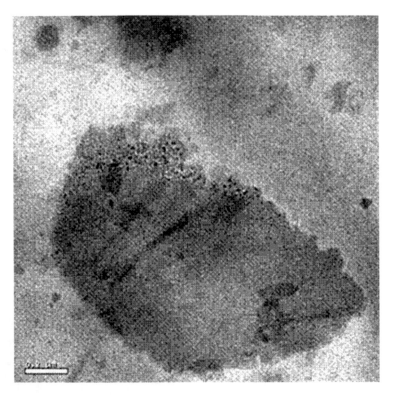

FIGURE 5.18   TEM image of HeLa cell line treated with silver ions for 96 h.

the cell membrane becoming porous. The transport of silver ions inside the cells may be facilitated by silver ions acting as biomimetic vehicle for transport through ionic channels. Also electrostatic interactions are created between membrane and cations due to negatively charged nature of membrane surface. These interactions may favorably increase the rate of ion transport inside the cells. Based on this investigation, we conclude that both intracellular and extracellular synthesis of silver nanoparticles was observed in all the three cell lines.

5.4   CONCLUSIONS

The synthesis of silver nanoparticles might be due to the reduction process by proteins or carbohydrate groups in the cell. The silver nanoparticles might be capped by the biomolecules stabilizing the nanoparticle. Several

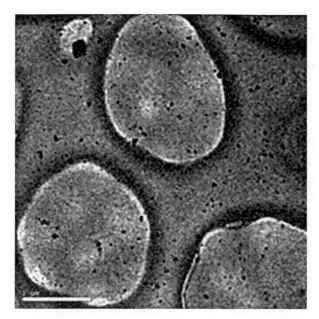

FIGURE 5.19   TEM image of HEK cell line treated with silver ions for 96 h.

other studies are in progress to find the location inside the cell, where the synthesis is occurring and to characterize the intra- and extracellular silver nanoparticles. This method could be a novel process of synthesis of silver nanoparticles using cell culture and will be applicable in the areas of drug targeting and delivery to the cancer cell.

5.5   SUMMARY

The cellular synthesis of nanoparticle is a green process and an alternative for a conventional process for the preparation of silver nanoparticles. In our research, focus has been given to the development of an efficient and eco-friendly viable process for the synthesis of silver nanoparticles using cancer and noncancerous cells, a cell culture that was isolated. The results of this investigation indicate that silver nanoparticles could be induced to be synthesized intracellularly and extracellularly using such cancerous and noncancerous mammalian cells.

The silver nanoparticles exhibited maximum absorbance at 415 nm in UV-visible spectroscopy. The X-ray diffraction (XRD) confirms the

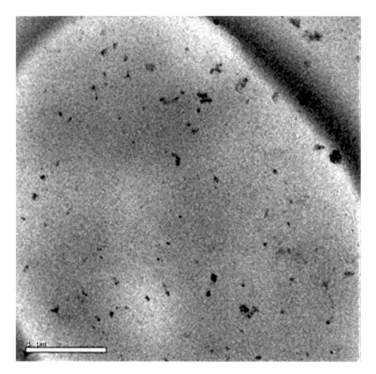

FIGURE 5.20   TEM image of HEK cell line treated with silver ions for 96 Hours.

characteristic of the crystal lattice of silver nanoparticles by observing three peaks at 38 which is due to reflection from (111), peak at 44 is due to reflection from (200), and peak at 65 is due to reflection from (220). TEM images showed the formation of stable silver nanoparticles in the cell lines.

KEYWORDS

- **Biosynthesis**
- **nanoparticles**
- **silver nanoparticles**
- **transmission electron microscopy**
- **XRD analysis**

REFERENCES

1. Bhainsa, K. C. Extracellular Biosynthesis of Silver Nanoparticle Using the Fungus *Aspergillus Fumigatus*. *Colloids Surf. B. Biointerfaces* **2006**, *47*(2), 160–164.
2. Burda, C; Chen, X; Narayanan, R; El-Sayed, M. A. Chemistry and Properties of Nanocrystals of Different Shapes. *Chem. Rev.* **2005**, *105*, 1025–1102.
3. Dameron, C. T.; Reese, R. N.; Mehra, R. K.; Kortan, A. R.; Caroll, P. J.; Steigerwald, M. L.; Brus, L. E.; Winge, D. R. Biosynthesis of Cadmium Sulphide Quantum Semi-conductor Crystallites. *Nature* **1989**, *338*, 596–597.
4. Gupta, A., Silver. S. The Silver as Biocide: Will Resistance Become a Problem? *Nat. Biotechnol.* **1998**, *16*, 888.
5. Hayat, M. A. *Colloidal gold: Principles, Methods, and Applications*; Academic Press, Inc.: New York, 1989; p 232.
6. Hirsch, L. R.; Stafford, R. J; Bankson, J. A; Sershen, S. R. Nano Shell-Mediated Near-Infrared Thermal Therapy of Tumors Under Magnetic Resonance Guidance. *PNAS* **2003**, *100*, 13549–13554.
7. Kim, K. W.; Wong, S. M. Biomimetic Synthesis of Cadmium Sulphide-Ferritin Nanocomposites. *Adv. Mater.* **1996**, *8*, 11.
8. Klaus, T; Joerger, R; Olsson, E; Granqvist, C. G. Silver-Based Crystalline Nanopar-ticles, Microbially Fabricated. *Proc. Natl. Acad. Sci. U S A* **1999**, *23*, 13611–13614.
9. Klaus-Joerger, T.; Joerger, R; Olsson, E; Granqvist, C. Bacteria as Workers in the Living Factory: Metal-Accumulating Bacteria and Their Potential for Materials Science. *Trends Biotechnol.* **2001**, *19*(1), 15–20.
10. Kowshik, M. Extracellular Synthesis of Silver Nanoparticles by a Silver-Tolerant Yeast Strain MKY3. *Nanotechnology*, **2003**, *14*, 95–100.
11. Labrenz, M.; Druschel, G. K.; Thomsen-Ebert, T.; Gilbert, B.; Susan, A. Formation of Sphalerite (ZnS) Deposits in Natural Biofilms of Sulfate-Reducing Bacteria. *Science* **2000** 290, 1744–1747.
12. Liau. S. Y; Read, D. C; Pugh, W. J; Furr, J. R; Russell, A. D. Interaction of silver nitrate with readily identifiable groups: relationship to the antibacterial action of silver ions. *Lett. Appl. Microbiol.* **1997**, *25*, 279–283.
13. Liz-Marzan, L. M. Nano-Metals: Formation and Color. *Mater. Today* **2004**, *7*, 26–31.
14. Mandal, S.; Phadtare, S.; Sastry, M. Interfacing Biology with Nanoparticles. *Curr. Appl. Phys.* **2005**, *5*, 127–218.
15. Morones, J. R. Interaction of Silver Nanoparticles with HIV- 1. *J. Nano-biotechnol.* **2005**, *3*, 6.
16. Mulvaney, P. Surface Plasmon Spectroscopy of Nanosized metal Particles. *Langmuir* **1996**, *12*, 788–800.
17. Nam, J. M.; Thaxton, C. S; Mirkin, C. A. Nanoparticle-Based Bio-Bar Codes for the Ultrasensitive Detection of Proteins. *Science* **2003**, *301*, 1884–1886.
18. Nomiya, K.; Yoshizawa, A.; Tsukagoshi, K.; Kasuga, N. C.; Hirakawa, S.; Watanabe, J. Synthesis and Structural Characterization of Silver (I), Aluminium (III) and Cobalt (II) Complexes with 4-Isopropyltropolone (hinokitiol) Showing Noteworthy Biolog-ical Activities: Action of Silver (I)-oxygen Bonding Complexes on the Antimicrobial Activities. *J. Inorg. Biochem.* **2004**, *98*, 46–60.

19. Pooley, F. D. Bacteria Accumulate Silver During Leaching of Sulphide ore Minerals. *Nature* **1982**, *296*, 642—643.
20. Prasanna, S. *Extracellular and Intracellular synthesis of silver nanoparticles*. M. Tech. Thesis, Anna University, Madras – India, **2006**; pp 5–30.
21. Ren, N. Synthesis of Silver Nanoparticles via Electrochemical Reduction on Compact Zeolite Film Modified Electrodes. *Chem. Commum.* **2002**, *7*(23), 2814–2815.
22. Richard P. Blackmore Magnetotactic Bacteria. *Ann. Rev. Microbiol.* **1982**, *36*, 217–238.
23. Rosemary, M.; Pradeep, T. Solvothermal Synthesis of Silver Nanoparticles from Thiolates. *J. Colloid Interface Sci.* **2003**, *268*(1), 81–84.
24. Sastry, M. Geranium Leaf Assisted Biosynthesis of Silver Nanoparticles. *Biotechnol. Prog.* **2003**, *19*(6), 1627–1631.
25. Schultz, S.; Smith, D. R.; Mock, J. J.; Schultz, D. A. Single-Target Molecule Detection with Nonbleaching Multicolor Optical Immunolabels. *Proc. Natl. Acad. Sci. U. S. A.* **2000**, *97*, 996–1001.
26. Sondi, I.; Salopek-Sondi, B. Silver Nanoparticles as Antimicrobial Agent: A Case Study on E. Coli as a Model for Gram-Negative Bacteria. *J Colloid Interface Sci.* **2004**, *275*,177–182.
27. Tkachenko, A. G; Xie, H; Coleman, D; Glomm, W; Ryan, J; Anderson, M. F; Franzen, S.; Feldheim, D. L; Multifunctional Gold Nanoparticle-Peptide Complexes for Nuclear Targeting. *J. Am. Chem. Soc.* **2003**, *125*, 4700–4701.
28. Yingwei, X.; Ruqiang, Y.; Honglai, L. Synthesis of Silver Nanoparticles in Reverse Micelles Stabilized by Natural bio Surfactant, *Colloids and Surf. Physicochem. Eng. Aspects* **2006**, *279*(1–3), 175–178.

CHAPTER 6

BIOSENSORS: A POTENTIAL TOOL FOR DETECTION OF MICROBIAL CONTAMINANTS FOR FOOD SAFETY

ANURAG JYOTI[1,*] AND RAJESH SINGH TOMAR[2]

[1]*Amity Institute of Biotechnology, Amity University Madhya Pradesh, Gwalior 474005, India*

[2]*Amity Institute of Biotechnology, Amity University Madhya Pradesh, Gwalior 474005, India*
E-mail: rstomar@amity.edu

*Corresponding author. E-mail: ajyoti@gwa.amity.edu; anurag.bt@gmail.com

CONTENTS

6.1   INTRODUCTION

Food is one of the most important necessities of life. The quality and safety of food have always been a prime importance for human beings, irrespective of developing and developed nations. These include proper handling, processing, and safe storage to prevent contamination from microorganisms. In general, the manufacturing processes are responsible for the food spoilage and upon consumption lead to foodborne diseases. Harmful microorganisms may enter at any level of food chain and web and reach the ultimate user. Food industries are always at alarm with the concern of pathogenic microorganisms.[3] A number of examples and cases have been documented for the withdrawal of food items from the market worldwide.[9]

Over the last 20 years, the world has suffered tremendous loss of human lives and money due to foodborne diseases. These have pulled significant attention from public. Several regulatory agencies related to health and environment is now formulating the guidelines for safety measures for proper food security and quality. In spite of potential safety measures and precautions in food industries, contamination of food items occurs. The identification of certain foodborne pathogens is the necessity of hour to manage foodborne diseases.

Methods for the detection of foodborne pathogens have been developed. Conventional culture-based methods detect pathogens based on their cultivability and often identify them based on their physical appearance. Culturing of pathogens on specific media often take 2–3 days for significant changes for identification. The culturing method is time taking and labor intensive. Therefore, its practical applications are often limited. Moreover, certain pathogens undergo in viable but non-culturable (VBNC) state, which cannot be detected by culture-based methods. In the further development of detection system, the antigen–antibody interactions have been explored for disease diagnosis. The surface antigens of pathogens are matched with available antibodies in the laboratory condition. Based on their interactions, the pathogen is confirmed.

With the development of molecular techniques, the gene-based detection of pathogens began. Polymerase chain reaction (PCR) is a powerful technique that amplifies certain fragments of the genome/gene for the identification of pathogens. A number of reports have been documented using this technique. The PCR-based method confirms the presence of pathogens, it is unable to quantify the same. As a result, the risk assessment for certain diseases are not possible. In order to assess the risk through foodborne pathogens, quantitative methods have been developed that not only detect the pathogens but also quantify them. Real-time PCR (RTi-PCR) has been a revolutionary technique for the quantification of pathogens even in low doses.[11] This method uses probe chemistries to quantify the pathogens from different food matrices. Although, the method is rapid and sensitive, still it faces challenges in terms of cost and sophistication.[8]

Newer techniques have been emerged including applications of nanoparticles for the colorimetric-based methods. These include the label-based and label-free-detection of pathogenic deoxyribonucleic acid (DNA). These assays are simple and cost-effective, but they detect the nucleic acids of pathogens, which is again not feasible for point-of-care diagnostics. Development of isothermal amplification-based platforms has shown promises for the on-site detection of pathogens in rapid manner. Various biosensors have been developed for the sensitive and quick detection of foodborne pathogens. Biosensors have capability to overcome these spaces. Biosensors consist of a specific molecular recognition probe targeting an analyte of interest and a means of converting that recognition event into a measurable signal. There is a growing interest in biosensors due to rapidity, high specificity, and simplicity. Different formats of biosensors (namely, optical, chemical, and electrical) show greater potential for the detection of pathogens in the food.

This chapter intends to focus on different formats available for the detection of pathogens in foods. It also aims to cover different methods employed for the generation of probe biomolecules used in developing biosensors.

6.2   FOODBORNE PATHOGENS: PREVALENT AND EMERGING

Several foodborne pathogens have emerged in last few decades. The frequent use of antibiotics has led to the development of resistance and emergence of new pathotypes. New species and serotypes are being developed due to the horizontal gene transfer. Among the foodborne microbial agents, bacterial pathogens are the best characterized and reported causative agents of food-borne illnesses. Potential pathogens causing foodborne illnesses include: *Campylobacter, Salmonella, Listeria monocytogenes*, and *Escherichia coli* O157:H7. These bacterial pathogens together constitute the greatest burden of foodborne illness.

6.2.1 SALMONELLA

Salmonella sp. is a facultative anaerobic, Gram's-negative foodborne pathogenic bacteria. The pathogen can colonize a wide range of hosts. Salmonella is responsible for the major and frequent foodborne outbreaks worldwide. Due to the low infectious dose, the pathogen can infect humans easily. Salmonella has the ability to enter the food chain at any level. The raw food materials are the main vehicles for Salmonella to enter the chain. Salmonella has the ability to grow, multiply, and survive for longer duration in unprocessed food. The pathogens survive even at low temperatures.[14]

Each year, approximately, 93.8 million human cases of gastroenteritis and 155,000 deaths occur due to Salmonella infection around the world.[13] The typhoid caused by *Salmonella enterica* serotype Typhi remains an important public health problem in developing countries. Further, Salmonellosis causes substantial medical and economic burdens worldwide. It has been demonstrated that countries in south Asia and particularly south-east Asia exhibit high burden of typhoid fever.[15] Countries like India, Indonesia, Bangladesh, and Pakistan have been identified as high-risk sites for infections caused by *Salmonella* sp.

Salmonella have a wide range of hosts and are commonly associated with food animal products.[7] Outbreaks of Salmonella infections have been reported due to consumption of eggs, cheese, ice cream premix, a variety of fresh sprouts, juice, fishes, and other fresh vegetables.[1] Eggs and poultry meat products are one of the most important sources of infection by *Salmonella* in humans.[6] *Salmonellae* can enter the food chain at any stage and has ability to multiply to harmful levels. The type of food plays a major role in the severity of illness. *Salmonellae* present in the fatty foods can pass through the acidic environment of stomach and become invasive in the intestine.

Salmonella chromosome bears the virulence factors referred as *Salmonella* pathogenicity islands.[21] Five Salmonella pathogenicity islands (SPIs) are present on the *Salmonella* genome, which are responsible for the pathogenesis. Typhimurium, Enteritidis, Newport, and Heidelberg are among the top four commonly found serotypes that infect human. These serotypes are also the most frequently isolated serotypes from food samples.[4]

6.2.2 ESCHERICHIA COLI

Escherichia coli are other highly successful gut colonizers in many host species. *E. coli* strains isolated from intestinal diseases have been grouped into at least six different diarrheagenic *E. coli* (DEC) groups based on the specific virulence factors and phenotypic traits; and these include: enteropathogenic *E. coli* (EPEC), enterotoxigenic *E. coli* (ETEC), entero-invasive *E. coli* (EIEC), enteroaggregative *E. coli* (EAggEC), diffusely adherent *E. coli* (DAEC), and verocytotoxin producing *E. coli* (VTEC) or Shiga toxin-producing *E. coli* (STEC).

Enterotoxigenic *E. coli* (ETEC), a potential pathovar of *E. coli* is regarded as a major cause of diarrhea worldwide in humans, mainly affecting children and travelers.[17] The contamination of drinking or recreational waters with ETEC has been associated with waterborne disease outbreaks. Diarrhea due to ETEC is caused by the consumption of contaminated water and food.[18] In the case of improper sanitation and hygiene, the ETEC is a major cause of diarrhea. In the developing countries, surface waters are the potential reservoirs of ETEC and transmission can occur while bathing and/or using water for food preparation.[18] Further, these forms of transmission lead to the infection of local populations and international travelers visiting these areas, hence often referred to as traveler's diarrhea. A few studies report the prevalence of ETEC in surface waters and in macrophytes.[23]

Food matrices are the potential carriers of pathogenic bacteria and are one of the most well investigated and monitored causes of intestinal infectious disease. Recently, the outbreaks of gastroenteritis occurred in Denmark due to consumption of contaminated lettuce. Investigations showed that the outbreaks were caused by ETEC and by norovirus of several genotypes.[5]

The pathogenesis of ETEC secretory diarrhea involves the colonization of small intestine epithelial cells by means of filamentous adhesins known as colonization factors (CFs) followed by the production of at least one out of two enterotoxin types, the heat-labile toxin (LT), and/or the heat-stable toxin (ST). Heat-labile toxins produced by ETEC strains are a heterogeneous group of toxins. Two major LT families have been identified, LT-I and LT-II. Heat-labile toxin-II is rarely found among human-derived ETEC strains. The enterotoxin gene *LT1* commonly present in strains associated with the human illness has been observed abundantly in ETEC recovered from food matrices.

Enteropathogenic *E. coli* (EPEC) is also considered a potential food-borne pathogen, though surveillance for this DEC type is generally poor. Although the true association with food vehicles is unclear, sporadic infections with this type continue to be reported with chicken and beef as common sources. Both ETEC and EAggEC may also be associated with foodborne outbreaks albeit infrequently. Most of the current knowledge on trends and persistence of foodborne *E. coli* infections is derived from studies on VTEC. As indicated previously, in the early outbreaks the sources of VTEC were most often found to be contaminated beef meat but today almost any vehicle in contact with ruminant feces is potentially a source including vegetables, sprouts, fruits, meat products (such as dry fermented sausages), juices and milk (both pasteurized and unpasteurized) as well as fecally-contaminated drinking, recreational and bathing waters, and novel transmission routes for outbreaks continue to arise.

6.3   CONVENTIONAL METHODS FOR DETECTION OF PATHOGENIC BACTERIA PREVALENT IN FOOD

Conventional methods for the detection and identification of microbial pathogenic agents mainly rely on specific microbiological and biochemical identification. Conventional methods use the culture-based methods: first involve counting of bacteria; second immunology-based methods involve antigen–antibody interactions; and the third PCR method which involves DNA analysis. While these methods can be sensitive, inexpensive, and give both qualitative and quantitative information of the tested microorganisms, they are greatly restricted by assay time, and also initial enrichment is needed in order to detect pathogens which typically occur in low numbers in food.

Classical culture-based method is the oldest bacterial detection technique and remains the standard detection methodology. This method has been widely adopted by the laboratories and is established as a gold standard for the identification and detection of pathogenic bacteria.

6.3.1   CULTURE-BASED METHODS

The traditional method of detection of ETEC often involves cultivation of bacteria in selective media. This includes the cultivation in MacConkey or Eosin Methylene Blue (EMB) agar followed by overnight incubation. The characteristic colonies are typical pink to red color for *E. coli*. Further confirmation is done by Indole, Methyl red, Voges-Proskauer, and Citrate (IMViC) test which examines the ability to produce indole and sufficient acid to change the color of a methyl red indicator.[24] These assays are laborious and time consuming.

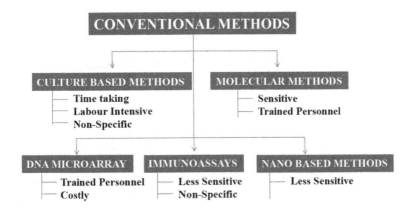

FIGURE 6.1   Conventional methods used for pathogen detection and their limitations.

Enterohemorrhagic *E. coli* (EHEC) is exploited in sorbitol MacConkey agar (SMAC). The International Organization for Standardization (ISO) protocol (ISO 16654) recommends that addition of Cefixime and potassium tellurite to SMAC (CT-SMAC) may increase the selectivity in samples. *E. coli* O157 generally produces colorless colonies when cultured on this media, thus distinguishing it from other EHEC serogroups. EHEC O157:H7 colonies are confirmed with biochemical tests and immunoassays having the O157 somatic antigen and H7 flagellar antigen (Figure 6.1).

Salmonellae serovars are identified and detected on the basis of colony appearance on chromogenic and other selective agar media, followed by confirmation using classical biochemical and serological testing. The biochemical tests include fermentation of glucose, negative urease reaction, lysine decarboxylase activity, and H_2S production. In the European Union (EU), the reference detection methods are published by the ISO. The procedure for the detection of foodborne *Salmonella* includes nonselective enrichment in buffered peptone water (BPW) broth for 16–20 h followed by selective enrichment cultivation in selective media and incubation on two different selective agar plates for isolation of colonies. The colonies are then identified by means of biochemical tests. This procedure takes at least 72 h to complete. In Bacteriological Analytical Manual (BAM), published from the United States, the procedure for the detection of *Salmonella* includes nonselective enrichment in nutrient broth (NB) for 16 h and followed by another 16-h selective enrichment cultivation in either Rappaport–Vassiliadis (RV) or tetrathionate brilliant green (TBG) broth.

The colonies are isolated by using selective agar plates and identified biochemically.

6.3.1.1 IMMUNOASSAYS

Enzyme-linked immunosorbent assay (ELISA) has been widely used for detection of LT using microtiter GM1 ganglioside methods. It requires culturing of the bacteria before testing for the presence of enterotoxins. This technique has also been further evolved into the inhibition-ELISA for detection of ST. Both ELISA and inhibition ELISA, based on monoclonal antibodies against LT and ST, are today used in many laboratories for detection and identification of ETEC. Although, the immunoassays, based upon the antigen-antibody interactions are established techniques, they often lack specificity. Serotyping has also been used to identify and characterize ETEC strains. The ETEC has more than 78 O groups and 34 H groups identified; therefore the determination of O serogroups associated with the lipopolysaccharides in the cell wall and H serogroups of the flagella is difficult.[17] A huge number of combinations of O and H groups make serotyping less suitable for identification of ETEC.

In general, the immunoassays use antisera for detection of flagellar (H) and somatic (O) antigens. Isolates with a typical biochemical profile, which agglutinate with both H and O antisera, are identified as *Salmonella sp.*[12] Further, the positive isolates are often confirmed by serotyping and using techniques such a phage typing and pulsed-field gel electrophoresis (PFGE). The immunoassays are capable of detecting 10^4–10^6 cells of *Salmonella* in foods per assay. The antibodies may cross-react with antigens in closely related bacteria, while showing low reactivity with some *Salmonella* serotypes (Fig. 6.1).

6.3.2 MOLECULAR METHODS

The advancements in nucleic acids (DNA/RNA) based methods to detect bacteria offer better sensitivity and selectivity over conventional microbiological techniques. The potential benefit of nucleic acid-based methods is high throughput, precise and reliable outputs with reduced time. The different methods that have been used frequently to detect water borne microorganism are mentioned in the following sections.

6.3.2.1   POLYMERASE CHAIN REACTION AND QUANTITATIVE POLYMERASE CHAIN REACTION (QPCRS)

The PCR and qPCRs are powerful molecular techniques frequently used to amplify the desired gene sequence. PCR was able to detect and amplify the targeted nucleic acid. However, its application to quantitative analysis was restricted due to its end point detection as well as low sensitivity. Further, at the end point in PCR the concentration of DNA/RNA is not proportionate to the initial concentration of template DNA, due to limitations of PCR and associated biases. qPCRs, an advance molecular technique, has capability of amplification and simultaneous detection in real time mode. The qPCR detection depends on the florescence generated as products accumulate. The fluorescence level is directly proportional to the quantity of target amplicons formed. There are several qPCR detection chemistries (for example, intercalating dyes, hydrolysis probes, and hybridization probes) available that have been widely used for the detection of water borne pathogens. The absolute quantification in qPCR has been used for quantitative detection of pathogens in environmental as well as clinical samples (Fig. 6.1). qPCR in combination with ethidium monoazide and propidium monoazide dye have been used for discriminating viable and nonviable pathogens.

6.3.2.2   NUCLEIC ACID MICROARRAYS

Nucleic acid microarray technology is fabricated on the ability of nucleic acids and complementary sequences to hybridize effectively to their target sequences. The oligonucleotide probes targeting specifically desired genes are made to attach to chemically treated glass slide surface. The DNA/RNA, extracted from the sample of interest, is incubated with the slide under defined conditions to facilitate hybridization. The hybridized materials labeled with a radioactive/fluorescent group produce radiation/fluorescence in the presence of the specific target sequence. The intensity of the radiation/fluorescence determines the concentration of the sequence.

This method has been implemented to detect specific microorganism within large array of microbes from environmental samples. The technique has also been utilized to detect alterations in gene expression profiling of particular microorganisms in response to any chemical and environmental stress (Fig. 6.1).

6.3.3 NANOTECHNOLOGY-BASED APPROACHES

6.3.3.1 NANOPARTICLE-BASED PROBES
FOR DETECTION OF DNA

The potential applications of nanotechnology are becoming increasingly important for the development of ultrasensitive DNA detection systems. The molecular recognition is fundamental for the development of diagnostic tools. Various organic molecules, possessing unique properties, have been used to achieve the recognition of different targets. Gold nanoparticles (GNPs) have been explored for bio-diagnostics due to unique optical properties (i.e., surface plasma resonance absorption and resonance light scattering), a variety of surface coatings and great biocompatibility.[20] GNP-based DNA detection has high sensitivity as compared to conventional fluorescence based assays due to the extremely high molar absorptivity of GNPs (1000 times higher than that of organic dyes).[27]

The optical properties of GNPs are governed by collective oscillation of electrons at surfaces known as surface plasmon resonance (SPR). The resonance frequency of this oscillation lies in the visible region of the electromagnetic spectrum. The GNPs have a high surface to volume ratio, therefore the surface electrons are sensitive to change in the dielectric (refractive index) of the medium. Any changes to the environment of these particles (surface modification, aggregation, medium refractive index, etc.) lead to colorimetric changes of the dispersions.[19] The aggregation behavior of GNPs has been widely explored for different applications. This has further facilitated the application in bio-detection via numerous detection methods. The versatile surface chemistry of GNPs can be achieved by attaching various bio-functional groups, such as nucleic acids, sugars, proteins via the strong affinity of gold surface with thiol ligands.

The colloidal solution of GNPs is monodisperse red and exhibit a narrow surface plasmon absorption band centered on 520 nm (depends upon particle size) in the UV-visible spectrum. In contrast, when the particles aggregate, the solution appears purple/blue, corresponding to the characteristic red shift in the SPR towards higher wavelength. A number of factors such as size and shape of the nanoparticle, refractive index of the surrounding media and interparticle distance are taken into account for use in colorimetric detection of DNA. The GNPs obtained by citrate reduction present in solution are charged particles and they are sensitive to changes in solution dielectrics. Hence, with the addition of NaCl the surface charge is shielded leading to a decrease in interparticle distance and particle aggregation.[10]

6.3.3.1.1 *Principle*

The use of GNP-based colorimetric methods for detection of DNA has been widely reported and reviewed.[19] Bio-functionalization of GNPs with the thiol modified single stranded DNA (ssDNA) also called GNP probe, has been synthesized for the DNA detection. The ssDNA probe strand is designed to be complementary to a target of interest and is attached to the GNPs through chemisorption of the thiol group onto the surface of the GNPs. After hybridization with the target, the GNP probes come in the close proximity, which leads to change in color.

Rosi and Mirkin reported the colorimetric detection of DNA targets based on the cross-linking mechanism use of GNP probes.[19] The two different batches of probes are designed to target the DNA. Thus, upon the addition of target DNA, a polymeric network of GNP probes is generated due to aggregation, turning the solution from red to blue. This aggregation mechanism is mainly applied to detect small sized targets. The GNP aggregation induced by interparticle cross-linking is a relatively slower process. The relatively slow aggregation is due to the nature of the interparticle cross-linking aggregation mechanism (Fig. 6.1). In general, the aggregation is driven by random collisions between nanoparticles with relatively slow Brownian motion.[19]

6.3.4 *BIOSENSORS FOR PATHOGEN DETECTION*

In recent years, the biosensors have emerged for the detection of pathogens. Biosensor is an analytical device which constitutes a bioreceptor (for recognition of target analytes) and a transducer which converts the biological interactions into a measurable electrical signal.[26] Different platforms of biosensors have been developed based on the mechanism of detection.

Optical biosensors often work by measuring the change in optical properties of the solution upon biological interactions of target and bioreceptor. Electrochemical biosensors detect target pathogens by measuring significant changes in conductance, resistance, or capacitance of the surface.[2,25]

Biosensor-based methods are advantageous over other conventional methods. These are integration and miniaturization of scattered detection components. The integrated sensors reduce time, labor and are simple in operation.[22] Management of food-borne disease outbreaks demands constant improvements in the detection technologies at cheap cost. The regulatory agencies formulate the guidelines for rapid and sensitive identification of pathogen(s).

The concept of "Lab-on-chip" has further set standards as these devices work on the integration of sensors and microfluidic systems in a miniaturized set up for real time monitoring of samples.[16] It integrates several components on a single chip and reduces sample and reagent consumption. This has fast detection times and low limits of detection.

Different platforms of biosensors exist which include, enzyme-based biosensors, DNA-based biosensors, and cell-based biosensors. Recently, paper-based biosensors have emerged. The capillary force generated in paper allows the transportation of samples/liquids. This eases in the movement and detection of target by immobilized probes. These paper-based biosensors often compromise with the sensitivity. Nanomaterials having outstanding properties are alternative to conventional methods. These materials show greater affinity and conjugate with biomolecules with ease. The conjugation of nanomaterials with biomolecules leads to the development of nanobiosensors. This has emerged into many devices with minimal noise. Colloidal gold solution is one of the most studied nanomaterials available for biosensors.

6.3.5 FEATURES OF A GOOD BIOSENSOR

- Selectivity: The biosensor must inclusive for its target and exclusive for nontarget analyte(s). The device should have ability to differentiate the closely related biomolecules or strains of a bacterium.
- Sensitivity: A biosensor should be sensitive enough to detect and measure the range of a given target analyte. In case of a pathogenic bacteria, the sensor should be enough sensitive to detect the infectious dose of pathogen.
- Signal reproducibility: The biosensor should show same results when samples having same concentrations are analyzed several times.

- Rapidity: A biosensor should respond quickly to the target(s) so that the management of particular pathogen and disease caused by it can be done efficiently.

6.4 SUMMARY

Globally the food industry is one of the largest industries. In general, the quality of food gets deteriorated due to microbial contamination. It not only poses major health threats but also causes major losses to food industries and food-borne illness. The quality control towards food safety is one of the major responsibilities for food industries to ensure high-quality standards and minimize the risk for the consumer. Existing methods for pathogen detection face challenges of inadequate monitoring in terms of specificity, rapidity and simplicity. Hence, there is an urgent need to overview and develop novel sensing formats for early detection of microbial contamination in food.

One promising area is the field of nanotechnology, where nanoparticle-based assays can be developed for specific detection of bioanalytes in food samples. Biosensors have capability to overcome these lacunae. Biosensors consist of a specific molecular recognition probe targeting an analyte of interest and a means of converting that recognition event into a measurable signal. There is a growing interest in biosensors due to rapidity, high specificity, and simplicity. Different formats of biosensors (namely, optical, chemical, and electrical) show greater potential for the detection of pathogens in the food. The present chapter intends to focus on different formats available for detection of pathogens in foods. It also aims to cover the different methods employed for the generation of probe biomolecules used in developing biosensors.

KEYWORDS

- Biosensors
- ELISA
- *Escherichia coli*
- gold nanoparticles
- microarrays

- pathogen detection
- point-of-care diagnostics
- polymerase chain reaction
- quantitative polymerase chain reaction
- *Salmonellae*

REFERENCES

1. Antony, B.; Dias, M.; Shetty, A. K.; Rekha, B. Food Poisoning due to *Salmonella Enterica* Serotype Weltevreden in Mangalore. *Indian J. Med. Microbiol.* **2009**, *27*(3), 257–258.

2. Bridle, H.; Desmulliez, M. Biosensors for the Detection of Waterborne Pathogens. In *Waterborne Pathogens. Detection Methods and Applications;* Bridle, H., Ed.; Elsevier B.V.: London, UK, 2014; p 401.

3. Cappitelli, F.; Polo, A.; Villa, F. Biofilm Formation in Food Processing Environments is Still Poorly Understood and Controlled. *Food Eng. Rev.* **2014**, *6*(1), 1–2.

4. CDC. *Salmonella* Suveillance Summary for 2002. CDC, US Department of Health and Human Services; 2003.

5. Ethelberg, S.; Lisby, M.; Böttiger, B.; Schultz, A. C.; Villif, A.; Jensen, T.; Olsen, K. E.; Scheutz, F.; Kjelsø, C.; Müller, L. Outbreaks of Gastroenteritis Linked to Lettuce. *Eurosurveillance* **2010**, *15*(6), 19484.

6. Hald, T.; Vose, D.; Wegener, H. C.; Koupeev, T. A Bayesian Approach to Quantify the Contribution of Animal-Food Sources to Human Salmonellosis. *Risk Anal.* **2004**, *24*(1), 255–269.

7. Hoelzer, K.; Moreno, S. A. I.; Wiedmann, M. Animal Contact as a Source of Human Non-Typhoidal Salmonellosis. *Vet. Res.* **2011**, *42*(1), 34.

8. Jyoti, A.; Tomar, R. S. Nanosensors for the Detection of Pathogenic Bacteria. *Nanosci. Food Agric.* **2016**, *1*(20), 129–150.

9. Jyoti, A.; Agarwal, M.; Tomar, R. S. Culture-Free Detection of Enterotoxigenic *Escherichia coli* in Food by Polymerase Chain Reaction. *Int. J. Res. Appl. Sci. Eng. Technol.* **2015**, *3*(4), 1060–1064.

10. Jyoti, A.; Pandey, P.; Singh, S. P.; Jain, S. K.; Shanker, R. Colorimetric Detection of Nucleic Acid Signature of Shiga Toxin Producing *Escherichia coli* using Gold Nanoparticles. *J. Nanosci. Nanotechnol.* **2010**, *10*(7), 4154–4158.

11. Kubista, M.; Andrade, J. M.; Bengtsson, M.; Forootan, A.; Jonak, J.; Lind, K.; Sindelka, R.; Sjoback, R.; Sjogreen, B.; Strombom, L.; Stahlberg, A.; Zoric, N. The Real-Time Polymerase Chain Reaction. *Mol. Aspects Med.* **2006**, *27*(2–3), 95–125.

12. Magliulo, M.; Simoni, P.; Guardigli, M.; Michelini, E.; Luciani, M.; Lelli, R.; Roda, A. A Rapid Multiplexed Chemiluminescent Immunoassay for the Detection of *Escherichia coli* O157:H7, *Yersinia enterocolitica, Salmonella typhimurium,* and *Listeria monocytogenes* Pathogen Bacteria. *J. Agric. Food Chem.* **2007**, *55*(13), 4933–4939.

13. Majowicz, S. E.; Musto, J.; Scallan, E.; Angulo, F. J.; Kirk, M.; O'Brien, S. J.; Jones, T. F.; Fazil, A.; Hoekstra, R. M. The Global Burden of Nontyphoidal Salmonella Gastroenteritis. *Clin. Infect. Dis.* **2010**, *50*(6), 882–889.

14. Malorny, B.; Löfström, C.; Wagner, M.; Krämer, N.; Hoorfar, J. Enumeration of *Salmonella* Bacteria in Food and Feed Samples by Real-Time PCR for Quantitative Microbial Risk Assessment. *Appl. Environ. Microbiol.* **2008**, *74*(5), 1299–1304.

15. Ochiai, R. L.; Acosta, C. J.; Holliday, M. C. D.; Baiqing, D.; Bhattacharya, S. K.; Agtini, M. D.; Bhutta, Z. A.; Canh, D. G.; Alim, M.; Shin, S.; Wain, J.; Page, A. L.; Albert, M. J.; Farrar, J.; Elyazeed, R. A.; Pang, T.; Galindo, C. M.; Seidlein, L.; Clemens, J. D. (The Domi Typhoid Study Group). Study of Typhoid Fever in Five

Asian Countries: Disease Burden and Implications for Controls. *Bull. World Health Organ.* **2008**, *86*(4), 260–268.

16. Pires, N. M.; Dong, T.; Hanke, U.; Hoivik, N. Recent Developments in Optical Detection Technologies in Lab-on-a Chip Devices for Biosensing Applications. *Sensors (Basel)* **2014**, *14*(8), 15458–15479.

17. Qadri, F.; Svennerholm, A. M.; Faruque, A. S.; Sack, R. B. Enterotoxigenic *Escherichia coli* in Developing Countries: Epidemiology, Microbiology, Clinical Features, Treatment, and Prevention. *Clin. Microbiol. Rev.* **2005**, *18*(3), 465–483.

18. Ram, S.; Vajpayee, P.; Shanker, R. Rapid Culture-Independent Quantitative Detection of Enterotoxigenic *Escherichia coli* in Surface Waters by Real-Time PCR with Molecular Beacon. *Environ. Sci. Technol.* **2008**, *42*(12), 4577–4582.

19. Rosi, N. L.; Mirkin, C. A. Nanostructures in Biodiagnostics. *Chem. Rev.* **2005**, *105*(4), 1547–1562.

20. Sato, K.; Hosokawa, K.; Maeda, M. Colorimetric Biosensors Based on DNA-Nanoparticle Conjugates. *Anal. Sci.* **2007**, *23*(1), 17–20.

21. Shea, J. E.; Hensel, M.; Gleeson, C.; Holden, D. W. Identification of a Virulence Locus Encoding a Second Type III Secretion System in *Salmonella* Typhimurium. *Proc. Natl. Acad. Sci. USA.* **1996**, *93*(6), 2593–2597.

22. Singh, A.; Poshtiban, S.; Evoy, S. Recent Advances in Bacteriophage Based Biosensors for Food-Borne Pathogen Detection. *Sensors*, **2013**, *13*(2), 1763–1786.

23. Singh, G.; Vajpayee, P., Ram, S.; Shanker, R. Environmental Reservoirs for Enterotoxigenic *Escherichia coli* in South Asian Gangetic Riverine System. *Environ. Sci. Technol.* **2010**, *44*(16), 6475–6480.

24. Todar, K. Pathogenic *Escherichia coli*. In *Todar's Online Textbook on Bacteriology;* University of Wisconsin, Dept. of Bacteriology: Madison, 2008; p 579.

25. Vidal, J. C.; Bonel, L.; Ezquerra, A.; Hernandez, S.; Bertolin, J. R.; Cubel, C.; Castillo, J. R. Electrochemical Affinity Biosensors for Detection of Mycotoxins: A Review. *Biosens. Bioelectron.* **2013**, *49*(15), 146–158.

26. Zhao, W.; Chiuman, W.; Lam, J.; McManus, S. A.; Chen, W.; Cui, Y.; Pelton, R.; Brook, M. A.; Li, Y. DNA Aptamer Folding on Gold Nanoparticles: From Colloid Chemistry to Biosensors. *J. Am. Chem. Soc.* **2008**, *130*(11), 3610–3618.

CHAPTER 7

MODELING AND FERMENTATATION ASPECTS OF PULLULAN PRODUCTION FROM JAGGERY

G. V. S. RAMA KRISHNA

Department of Biotechnology, School of Life Sciences, K. L. University, Green Fields, Vaddeswaram 522502, District Guntur, Andhra Pradesh, India

Corresponding author. E-mail: krishna.ganduri@kluniversity.in

CONTENTS

7.1 INTRODUCTION

Over the past few decades, the number of polysaccharides produced by microbial fermentation has been gradually increasing. The first microbial polysaccharide discovered was dextran, in 1940. Later, other complex polysaccharides such as Xanthan, Gellan, Curdlan, Emulsan, Sclero-glucan, Succinoglycon, Lentinan, and Pullulan were gaining commercial importance. The biotechnological production of biopolymers may occur in intracellular or extracellular and pose severe consequences in upstream and downstream processing to obtain them in a purified state. Mostly, exopolysaccharides (EPSs) are produced by microbes either as an extracellular or cell-surface-attached material in amorphous slime forms,[66] which are categorized as (neutral) homopolysaccharides and (polyionic) heteropolysaccharides. One of such high priced ($25/Kg) biopolymer, pullulan is getting renewed focus because of its distinctive properties and excellent food, pharmaceutical, and biomedical applications.

Pullulan is a water soluble random coil glucan, consisting of regularly repeating copolymer of maltotriose trimer {1→6)-α-D-glucopyranosyl-(1→4)-α-D-glucopyranosyl-(1→4)-α-D-glucopyranosyl-(1→n}. This regular alteration of α-(1→4) and α-(1→6) bonds results in two distinctive properties of structural flexibility and enhanced solubility along with adhesive film/fiber forming, oxygen impermeability properties.[60] Pullulan has been considered as "Generally Regarded As Safe" (GRAS) status by United States Food and Drug Administration (USFDA) (Fig. 7.1). Pullulan is being widely used as an edible film, binding agent, flocculating agent, as plasma expander, entitles this polysaccharide as "wonder biopolymer." Pullulan production has been stable with major applications in food and pharma industries for number of years, but nowadays pullulan is being tested for capsule formation, edible packaging and environmental bioremediation.

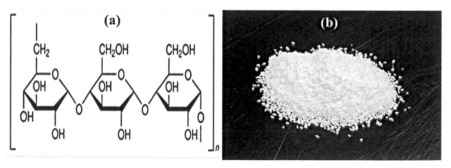

FIGURE 7.1   (a) Chemical structure, (b) whitish granular form of pullulan.

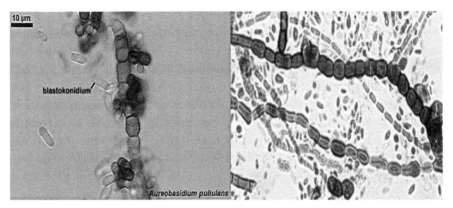

FIGURE 7.2 Microscopic view of *Aureobasidium pullulans* showing one to two-celled, darkly pigmented arthroconidia and hyaline, single-celled, ovoid-shaped conidia.

Pullulan has been produced in large amounts by *Aureobasidium pullulans* (De Bary), cosmopolitan yeast-like micro fungus, a saprophyte found in tropical and temperate environment with fluctuating moisture content in phyllospore and also present in damp indoor surfaces, food and feed substances (Fig. 7.2).[60] Multiple factors affect the biosynthesis of pullulan in fermentation media includes temperature, pH, viscosity, type and sources of carbon, nitrogen concentrations, high carbon: nitrogen ratio, oxygen supply, mineral salts, agitation rate, mixing control, and so forth. Phosphates and potassium are also important for improving buffering capacity of microbial fermentation broths. Pullulan can be produced in high quantities, when *A. pullulans* cultured in sucrose, glucose, fructose, maltose, starch, or maltooligosaccharides.[13,14]

Several researchers have exercised various agro-industrial wastes on growth of *A. pullulans* and pullulan production. Jaggery, a concentrated sugar cane juice with or without prior purification produced by cottage industry, yielded maximum production of pullulan in batch cultivation of *A. pullulans* CFR-77.[75]Statistical optimization of process parameters of any fermentation gives the idea of most significant variables which affect the yield and concentration of final bioproduct. Response surface methodology (RSM) is a statistical modeling and optimization approach to evaluate the interactive and synergistic effects of the given quantitative data from appropriate experiments to determine and draw solution to multivalent equations, thus providing an optimum solution to achieve the maximum output.[39]

Kinetic models describe the behavior of microorganisms under physical and chemical conditions such as profiles of pH, temperature, substrate, biomass, and product. In order to develop these models, all the process variables have to be measured and modeled.A very useful logistic type of unstructured kinetic model for pullulan fermentation was developed for *A. pullulans* growth, limited substrate consumption and pullulan production.[14]Therefore, it is remarkably attractive to maximize production levels of pullulan using *A. pullulans* by varying carbon and nitrogen substrates under batch fermentation. Because of excellent properties possessed by pullulan, it has been a major research subject for many researchers worldwide.

This chapter discusses the batch fermentative production of pullulan using jaggery, as a carbon substrate, evaluation of kinetic parameters, and statistical optimization of process variables for effective production of pullulan.

7.2　LITERATURE REVIEW

Bio-based polymers have become closer to the reality of replacing conventional, synthetic polymer than ever before, thanks to progress in white biotechnology, the production of biopolymers such as proteins, microbial polysaccharides, lipids and special polymers from renewable, and natural resources. A variety of microbial EPSs possess diverse applications in food, chemical, energy, and pharmaceutical industries. Pullulan is one of the biopolymer that has been safely used in Japan as a food ingredient and as pharmaceutical bulking agent, over the last two decades.[49]

7.2.1　PULLULAN

7.2.1.1　HISTORICAL OUTLINE

Pullulan is synthesized as a water-soluble, neutral polysaccharide by different strains of polymorphic fungus *A. pullulans* (De Bary) Arnaud, formerly known as *Pullularia pullulans* (De Bary) Berkhout or *Dematium pullulans* (De Bary). Pullulan, a biopolymer, was first reported by Bauer.[4] Although *A. pullulans*, as pullulan producer strain, was first explained (as *D. pullulans*) by De Bary in 1866, yet the polysaccharide formation by *A. pullulans* was first observed by Bauer[4]; while Bernier[6] had isolated and characterized the produced polymer. However, pullulan name was first coined by Bender.[5] Commercial production of pullulan began in 1976 by M/s. Hayashibara Company Ltd., Okayama, Japan. Elemental analysis revealed that the chemical formula of pullulan was $(C_6H_{10}O_5)n$. This EPS was produced mostly by *A. pullulans* as an amorphous slime matter.[11,66]

7.2.1.2 PROPERTIES AND STRUCTURE

Pullulan chemical structures revealed the predominant linkages of α-(1→4) of α-glucans and were concluded by Bender.[5] Subsequent experiments confirmed that pullulan was essentially a linear glucan containing α-(1→4) and α-(1→6) linkages in 2:1 ratio from Infrared (IR), periodate oxidation, and methylation analysis (Fig. 7.1(a)). Pullulan possess distinctive physical traits because of unique linkage pattern of maltotriose units (Table 7.1). Owing to its unique linkage pattern in chemical structure, pullulan exhibits distinctive physical, chemical, and biological properties such as not having color, odor, taste, having high viscosity, solubility in water, insoluble in organic solvents, nonhygroscopic in nature, moldable, spinnable and good adhesive and binder, nontoxic, edible and biodegradability.

TABLE 7.1 Main Quality Features of Pullulan.

Parameter	Specification
Appearance	White or yellowish-white
Water solubility (25°C)	Easily soluble
Specific optical activity [α] D_2O	Min. + 160°
Polypeptides (%)	Max. 0.5
pH of solutions	5–7
Mineral residue-ash (sulfated, %)	Max. 3
Moisture (loss on drying, %)	Max. 6
Molecular weight (range, kDa)	100–250

7.2.1.3 APPLICATIONS

Due to excellent properties possessed by pullulan, it has been used in many different fields ranging from food, pharmaceutical, cosmetic, and other allied industries. Significant efforts have been made by several researchers towards modification of pullulan to improve its further applicability in diverse industries. Recent literature of modifications and applications of pullulan has been summarized in Table 7.2.

TABLE 7.2 Pullulan Derivatives and Their Current Potential Applications.

Pullulan product/derivatives	Current or potential applications	Reference
2-Nitroalkyl pullulan ester	EPS with amino functionalities	[26]
Anionically modified pullulan	Blood plasma substitute	[57]
Antibacterial film	Food preservation	[24]
Biodegradable pullulan	Plastic	[34]
Carbonation	Drug carriers	[9]
Carboxymethyl pullulan gel	Antibacterial release wound dressing	[33]
Food additive	Dietary and functional food	[42, 62]
Heparin-conjugated pullulan	Cell/tissue engineering	[13]
Methacrylate pullulan gel	Cell proliferation and cluster formation	[3]
NPcaps® capsules	Capsules	[29]
Palmitoyl and cholesteryl derivatives	Hydrophobic reactant	[41]
Poly(L-lactide)-grafted pullulan	Biodegradable polymer	[43]
Pullulan 6-hydroxyhexanoates/6-dilactates	Biomedical/tissue engineering	[19]
Pullulan additives	Blood plasma substitutes	[70]
Pullulan blend	Drug/gene delivery and imaging	[8, 51]
Pullulan cinnamates	Modeling of cell wall biogenesis	[28]
Pullulan gel	Electrophoresis	[40]
Siloxane pullulan	Water stable pullulan/silicone composite	[71]

7.2.2 MORPHOLOGY OF A. PULLULANS

Microbial synthesis of pullulan was first reported by ubiquitous fungi called *Aureobasidiu mpullulans*, isolated from moist and damped environments. Other than *A. pullulans*, many organisms that produce pullulan during their metabolism are: *Tremella mesenterica*, *Cyttaria hariotii*, *Cyttaria darwinii*, *Cryphonectria parasitica*, *Teloschistes flavicans*, and *Rhodototula bacarum*.[59] *A. pullulans*, popularly known as black yeast, is a most widespread polymorphic, saprophytic fungus associated with wide range of terrestrial and aquatic habitats, in temperate and tropical environments.[22] *A. pullulans* is a cosmopolitan, dematiaceous fungus grows moderate rapidly and matures (colony diameter is 1–3 cm) within 7 days of incubation, at 25°C on PDA. Colonies appear as flat, smooth, moist, mucoid to pasty, shiny and leathery with surface pale pink in color and develops into blackish brown. Most undesirable characteristic feature of this fungus is the production of dark, melanin-like pigment that appears dark green to black in color.

During the growth of *A. pullulans*, attempts have been made to identify the precise location and possible site of α-glucanpullulan production in different morphological forms using gold-conjugated pullulanase by transmission electron microscopy (TEM). Silver enhancement was used to enlarge colloidal gold particles has shown the multicelled chlamydospores, swollen cells, blastospores, germ tubes arising from swollen cells, and fungal hyphae.[10] In addition to pullulan, different strains of *A. pullulans* have also synthesized wide range of important metabolites, enzymes (amylase, proteinase, lipase, cellulase, xylanase, mannanase, and transferases), antibiotics, siderophores, single-cell protein (SCP), alkaloids, and so forth. Some of the *A. pullulans* exhibit potential antagonistic activity against phytopathogenic fungi, and thus used as biocontrol agents in postharvest diseases and also possess capability to degrade xenobiotic compounds.[22]

7.2.3   PULLULAN BIOSYNTHETIC PATHWAY

Aureobasidium pullulans synthesizes pullulan intracellularly at the cell wall membrane and is secreted out to the cell surface to form a loose, slimy layer.[58] For many years, mechanism of pullulan biosynthesis in *A. pullulans* was not fully understood. Duan[20] revealed the possible pathway for pullulan synthesis. A set of three key enzymes namely, α-phosphoglucosemutase, uridine diphospho glucose pyrophosphorylase (UDPG-pyrophosphorylase), and glucosyltransferase, needed to convert glucose into pullulan. *A. pullulans* also utilizes other sugars like sucrose, mannose, galactose, maltose, fructose, and even agricultural wastes such as carbon sources.[31] Hexokinase and isomerase presence is also important for *A. pullulans* to convert different carbon sources into the pullulan precursor, UDPG. The sequential steps of pullulan synthesis are given further.[13]

7.2.4   FERMENTATIVE PRODUCTION OF PULLULAN

Aureobasidium pullulans, pullulan producer strain is capable of growing on variety of substrates, including even with agricultural waste without chemical or enzymatic pretreatment due to the multiple enzyme systems available for saccharification of complex sugars.[31] Pullulan was synthesized from wide variety of carbon and nitrogen sources and were reported by various research groups. Table 7.3 summarizes, the types of *A. pullulans* strains used for pullulan production on different carbon substrates indicating their fermentation conditions. Pullulan was also produced from cocultures of pullulan producing strain *A. pullulans* and inulin-degrading strain Kluyveromyces *fragile*.[56]

In addition to carbon substrates, nitrogen-rich substrates also play important role in pullulan production. High concentration of nitrogen source in production medium contributed to the transition from yeast-like cells to chlamydospores in *A. pullulans*. Nitrogen depletion in medium triggered more pullulan by stimulating glycolysis and also 10:1 carbon:nitrogen ratio was the most favorable condition of pullulan production.[63] Fermentation conditions namely, pH, temperature, dissolved oxygen (DO), and so forth were also played vital influence on pullulan synthesis. A high concentration of yeast-like cells, only swollen cells and chlamydospores were responsible for pullulan formation.[10] Not only the cultivation parameter of fermentation, the bioreactor design is also influenced the pullulan release in to medium.

In order to overcome the suppression effect, fed-batch mode of operation was conducted by Shin.[56] There were not many reports by researchers for continuous fermentation of pullulan due to low dilution rate effect and septic problems.[38]

As media optimization influences pullulan production cost, medium and process variables are optimized with statistical design of experiments to achieve maximum pullulan yield under low-cost. For this purpose,

TABLE 7.3 Summaries of Carbon Substrates, Fermentation Conditions and Pullulan Yield.

Organism	Carbon substrate type/concentration	Fermentation conditions			Pullulan yield	Reference
		Temperature (°C)	pH	Time (h)		
Aureobasidium pullulans wild	Glucose (22 g/l) + Sucrose (20 g/l)	NA	NA	120	1.3 g/g	[5]
Pullularia pullulans 3092	Sucrose (5%)	25–27	5.0	100	14.8 g/l	[11]
				160	20.7 g/l	
	Maltose (2.5%)			100	4.9 g/l	
	Glucose (2.5%)				8.8 g/l	
	Fructose (2.5%)				6.8 g/l	
A. pullulans	Sucrose (10% w/v)	NA	NA	NA	58.0 g/l	[56]
Aureobasidium sp. strain NRRL Y-12,974	Corn fiber	28	NA	216	0.9±0.1 g/l	[32]
	Thin stillage				8.3±1.7 g/l	
	Corn condensed Distiller's solubles				4.5±0.2 g/l	
A. pullulans	Beet molasses	NA	NA	107	32 g/l	[50]
	Brewery wastes	NA	NA	72	6.0 g/l	
A. pullulans CFR-77	Jaggery	30	5	72	51.9 g/l	[75]
A. pullulans P56	Molasses	20	8.0	NA	8.05 g/l	[23]

TABLE 7.3 *(Continued).*

Organism	Carbon substrate type/concentration	Fermentation conditions			Pullulan yield	Reference
		Temperature (°C)	pH	Time (h)		
A. pullulans HP-2001	Soybean pomace	30	5.7	96	7.5 g/l	[53]
	Yeast extract		4.5		5.5 g/l	
	Glucose (10%)	30	5.81	72	36.87 g/l	
			3.2		32.12 g/l	
	Glucose (15%)		4.98	96	27.65 g/l	
A. pullulans P56	Synthetic medium	28	7.5	144	16.7 g/l	[23]
				96	6 g/l	
	Molasses medium			144	16.9 g/l	
A. pullulans ATCC 42023	Corn oil (0.4%)+ Boric acid (0.2 g/l)	28	NA	96	26.6 g/l	[21]
A. pullulans NRRL Y-6220	Soya bean oil	26	4.0	120	17.4 g/l	[52]
A. pullulans NRRL Y-2311			3.5	96	26.24 g/l	
A. pullulans NRM2	Sucrose	30	3	168	25.1 g/l	[47]
A. pullulans MTCC 1991	Cassava starch residue (CSR)	28	6.5	168	27.5 g/kg of S	[48]
	Wheat bran (WB)				19.5 g/kg of S	
	Rice bran (RB)				2.4 g/kg of S	

TABLE 7.3 (Continued).

Organism	Carbon substrate type/concentration	Fermentation conditions			Pullulan yield	Reference
		Temperature (°C)	pH	Time (h)		
A. pullulans ATCC 42023	7% CSL + 20% sucrose	28	2.2	120	65.3 g/l	[1]
	Cane molasses (10% sugars)		2.2		47.84 g/l	
	Glucose syrup (5% sugars)		3.3		33.21 g/l	
	Potato starchy waste (3%)		3.5		22.33 g/l	
	Hydrolyzed sweet whey (5% lactose)		2.3		12.4 g/l	
	Rice straw (4%) + 1% sucrose		4.0		9.36 g/l	
A. pullulans MTCC 2195	Cashew juice (110 g/l)	28	6.5	156	92.5 g/l	[68]
	Maize (50 g/l)			96	71.0 g/l	
	Cassava (50 g/l)				65.0 g/l	
	Bakery waste				27.0 g/l	
A. pullulans MTCC 2195	Coconut water	28	7	144	38.3 g/l	[67]
	Coconut milk (50 g/l)				58.0 g/l	
A. pullulans AP329	Sweet potato	28	5.5	96	29.43 g/l	[77]
A. pullulans CGMCC1234	Sucrose (50 g/l)	26	6.5	96	27.80%(w/w)	[76]

TABLE 7.3   Summaries of Carbon Substrates, Fermentation Conditions and Pullulan Yield (Continued).

Organism	Carbon substrate type/concentration	Fermentation conditions			Pullulan yield	Reference
		Temperature (°C)	pH	Time (h)		
A. pullulans ATCC 201253	Sucrose (75 g/l)	30	5.0	168	22.6 g/l	[14]
					32.9 g/l	
					23.1 g/l	
A. pullulans SK1002	Sucrose (50 g/l)	28	5.5	120	30.28 g/l	[27]
Rhodotorula bacarum	Glucose (80 g/l)	28	7	60	59 g/l	[15]
A. pullulans P56	Hydrolyzed potato starch waste	NA	NA	NA	19.2 g/l	[77]
A. pullulans ATCC 201253	Sucrose (75 g/l)	30	5	168	25.8 g/l	[13]
A. pullulans MTCC 1991	Full fat soya flour + soya extract	27	5.9	168	125.7 g/l	[55]
A. pullulans	DOJSC	28	6.0	120	83.98 g/l	[17]
RBF-4A3	Glucose	28	NA	96	70.43 g/l	
A. pullulans CJ001	Sucrose (50 g/l)	28	5.5	96	26.13 g/l	[12]
A. pullulans RBF 4A3	Dextrose	28	NA	96	65.34±0.21 g/l	[16]
A. pullulans MTCC 2195	Jack fruit seed powder	30	7	168	22.49 g/l	[54]
A. pullulans NCIM 1049	Jack fruit seed	NA	NA	NA	34.22 mg/g S	[65]
A. pullulans MTCC 2670	Wheat bran	NA	6	96	7.6 g/l	
	Rice bran		5	72	5.2 g/l	
	Coconut kernel		7	144	9.6 g/l	
	Palm kernel		6.5	168	16 g/l	

TABLE 7.3 Summaries of Carbon Substrates, Fermentation Conditions and Pullulan Yield (*Concluded*).

Organism	Carbon substrate type/concentration	Fermentation conditions			Pullulan yield	Reference
		Temperature (°C)	pH	Time (h)		
A. pullulans RBF 4A3	18% Jaggery, 3% DOJSC & 0.97% CSL	28	NA	72	66.25 g/l	[35]
A. pullulans MTCC 2670	Sucrose Asian palm kernel	30	NA	168	33.4 mg/g s 28.7 mg/g s	[64]
A. pullulans ATCC 42023	Sucrose	24	5.0	24	10.2 g/l	[38]

initially key (medium and process) variables were screened by fractional factorial design, called Plackett–Burman (PB) design.[46] Further, to evaluate interaction among variables for finding optimum conditions for desired response variable, RSM as statistical tool was used by many researchers to produce optimum pullulan from sweet potato,[44,74] from process conditions,[27] from Jatropha seed cake (JSC),[17] from jack fruit seed,[65] from Asian palm kernel,[64] from molasses,[61] from sweet potato.[44]

Many attempts have been done to develop a model which describes growth of pullulan producers and pullulan production in batch fermentation.[7,30,37,69] Mostly, the logistic function model was used to describe growth of *A. pullulans* based on assumption that growth of microbial population as a function of maximum population density.[37,45] Luedeking–Piret (LP) equation was used to calculate batch kinetics of sucrose utilization.[37] As pullulan was identified as a secondary metabolite (during late exponential phase) and produced more when growth limitation (during stationary phase), modified LP model was developed by Mohammad.[37] In other study by Thirumavalavan[68] used kinetic models developed by Dhanasekar[18] for the optimization of pullulan production by *A. pullulans* using different initial concentrations of cashew juice. Alemzadeh[2] also estimated kinetic parameters of *P. pullulans* fermentation using similar mathematical models.

7.3   KINETIC MODEL DEVELOPMENT: CELL GROWTH, SUBSTRATE CONSUMPTION, AND PULLULAN (AS PRODUCT) SYNTHESIS

The kinetic models for cell growth, substrate consumption, and pullulan (as product) synthesis in a batch system have been developed by many researchers.[7,30,37,69] Under optimal growth conditions, growth kinetic model of *A. pullulans* (X) (as per Malthus's law), in a batch fermentation is described below:[72]

$$\frac{dX}{dt} = \mu_{max}X\left(1 - \frac{X}{X_m}\right) \tag{7.1}$$

The logistic (L)-type model equation derived from the integration of above equation results:

$$X(t) = \frac{X_0 e^{\mu_{max}t}}{1 - \frac{X_0}{X_m}\left(1 - e^{\mu_{max}t}\right)} \tag{7.2a}$$

$$\ln\left[\frac{X_0 e^{\mu_{max}t}}{1 - \frac{X_0}{X_m}\left(1 - e^{\mu_{max}t}\right)}\right] = mt + C \quad \text{or} \quad Y = mt + C \tag{7.2b}$$

A plot of $\ln\left(\dfrac{X_t(X_m - X_0)}{X_0(X_m - X(t))}\right)$ versus t will yield maximum specific growth rate of biomass, μ_{max} as slope. The substrate utilization kinetics in microbial polysaccharide production can be taken from modified Luedeking–Piret (MLP) equation:

$$-\frac{dS}{dt} = r_S = \gamma\left(\frac{dX}{dt}\right) + \eta X \tag{7.3}$$

Logistic Incorporated Modified Luedeking–Piret (LIMLP) equation derived from integration of above equation results:

$$S(t) = S_0 - \gamma\left[\frac{X_0 e^{\mu_{max}t}}{1 - \left(\frac{X_0}{X_m}\right)\left(1 - e^{\mu_{max}t}\right)} - X_0\right]$$

$$+ \frac{\eta X_m}{\mu_{max}} \ln\left[1 - \left(\frac{X_0}{X_m}\right)\left(1 - e^{\mu_{max}t}\right)\right] \tag{7.4}$$

Non growth associated constant, η, in above equation can be calculated from stationary phase data (where $\dfrac{-dS}{dt} = 0$):

$$\eta = \frac{-\left(\dfrac{dS}{dt}\right)_{stationary\ phase}}{X_{max}} \tag{7.5}$$

And, a plot of

$$(S_0 - S(t)) + \frac{\eta X_m}{\mu_{max}} ln \left[1 - \left(\frac{X_0}{X_m} \right) \left(1 - e^{\mu_{max}t} \right) \right] \ vs. \left[\frac{X_0 e^{\mu_{max}t}}{1 - \left(\frac{X_0}{X_m} \right) \left(1 - e^{\mu_{max}t} \right)} - X_0 \right]$$

will yield growth associated constant, γ as slope. Product formation kinetics follows LP equation, as:

$$\frac{dP}{dt} = \alpha \frac{dX}{dt} + \beta X \qquad (7.6)$$

Logistic Incorporated Luedeking–Piret (LILP) equation derived from integration of above equation results:

$$P(t) = P_0 + \alpha \left[\frac{X_0 e^{\mu_{max}t}}{1 - \left(\frac{X_0}{X_m} \right) \left(1 - e^{\mu_{max}t} \right)} - X_0 \right]$$

$$+ \frac{\beta X_m}{\mu} ln \left[1 - \left(\frac{X_0}{X_m} \right) \left(1 - e^{\mu_{max}t} \right) \right] \qquad (7.7)$$

The β, nongrowth associated parameter can be determined from stationary phase data (where $\frac{dX}{dt} = 0$):

$$\beta = \frac{\left(\frac{dP}{dt} \right)_{stationary\ phase}}{X_{max}} \qquad (7.8)$$

A plot of

$$(P_t - P_o) + \frac{\beta X_m}{\mu} ln \left[1 - \left(\frac{X_0}{X_m} \right) \left(1 - e^{\mu_{max}t} \right) \right] \ vs. \left[\frac{X_0 e^{\mu_{max}t}}{1 - \left(\frac{X_0}{X_m} \right) \left(1 - e^{\mu_{max}t} \right)} - X_0 \right]$$

yields growth associated parameter, α as slope.

In this study, Equations (7.2), (7.4), and (7.7) were used to simulate the experimental data obtained in the shake flask batch fermentations with initial sucrose and jaggery concentrations of 50, 75, and 100 g/l. The software Microsoft Excel 2010 was employed to estimate the values of modeled kinetic parameters.[72]

7.4 MATERIALS AND METHODS

7.4.1 MATERIALS USED

Jaggery was purchased from the local market and was found to be mainly composed of sucrose. The composition of Standard Cultivation Medium (SCM) used in the shake flask fermentation (in g/l) is sucrose, 50.0; yeast extract, 3.0; KH_2PO_4, 5.0; KCl, 0.5; $MgSO_47H_2O$, 0.2; NaCl, 1.0, and distilled water 1 L. All the chemicals were purchased from Qualigens Chemicals and the biochemicals are purchased from M/s. HiMedia chemicals Ltd. Standard pullulan (*Kopulan*) was purchased from M/s. Kumar Organics Pvt. Ltd., Bengaluru, and was used in structural characterization. All ingredients used here were of analytical grade.[73] Media components (Luria Bertani Agar and Potato Dextrose Agar (PDA)) used in this study were obtained from M/s. HiMedia Chemical Ltd. (India).

7.4.2 MICROORGANISM AND INOCULUM DEVELOPMENT

Microorganism, *A. pullulans* MTCC 2195, was obtained from Microbial Type Culture Collection and Gene Bank (MTCC), IMTECH, Chandigarh and maintained on potato dextrose agar (PDA) slants at 4°C and subculture prior to each experimental run. A loopfull of freshly grown cultures from PDA agar slants were transferred to a 250 ml conical flask containing 50 ml standard cultivation media. Media pH (initially) was adjusted to 5.0 (after autoclaving at 121°C, for 15 min.) and incubated at 30°C for 48h on a rotary shaker at 150 rpm. This resulted suspension (at 5% v/v) was then used as inoculum for jaggery medium fermentations.[73]

7.4.3   SHAKE FLASK FERMENTATION

Shake flask fermentations were carried out with standard cultivation medium components at initial sucrose concentrations (g/l) of 50, 75, and 100, respectively. Further, the cultivation media (JCM) was made by replacing sucrose with jaggery (based on weight but not on percentage of sucrose content) for the same concentrations (g/l) of 50, 75, and 100. Both SCM and JCM in 100 ml aliquots were distributed in 500 ml Erlenmeyer flasks and autoclaved. These sterilized media were inoculated, 5% (v/v) aseptically and incubated for 172 h at 30°C and 150 rpm on a rotary shaker. Fermentation broth samples were collected, aseptically, at irregular intervals and determined the concentrations of dry cell biomass, pullulan and residual sugar.[73]

7.4.3.1   EFFECT OF INITIAL PH ON FERMENTATION

In order to study, the influence of pH on the shake flask fermentation for the pullulan production, the pH of *jaggery* in cultivation medium (after autoclaving) was adjusted to 3.0–7.0, respectively, using either 1N HCl or 1N NaOH and left uncontrolled during the fermentation. A 100 ml sterile media in 500 ml Erlenmeyer flasks by inoculating 5% (v/v) inoculum were incubated for 156 h at 30°C and 150 rpm on a rotary shaker. The samples were withdrawn for every 12 h and analyzed for cell biomass, Pullulan and residual sugar contents.[73]

7.4.3.2   ESTIMATION OF DRY CELL BIOMASS AND PULLULAN

At specific intervals of time, 2 ml of broth volume from each flask was centrifuged at 10,000 rpm for 20 min at ambient temperature to separate the cell biomass (pellet) from supernatant liquid. The collected cell biomass was washed twice with saline and distilled water and dried to constant weight in an oven at 90°C. The dry cell biomass weight was expressed in g/l. The polysaccharide, pullulan, was precipitated (kept at 4°C for 12 h) using cell-free supernatant liquid by adding cold ethanol (in the ratio of 1:2 v/v). The precipitate obtained was filtered through a pre-weighed Whatman No.1 filter paper and dried to constant weight at 80°C. The dry weight of pullulan and yield of pullulan were expressed in g/l and gram EPS per 100 g of sugar consumed.[73]

7.4.3.3   ESTIMATION OF RESIDUAL SUGAR CONCENTRATION

The residual sugar content in the cell-free fermentation broth was measured by the Miller's method[36] using double beam ELICO SL-159 UV-visible spectrophotometer.[73]

7.4.3.4   CHARACTERIZATION OF PULLULAN BY FTIR SPECTROSCOPY

The precipitated pullulan from each flask, at the end of fermentation was characterized to compare the quality of pullulan obtained from sucrose and jaggery media. The structural characterization of pullulan was carried out using Fourier-transform Infrared (FTIR) spectroscopy and IR spectra were recorded with Shimadzu IRTracer-100 FTIR spectrophotometer.

7.5   RESULTS AND DISCUSSION

In the present study, an attempt has been made to use jaggery as novel substrate and other process parameters like pH was optimized to produce pullulan at different initial concentrations of jaggery, using *A. pullulans* (MTCC 2195).[73] The culture was procured from microbial type culture collection and gene bank (MTCC), Institute of Microbial Technology, Chandigarh and the lyophilized culture was revived on malt yeast agar medium and further subcultured on PDA medium on petri plates.[73]

7.5.1   EFFECT OF INITIAL PH ON FERMENTATIVE PRODUCTION OF PULLULAN IN JAGGERY MEDIUM

Effects of environmental variables like pH of the cultivation medium enhanced the yields of pullulan formation by *A. pullulans* and were studied on wide variety of carbon substrates (Table 7.3). Changes in initial pH of fermentation medium affects the growth rate of *A. pullulans* (MTCC 2195) using coconut by-products, bakery waste, cassava and maize powders, cashew fruit juice, according to Thirumavalavan.[67,68]

In another study, Vijayendra SVN et al. (2001) varied the initial pH from 2 to 7 for the study of *A. pullulans* CFR-77 growth and reported that maximum pullulan production was achieved at an initial pH of 5.0. Here, we attempted the change in initial pH (3.0–7.0) effect on the kinetics of *A. pullulans* MTCC 2195 in jaggery medium. Pullulan content released into the medium increases with pH up to 5.0 and then decreased. The maximum concentration (9.88 g/l) of pullulan EPS was obtained at pH 5.0.[73]

7.5.2 STUDY OF JAGGERY AS A CARBON SOURCE ON GROWTH OF A. PULLULANS AND PRODUCTION OF PULLULAN

Preliminary checking of jaggery in replace of sucrose to study the growth of *A. pullulans* (MTCC 2195) in the cultivation media with agar on petri plates was performed. Several workers have exercised the wide variety of carbon substrates, ranging from agro-industrial residues and wastes for the economic production of pullulan (Table 7.3). The earlier investigations[35,72] on jaggery reveal that jaggery could be used as a good carbon source, because of its high sucrose percentage, for the growth of *A. pullulans* and towards cost-effective production of pullulan. Vijayendra[75] reported that the yield of pullulan by *A. pullulans* depends on initial sugar concentration in the cultivation medium. Therefore, in the present study, the increasing initial levels of jaggery concentrations, 50 g/l, 75 g/l, and 100 g/l were employed to determine the maximum pullulan content using *A. pullulans* (MTCC 2195).

The profiles of pullulan concentration at specific time points indicate that a considerable amount of sugar in the media was depleted in the early hours of fermentation and results a quick growth of *A. pullulans,* and there was a concomitant increase in the pullulan production with decline in the residual sugar content in the media. The maximum pullulan concentrations obtained from jaggery medium containing 50, 75, and 100 g/l of sugar concentration were 10.8, 14.8 and 18.6 g/l, respectively.[73]

The time course profiles of substrate, biomass, and pullulan, in case of sucrose containing media with the similar initial sucrose concentrations, indicate pullulan concentration at specific time points in the above profiles.[73] Kinetic parameters were calculated from experimental data using jaggery and sucrose as carbon substrates for the production of pullulan in this study.[74]

Further, to test the appropriateness of fermentation data, kinetic models were developed and fitted the results of comprehensive biomass growth and pullulan formation.[74]Among these models, Logistic (L), Luedeking-Piret (LP), Logistic incorporated Luedeking–Piret (LILP), Modified Luedeking–Piret (MLP), and LMLP have already been tested to describe biomass, pullulan and sugar profiles in the batch cultivation of *A. pullulans*.[14,68]

7.5.3   STRUCTURAL CHARACTERIZATION OF PULLULAN BY FTIR SPECTROSCOPY

Structural characterization is used to examine the possible functional groups for commercial pullulan and synthesized pullulan from batch cultivation; and it was done by FTIR spectrophotometer and spectra.[73] The top spectrum shows the absorption peaks of synthesized pullulan from this study and bottom spectrum indicates that of standard pullulan.[73] A strong absorption peaks at ranges of 3300and 1640 cm^{-1} confirms repeating units of –OH and O–C–O, respectively. In the specific area (1500–650 cm^{-1}) which is a characteristic for the pullulan molecule as a whole. Another strong absorption at 1000 cm^{-1} characterizes the C–O group and peaks at 900 and 650 cm^{-1} proves the presence of α-1, 6- and α-D-glucopyranoside units, respectively.[73]

7.5.4   STATISTICAL OPTIMIZATION OF FERMENTATION CONDITIONS

Total of 20 shake flask fermentation experiments were conducted with combinations of initial jaggery concentration, initial pH and total fermentation time by *A. pullulans* MTCC 2195. Central composite design (CCD) of RSM of Design Expert-10 was used. The experimental design and the results obtained from experiments are shown in Table 7.4. High-degree polynomial models (namely, linear, interactive (two factorial), quadratic and cubic models) were tested to fit into the experimental data to analyze the actual relationship between response (pullulan concentration, g/l) and the variables, sequential model sum of squares, lack of fit tests and model summary statistics were also used to study the adequacy of models among various models. The results obtained are represented in Table 7.5.

The quadratic model based on sequential model sum of squares was found to be significant (*p*-value <0.0001). Lack of fit tests values from

TABLE 7.4 Experimental Designs used in Response Surface Methodology (RSM) Studies to Understand the Interactions among Process Conditions for Pullulan Maximization.

Standard order	Run numbers	A	B	C	Response	
		Initial jiggery concentration (g/l)	Initial pH	Total fermentation time (h)	Observed pullulan (g/l)	Predicted pullulan (g/l)
1	13	25 (−1)[a]	3 (−1)	96 (−1)	9.96 ± 0.14[b]	9.80
2	10	75 (+1)	3 (−1)	96 (−1)	11.3 ± 0.08	11.25
3	9	25 (−1)	7 (+1)	96 (−1)	9.98 ± 0.04	9.99
4	7	75 (+1)	7 (+1)	96 (−1)	11.19 ± 0.05	11.11
5	11	25 (−1)	3 (−1)	144 (+1)	11.45 ± 0.1	11.52
6	6	75 (+1)	3 (−1)	144 (+1)	12.89 ± 0.08	12.87
7	18	25 (−1)	7 (+1)	144 (+1)	11.76 ± 0.08	11.80
8	14	75 (+1)	7 (+1)	144 (+1)	12.68 ± 0.19	12.83
9	12	25 (−1)	5 (0)	120 (0)	13.04 ± 0.08	13.08
10	2	75 (+1)	5 (0)	120 (0)	14.33 ± 0.05	14.32
11	19	50 (0)	3 (−1)	120 (0)	14.89 ± 0.06	15.05
12	8	50 (0)	7 (+1)	120 (0)	15.24 ± 0.11	15.12
13	1	50 (0)	5 (0)	96 (−1)	12.66 ± 0.12	12.93
14	3	50 (0)	5 (0)	144 (+1)	14.89 ± 0.14	14.65
15	17	50 (0)	5 (0)	120 (0)	15.6 ± 0.03	15.59
16	20	50 (0)	5 (0)	120 (0)	15.6 ± 0.03	15.59
17	15	50 (0)	5 (0)	120 (0)	15.6 ± 0.03	15.59
18	5	50 (0)	5 (0)	120 (0)	15.6 ± 0.03	15.59
19	4	50 (0)	5 (0)	120 (0)	15.6 ± 0.03	15.59
20	16	50 (0)	5 (0)	120 (0)	15.6 ± 0.03	15.59

[a]levels (−1): Low, (0): Middle, (+1): High used in RSM table.
[b]Mean ± S.D. from three replicate experiments.

TABLE 7.5   Analysis of Variance (ANOVA) for all Terms in Quadratic Model.

Source of variation	Sum of squares (SS)	Degree of freedom (df)	Mean squares (MS)	F-Value	p-value probability > F
Model	78.09	9	8.68	366.84	< 0.0001 (*significant*)
A-jaggery	3.84	1	3.84	162.51	< 0.0001
B-pH	0.013	1	0.013	0.55	0.4762
C-Time	7.36	1	7.36	311.23	< 0.0001
AB	0.053	1	0.053	2.23	0.1660
AC	4.512×10^{-3}	1	4.512×10^{-3}	0.19	0.6716
BC	4.513×10^{-3}	1	4.513×10^{-3}	0.19	0.6716
A^2	9.79	1	9.79	414.10	< 0.0001
B^2	0.71	1	0.71	29.92	0.0003
C^2	8.88	1	8.88	375.54	< 0.0001
Residual	0.24	10	0.024		
Lack of Fit	0.24	5	0.047		
Pure Error	0.000	5	0.000		
Total	**78.33**	**19**			

sequential model fitting for the quadratic model did not show significant lack of fit. Adjusted R^2 and predicted R^2 for this quadratic model gave the best model. The quadratic polynomial equation for predicting the pullulan concentration (response) is given below:

$$\text{Pullulan} = + 15.59 + 0.62 \times A + 0.036 \times B + 0.86 \times C - 0.081 \times AB - 0.024 \times AC + 0.024 \times BC - 1.89 \times A^2 - 0.51 \times B^2 - 1.8 \times C^2 \quad (7.9)$$

The predicted model was highly significant, as manifested by the F- value and the probability (p) value (*Probability* $> F$ = <0.0001). The goodness of fit was manifested by determination coefficient (R^2) and that was 0.997 indicating variations of 99.7% inthe response model. Generally, $R^2 > 0.9$ is well considered for any regression model.[25] Adjusted determination coefficient ($R^2_{Adj.}$ = 0.9943) was also very high that confirmed the significance of model and more useful to assess the direct interaction and quadratic effects in optimizing the parameters for increasing the pullulan production.

The optimum of location, obtained from quadratic model, for achieving maximal pullulan production was at A = 50 g/l, B = 5.0 and C = 120 h and the predicted optimum pullulan concentration corresponding to these values was 15.59 g/l (experiment value = 15.6 g/l). Goodness of fit for maximum pullulan concentration was also confirmed by performing additional triplicate of experiments that yielded an average maximum pullulan concentration of 15.54 g/l. The pareto plot showed a satisfactory correlation between the actual and predicted values, with the points clustured around the diagonal line, which indicates the good fit of the model (Fig. 7.3a).

The 3D response plots (Fig. 7.3(b, c, d)) were used to study the interaction of fermenation conditions and the optimum levels that have most significant effects on pullulan production. Minimum response of pullulan production (9.96 g/l) occurred when jaggery concentration and initial pH were at their lowest level. Pullulan concentration has increased considerably as jaggery concentration increased, that indicated that jaggery concentration for pullulan production has a significant effect on the responses (Fig. 7.3(b)). As the concentration of jaggery was between 40 and 60 g/l, the responses were maximal nearly at the middle of initial pH (Fig. 7.3b).

Figure 7.3(c) shows the effects of initial jaggery concentration and time on pullulan production. From this contour and 3D plot, it is evident that pullulan production was very much influenced by time and jaggery concentrations and very good interaction between them for pullulan production. From Figure 7.3(d), it is also clear that the there was considerable effect on pullulan production by the total fermentation time and initial pH parameters. Response surface optimzation using CCD supported 15.59 g/l of pullulan concentration on 50 g/l of initial jaggery concentration, 5.0 of initial pH and 120 h of total fermentation time by *A. pullulans* (MTCC 2195). From the other studies, optimized pullulan concentration was 45 g/l for molasses,[61] 18.76 g/l for jack fruit seed powder[54] by *A. pullulans* (MTCC 2195).

7.5.5   AUREOBASIDIUM PULLULANS (MTCC 2195) GROWTH

The lag phase of *A. pullulans* (MTCC 2195) in fermentation was very short as the cells were already adapted before they were used for pullulan production. *A. pullulans* started to form pullulan instantly as the cells

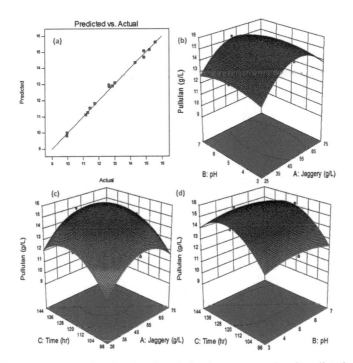

FIGURE 7.3   (a) Pareto plot showing the relation between actual and predicted values for pullulan maximization; (b), (c) and (d) 3D response surface plots for pullulan production using their process variables.

entered the logarithmic phase and therefore *A. pullulans* growth and pullulan production took place simultaneously. In order to validate the developed model, it is necessary to study the cell growth as function of time (i.e., Logistic growth). Sigmoidal curves are useful in describing the growth of organisms. The effect of initial substrate concentration changes on growth related parameters was done using 50, 75, and 100 g/l of jaggery and sucrose in batch fermentation for about 172 h. Other conditions of fermentation were kept at same values.

From experiments, maximum cell concentrations (X_m) were considered for the initial jaggery and sucrose concentrations of 50, 75, and 100 g/l, respectively. Upon linear fitting the experimental data into Eq. (7.2a and 7.2b), Logistic (L) model equation parameters, maximum specific growth rate (μ_{max}), and initial biomass concentrations (X_0) for increased concentrations of jaggery were determined. The resulting R^2 values and calculated values are summarized in Table 7.6. The Figures 7.4 (a) and (b) show

TABLE 7.6   Kinetic Parameters of Logistic (L) Model for *Aureobasidium pullulans* Growth on Jaggery and Sucrose.

Parameters	Initial jiggery concentration (g/l)			Initial sucrose concentration (g/l)		
	50	75	100	50	75	100
μ_{max}, h^{-1}	0.0706	0.0545	0.0679	0.048	0.0611	0.0586
R^2	0.83	0.81	0.93	0.77	0.92	0.91
X_0, g/l	2.28	0.667	0.636	4.29	0.547	1.0026
X_m, g/l	45.24	48.36	73.02	40	56	60.2

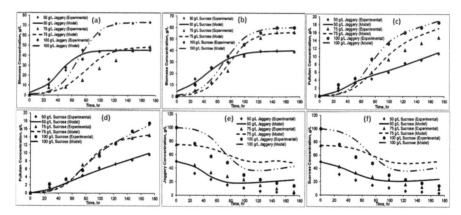

FIGURE 7.4   Profiles of (a), (b) *A. pullulans growth*, (c), (d) pullulan formation and (e), (f) substrate consumption based on L, LILP, and LIMLP modeled data.

the comparison of experimental data and model predictions for *A. pullulans* growth from increasing concentrations of jaggery and sucrose, respectively.

7.5.6 PULLULAN PRODUCTION

Comparisons of pullulan production profile with LILP model are shown by plotting both the experimental data and the predicted values from the models in Figure 7.4(c) and (d). The Figure 7.4(c) represents increasing concentrations of pullulan with increasing jaggery (as the initial substrate) concentrations (50, 75, and 100 g/l). Similar curves are also observed in case of increasing sucrose (as initial substrate) concentrations (50, 75, and 100 g/l), in Figure 7.4(d). The LILP model did properly fit the experimental data and the estimated kinetic parameters with R^2 values are listed in Table 7.7.

TABLE 7.7 Kinetic Parameters of Logistic Incorporated Luedeking–Piret (LILP) Model for Pullulan Production on Jaggery and Sucrose.

Parameter	Initial jiggery concentration (g/l)			Initial sucrose concentration (g/l)		
	50	75	100	50	75	100
α, g.P/g.X	0.069	0.2513	0.1895	0.0849	0.2227	0.1449
R^2	0.81	0.83	0.98	0.84	0.91	0.94
β, g.P/ (g.X.hr)	0.0014	0.00115	0.00071	0.0014	0.00042	0.0013

7.5.7 SUBSTRATE (JAGGERY AND SUCROSE) CONSUMPTION

To study the substrate consumption in EPS production, a LIMLP model was used. Subjective comparisons of actual substrate utilization by *A. pullulans* towards pullulan production with LIMLP model were shown by plotting both experimental and predicted data from model. The Figure 7.4(e) and (f) demonstrated the reasonable fit of experimental data with predicted values of model. Table 7.8 gives the comparison of estimated kinetic parameters of LIMLP models for initial jaggery and sucrose concentration of 50, 75, and 100 g/l.

TABLE 7.8 Kinetic Parameters of Logistic Incorporated Modified Luedeking–Piret (LIMLP) Model for Pullulan Production on Jaggery and Sucrose.

Parameter	Initial jaggery concentration (g/l)			Initial sucrose concentration (g/l)		
	50	75	100	50	75	100
γ, g.S/g.X	0.8534	0.7817	1.0233	0.992	0.6138	1.2776
R^2	0.96	0.61	0.82	0.965	0.66	0.88
η, g.S/ (g.X.hr)	0.00168	0.0038	0.0023	0.0019	0.00175	0.00279

7.6 SUMMARY

Microbial polysaccharides are very important biomaterials produced in nature and need additional research. The research study in this chapter focused on the suitability of using jaggery as an effective and cheaper carbon substrate for the maximum production of EPS, pullulan by *A. pullulans* (MTCC 2195). The effect of initial jaggery concentrations and initial pH in fermentation medium of batch cultivation for biopolymer production was investigated. Change in the initial pH (from 3.0 to 7.0) of media containing jaggery was varied to study the influence of pH on fungal fermentation. The maximum pullulan yield obtained was at a pH of 5.0.

An increase in the initial concentrations (50, 75, and 100 g/l) of jaggery in the media was used to study the effect of initial substrate concentration on *A. pullulans* fermentation. The maximum pullulan yield was estimated as 21.6, 19.7, and 18.6 g per 100 g of jaggery. The data—concentrations (g/l) of *A. pullulans* growth, jaggery utilization, pullulan formation as a function of time—were obtained from shake flask, batch fermentations were compared with the data obtained from batch fermentations with sucrose (replacing jaggery) media. Further, rates of jaggery utilization and pullulan production were high, when the initial substrate concentration (50 g/l) was comparatively low.

Structural characterization using FTIR spectroscopic analysis was done to examine the possible functional groups present in synthesized pullulan and compared with that of commercial pullulan. Therefore results in this chapter indicate that jaggery (an agro-industrial residue) is an alternative carbon substrate source for melanin-free pullulan production by *A. pullulans* (MTCC 2195). Statistical fermentation medium optimization could overcome the limitations of classical empirical methods like one factor at a time (OFAT). In this study, RSM of statistical design was used as a tool for optimization of fermentation conditions such as initial jaggery concentration (g/l), initial pH, and total fermentation time (h) for the pullulan production by the strain *A. pullulans* (MTCC 2195) in shaking flask cultures.

Central composite design of RSM was proposed to study the combined effects of culture medium components. The existence of very good interactions between independent variables with pullulan response was observed. The optimal fermentation conditions stimulated the maximal pullulan production were as follows: initial jaggery concentration, 50 g/l; initial pH: 5.0; and total fermentation time: 120 h. Under these optimized conditions, the predicted pullulan concentration was 15.59 g/l. Validation experiments were performed to verify the accuracy of the models and the results showed that the experimental values have good agreement with the predicted values. Results of statistical analysis have shown probability value ($p < 0.0001$) and determination coefficient ($R^2 = 0.997$) from regression model. Hence, this concludes the RSM was a promising method for optimization of pullulan production.

In this study, mathematical approach of batch and continuous fermentation of *A. pullulans* for pullulan production was developed. These unstructured models of *A. pullulans* growth, sucrose consumption, pullulan production in a batch system could be used for modeling the fermentation kinetics. Set of ordinary differential equations were developed as Logistic (L) model for biomass growth, LILP model for product formation, and LIMLP model for substrate utilization to understand the dynamic behavior of batch fermentation. Further, the similar sets of equations for a steady-state continuous fermentation were also developed. However, a detailed model presented in this study, for pullulan fermentation in a continuous system was not reported yet, till date.

Research work in this study also estimated the growth and nongrowth kinetic parameters of unstructured, mathematical models of batch shake flask pullulan fermentation by *A. Pullulans* (MTCC 2195) on jaggery and sucrose (as substrates) and validated with experimental results. The data obtained from increased concentrations (50, 75, and 100 g/l) of both jaggery and sucrose were used to fit the models of L, LILP, and LIMLP. Comparisons of kinetic parameters of Logistic (μ_{max}, X_0, and X_m), LILP (α, β) and LIMLP (γ, η) models adequately fit the experimental results with predicted data. All these parameters were predicted accurately with reasonable R^2 values. A good concurrence of the data was shown in *A. pullulans* growth, pullulan synthesis and jaggery and sucrose utilization profiles. Estimated values of kinetic parameters were also compared with literature. So, the information obtained in this study would be helpful for further developments in scaled-up productions of Pullulan.

KEYWORDS

- *Aureobasidium pullulans*
- carbon substrate
- central composite design
- exopolysaccharide
- Fourier-transform infrared
- GRAS
- jaggery
- Luedeking–Piret
- logistic
- Malthus law
- MTCC
- pullulan
- putative
- recovery
- regression coefficient
- response surface methodology

REFERENCES

1. Abdel Hafez, A. M.; Abdelhady, H. M.; Sharaf, M. S.; El-Tayeb, T. S. Bioconversion of Various Industrial by Products and Agricultural Wastes into Pullulan. *J. Appl. Sci. Res.* **2007**, *3*(11), 1416–1425.

2. Alemzadeh, I. The Study on Microbial Polymers: Pullulan and PHB. *Iran. J. Chem. Chem. Eng.* **2009**, *28*(1), 13–21.

3. Bae, H.; Ahari, A. F.; Shin, H.; Nichol, J. W.; Hutson, C. B.; Masaeli, M.; Khadem-hosseini, A. Cell-Laden Microengineered Pullulan Methacrylate Hydrogels Promote Cell Proliferation and 3D Cluster Formation. *Soft Matter* **2011**, *7*(5), 1903–1911.

4. Bauer, R. Physiology of Dematium Pullulans de Bary. *Zentralbl Bacteriol Parasitenkd Infektionskr Hyg Abt 2* **1938**, *98*, 133–167.

5. Bender, H.; Lehmann, J.; Wallenfels, K. Pullulan, in extracelluläres Glucan von *Pullularia pullulans*. *Biochim. Biophys. Acta* **1959**, *36*(2), 309–316.

6. Bernier, B. The Production of Polysaccharides by Fungi Active in the Decomposition of Wood and Forest Litter. *Can. J. Microbiol.* **1958**, *4*(3), 195–204.

7. Boa, J. M.; LeDuy, A. Pullulan from Peat Hydrolyzate Fermentation Kinetics. *Biotechnol. Bioeng.* **1987**, *30*(4), 463–470.

8. Boridy, S.; Takahashi, H.; Akiyoshi, K.; Maysinger, D. The Binding of Pullulan Modified Cholesteryl Nanogelsto Aβ Oligomers and their Suppression of Cytotoxicity. *Biomaterials* **2009**, *30*(29), 5583–5591.

9. Bruneel, D.; Schacht, E. Chemical Modification of Pullulan: 1. Periodate Oxidation. *Polymer* **1993**, *34*(12), 2628–2632.

10. Campbell, B. S.; Siddique, A. B. M.; McDougall, B. M.; Seviour, R. J. Which Morphological Forms of the Fungus *Aureobasidium pullulans* are Responsible for Pullulan Production? *FEMS Microbiol. Lett.* **2004**, *232*(2), 225–228.

11. Catley, B. J.; Ramsay, A.; Servis, C. Observations on the Structure of the Fungal Extracellular Polysaccharide, Pullulan. *Carbohydr. Res.* **1986**, *153*(1), 79–86.

12. Chen, J.; Wu, S.; Pan, S. Optimization of Medium for Pullulan Production using a Novel Strain of *Aureobasidium pullulans* Isolated from Sea Mud Through Response Surface Methodology. *Carbohydr. Polym.* **2012**, *87*(1), 771–774.

13. Cheng, K. C.; Demirci, A.; Catchmark, J. M. Pullulan: Biosynthesis, Production, and Applications. *Appl. Microbiol. Biotechnol.* **2011**, *92*(1), 29–44.

14. Cheng, K. C.; Demirci, A.; Catchmark, J. M.; Puri, V. M. Modeling of Pullulan Fermentation by Using a Color Variant Strain of *Aureobasidium pullulans*. *J. Food Eng.* **2010**, *98*(3), 353–359.

15. Chi, Z.; Wang, F.; Chi, Z.; Yue, L.; Liu, G.; Zhang, T. Bioproducts from *Aureobasidium pullulans*, a Biotechnologically Important Yeast. *Appl. Microbiol. Biotechnol.* **2009**, *82*(5), 793–804.

16. Choudhury, A. R.; Bhattacharjee, P.; Prasad, G. S. Development of Suitable Solvent System for Downstream Processing of Biopolymer Pullulan Using Response Surface Methodology. *PLoS One* **2013**, *8*(10), e77071.

17. Choudhury, A. R.; Sharma, N.; Prasad, G. S. Deoiled Jatropha Seed Cake is a Useful Nutrient for Pullulan Production. *Microb. Cell Fact.* **2012**, *11*(1), 1.

18. Dhanasekar, R.; Viruthagiri, T.; Sabarathinam, P. L. Poly (3-Hydroxy Butyrate) Synthesis from a Mutant Strain *Azotobacter Vinelandii* Utilizing Glucose in a Batch Reactor. *Biochem. Eng. J.* **2003**, *16*(1), 1–8.

19. Donabedian, D. H.; McCarthy, S. P. Acylation of Pullulan by Ring-Opening of Lactones. *Macromolecules* **1998**, *31*(4), 1032–1039.
20. Duan, X.; Chi, Z.; Wang, L.; Wang, X. Influence of Different Sugars on Pullulan Production and Activities of α-Phosphoglucose Mutase, UDPG-Pyrophosphorylase and Glucosyltransferase involved in Pullulan Synthesis in *Aureobasidium Pullulans* Y68.*Carbohydr. Polym.* **2008**, *73*(4), 587–593.
21. El-Tayeb, T. S.; Khodair, T. A. Enhanced Production of Some Microbial Exo-Polysaccharides by Various Stimulating Agents in Batch Culture. *Res. J. Agric. Biol. Sci.* **2006**, *2*(6), 483–492.
22. Gaur, R.; Singh, R.; Tiwari, S.; Yadav, S. K.; Daramwal, N. S. Optimization of Physico-Chemical and Nutritional Parameters for a Novel Pullulan-Producing Fungus, *Eurotium Chevalieri*. *J. Appl. Microbiol*.**2010**, *109*(3), 1035–1043.
23. Göksungur, Y.; Uçan, A.; GÜvenÇ, U. Production of Pullulan from Beet Molasses and Synthetic Medium by *Aureobasidium pullulans*. *Turk. J. Biol.* **2004**, *28*(1), 23–30.
24. Gounga, M. E.; Xu, S. Y.; Wang, Z.; Yang, W. G. Effect of Whey Protein Isolate–Pullulan Edible Coatings on the Quality and Shelf Life of Freshly Roasted and Freeze-Dried Chinese Chestnut. *J. Food Sci.* **2008**, *73*(4), E155–E161.
25. Haaland, P. D. *Experimental Design in Biotechnology;* CRC press: USA, 1989; 105, p 261.
26. Heeres, A.; Spoelma, F. F.; Van Doren, H. A.; Gotlieb, K. F.; Bleeker, I. P.; Kellogg, R. M. Synthesis and Reduction of 2-Nitroalkyl Polysaccharide Ethers. *Carbohydr. Polym.* **2000**, *42*(1), 33–43.
27. Jiang, L. Optimization of Fermentation Conditions for Pullulan Production by *Aureobasidium pullulans* Using Response Surface Methodology. *Carbohydr. Polym.* **2010**, *79*(2), 414–417.
28. Kaya, A.; Du, X.; Liu, Z.; Lu, J. W.; Morris, J. R.; Glasser, W. G.; Esker, A. R. Surface Plasmon Resonance Studies of Pullulan and Pullulan Cinnamate Adsorption onto Cellulose. *Biomacromolecules* **2009**, *10*(9), 2451–2459.
29. Kimoto, T.; Shibuya, T.; Shiobara, S. Safety Studies of a Novel Starch, Pullulan: Chronic Toxicity in Rats and Bacterial Mutagenicity. *Food Chem. Toxicol.* **1997**, *35*(3), 323–329.
30. Klimek, J.; Ollis, D. F. Extracellular Microbial Polysaccharides: Kinetics of *Pseudomonas* sp., *Azotobacter vinelandii*, and *Aureobasidium pullulans* batch fermentations. *Biotechnol. Bioeng.* **1980**, *22*(11), 2321–2342.
31. Leathers, T. D. Biotechnological Production and Applications of Pullulan. *Appl. Microbiol. Biotechnol.* **2003**, *62*(5–6), 468–473.
32. Leathers, T. D.; Gupta, S. C. Production of Pullulan from Fuel Ethanol Byproducts by *Aureobasidium* sp. Strain NRRl Y-12,974. *Biotechnol. Lett.* **1994**, *16*(11), 1163–1166.
33. Li, H.; Yang, J.; Hu, X.; Liang, J.; Fan, Y.; Zhang, X. Superabsorbent Polysaccharide Hydrogels Based on Pullulan Derivate as Antibacterial Release Wound Dressing. *J. Biomed. Mater. Res. Part A* **2011**, *98*(1), 31–39.
34. Mayer, J. M. Polysaccharides, Modified Polysaccharides and Polysaccharide Blends for Biodegradable Materials. In *Novel Biodegradable Microbial Polymers;* Mayer, J. M. Ed.; Kluwer Academic Publishers: Netherlands, 1990; pp 465–467.
35. Mehta, A.; Prasad, G. S.; Choudhury, A. R. Cost Effective Production of Pullulan from Agri-Industrial Residues Using Response Surface Methodology. *Int. J. Biol. Macromol.* **2014**, *64*, 252–256.

36. Miller, G. L. Use of Dinitrosalicylic Acid Reagent for Determination of Reducing Sugar. *Anal. Chem.* **1959**, *31*(3), 426–428.
37. Mohammad, F. H. A.; Badr-Eldin, S. M.; El-Tayeb, O. M.; El-Rahman, O. A. Polysaccharide Production by *Aureobasidium pullulans* III. The Influence of Initial Sucrose Concentration on Batch Kinetics. *Biomass Bioenergy* **1995**, *8*(2), 121–129.
38. Moubasher, H.; Wahsh, S. Pullulan Production from *Aureobasidium pullulans* by Continuous Culture. *Basic Res. J. Microbiol.* **2014**, *1*, 11–15.
39. Muthuvelayudham, R.; Viruthagiri, T. Application of Central Composite Design Based Response Surface Methodology in Parameter Optimization and on Cellulase Production Using Agricultural Waste. *Int. J. Chem. Biol. Eng.* **2010**, *3*(2), 97–104.
40. Nakatani, M.; Shibukawa, A.; Nakagawa, T. Separation Mechanism of Pullulan Solution-Filled Capillary Electrophoresis of Sodium Dodecyl Sulfate Proteins. *Electrophoresis* **1996**, *17*(10), 1584–1586.
41. Nishikawa, T.; Akiyoshi, K.; Sunamoto, J. Supramolecular Assembly Between Nanoparticles of Hydrophobized Polysaccharide and Soluble Protein Complexation Between The Self-Aggregate of Cholesterol-Bearing Pullulan and alpha-chymotrypsin. *Macromolecules* **1994**, *27*(26), 7654–7659.
42. Oku, T.; Yamada, K.; Hosoya, N. Effects of Pullulan and Cellulose on the Gastrointestinal Tract of Rats. *J. Jpn. Soc. Food Nutr. (Japan)* **1979**, *32*(4), 235–241.
43. Ouchi, T.; Minari, T.; Ohya, Y. Synthesis of Poly (L-lactide) -Grafted Pullulan Through Coupling Reaction Between Amino Group End-Capped Poly (L-lactide) and Carboxymethyl Pullulan and its Aggregation Behavior in Water. *J. Polym. Sci. Part A: Polym. Chem.* **2004**, *42*(21), 5482–5487.
44. Padmanaban, S.; Balaji, N.; Muthukumaran, C.; Tamilarasan, K. Statistical Optimization of Process Parameters for Exopolysaccharide Production by *Aureobasidium pullulans* Using Sweet Potato Based Medium. *3 Biotech.* **2015**, *5*(6), 1067–1073.
45. Pearl, R.; Reed, L. J. On the Rate of Growth of the Population of the United States Since 1790 and its Mathematical Representation. *Proc. Natl. Acad. Sci.* **1920**, *6*(6), 275–288.
46. Plackett, R. L.; Burman, J. P. The Design of Optimum Multifactorial Experiments. *Biometrika* **1946**, *33*(4), 305–325.
47. Prasongsuk, S.; Berhow, M. A.; Dunlap, C. A.; Weisleder, D.; Leathers, T. D.; Eveleigh, D. E.; Punnapayak, H. Pullulan Production by Tropical Isolates of *Aureobasidium pullulans. J. Ind. Microbiol. Biotechnol.* **2007**, *34*(1), 55–61.
48. Ray, R. C.; Moorthy, S. N. Exopolysaccharide (Pullulan) Production from Cassava Starch Residue by *Aureobasidium pullulans* Strain MTTC 1991. *J. Sci. Ind. Res.* **2007**, *66*, 252–255.
49. Rekha, M. R.; Sharma, C. P. Pullulan as a Promising Biomaterial for Biomedical Applications: A Perspective. *Trends Biomater. Artif. Organs.* **2007**, *20*(2), 116–121.
50. Roukas, T.; Liakopoulou-Kyriakides, M. Production of Pullulan from Beet Molasses by *Aureobasidium pullulans* in a Stirred Tank Fermenter. *J. Food Eng.* **1999**, *40*(1), 89–94.
51. San Juan, A.; Bala, M.; Hlawaty, H.; Portes, P.; Vranckx, R.; Feldman, L. J.; Letourneur, D. Development of a Functionalized Polymer for Stent Coating in the Arterial Delivery of Small Interfering RNA. *Biomacromolecules* **2009**, *10*(11), 3074–3080.

52. Sena, R. F.; Costelli, M. C.; Gibson, L. H.; Coughlin, R. W. Enhanced Production of Pullulan by Two Strains of *A. pullulans* with Different Concentrations of Soybean Oil in Sucrose Solution in Batch Fermentations. *Braz. J. Chem. Eng.* **2006**, *23*(4), 507–515.

53. Seo, H. P.; Son, C. W.; Chung, C. H.; Jung, D. I.; Kim, S. K.; Gross, R. A.; Lee, J. W. Production of High Molecular Weight Pullulan by *Aureobasidium pullulans* HP-2001 with Soybean Pomace as a Nitrogen Source. *Bioresour. Technol.* **2004**, *95*(3), 293–299.

54. Sharmila, G.; Muthukumaran, C.; Nayan, G.; Nidhi, B. Extracellular Biopolymer Production by *Aureobasidium pullulans* MTCC 2195 Using Jackfruit Seed Powder. *J. Polym. Environ.* **2013**, *21*(2), 487–494.

55. Sheoran, S. K.; Dubey, K. K.; Tiwari, D. P.; Singh, B. P. Directive Production of Pullulan by Altering Cheap Source of Carbons and Nitrogen at 5 L Bioreactor Level. *ISRN Chem. Eng.* **2012**, *2012*, 1–5.

56. Shin, Y. C.; Kim, Y. H.; Lee, H. S.; Kim, Y. N.; Byun, S. M. Production of Pullulan by a Fed-Batch Fermentation. *Biotechnol. Lett.* **1987**, *9*(9), 621–624.

57. Shingel, K. I.; Petrov, P. T. Behavior of γ-Ray-Irradiated Pullulan in Aqueous Solutions of Cationic (Cetyltrimethylammonium Hydroxide) and Anionic (Sodium Dodecyl Sulfate) Surfactants. *Colloid Polym. Sci.* **2002**, *280*(2), 176–182.

58. Simon, L. Caye-Vaugien, C. Bouchonneau, M. Relation Between Pullulan Production, Morphological State and Growth Conditions in *Aureobasidium pullulans*: New Observations. *Microbiology* **1993**, *139*(5), 979–985.

59. Singh, R. S.; Saini, G. K. Pullulan-Hyperproducing Color Variant Strain of *Aureobasidium pullulans* FB-1 Newly Isolated from Phylloplane of *Ficus* sp. *Bioresour. Technol.* **2008**, *99*(9), 3896–3899.

60. Singh, R. S.; Saini, G. K.; Kennedy, J. F. Pullulan: Microbial Sources, Production and Applications. *Carbohydr. Polym.* **2008**, *73*(4), 515–531.

61. Srikanth, S.; Swathi, M.; Tejaswini, M.; Sharmila, G.; Muthukumaran, C.; Jaganathan, M. K.; Tamilarasan, K. Statistical Optimization of Molasses Based Exopolysaccharide and Biomass Production by *Aureobasidium pullulans* MTCC 2195. *Biocatal. Agric. Biotechnol.* **2014**, *3*(3), 7–12.

62. Sugawa-Katayama, Y.; Kondou, F.; Mandai, T.; Yoneyama, M. Effects of Pullulan, Polydextroseand Pectin on Cecal Microflora. *Oyo Toshitsu Kagaku* **1994**, *41*, 413–418.

63. Sugumaran, K. R.; Gowthami, E.; Swathi, B.; Elakkiya, S.; Srivastava, S. N.; Ravikumar, R.; Ponnusami, V. Production of Pullulan by *Aureobasidium pullulans* from Asian Palm Kernel: A Novel Substrate. *Carbohydr. Polym.* **2013**, *92*(1), 697–703.

64. Sugumaran, K. R.; Ponnusami, V. Downstream Processing Studies for Pullulan Recovery in Solid State Fermentation Using Asian Palmyra Palm Kernel-Inexpensive Substrate. *Biotechnol. Indian J.* **2014**, *9*(2), 79–82.

65. Sugumaran, K. R.; Sindhu, R. V.; Sukanya, S.; Aiswarya, N.; Ponnusami, V. Statistical Studies on High Molecular Weight Pullulan Production in Solid State Fermentation Using Jack Fruit Seed. *Carbohydr. Polym.* **2013**, *98*(1), 854–860.

66. Sutherland, I. W. Novel and Established Applications of Microbial Polysaccharides. *Trends Biotechnol.* **1998**, *16*(1), 41–46.

67. Thirumavalavan, K.; Manikkadan, T. R.; Dhanasekar, R. Pullulan Production from Coconut by-Products by *Aureobasidium pullulans*. *Afr. J. Biotechnol.* **2009**, *8*(2), 254–258.
68. Thirumavalavan, K.; Manikkandan, T. R.; Dhanasekar, R. Batch Fermentation Kinetics of Pullulan from *Aureobasidium pullulans* Using Low Cost Substrates. *Biotechnology* **2008**, *7*(2), 317–322.
69. Thomson, N.; Ollis, D. F. Extracellular Microbial Polysaccharides. II. Evolution of Broth Power–Law Parameters for Xanthan and Pullulan Batch Fermentation. *Biotechnol. Bioeng.* **1980**, *22*(4), 875–883.
70. Tsujisaka, Y.; Mitsuhashi, M. *Pullulan*. In *Industrial Gums—Polysaccharides and Their Derivatives*; Academic Press: San Diego, CA, **1993**; pp 447–560.
71. Uchida, S.; Yamamoto, A.; Fukui, I.; Endo, M.; Umezawa, H.; Nagura, S.; Kubota, T. U.S. Patent5,583,244, 1996; U.S. Patent and Trademark Office: Washington, DC.
72. Venkata, S. R. K. G.; Podha, S. Unstructured Modeling of *Aureobasidium pullulans* Fermentation for Pullulan Production—A Mathematical Approach. *Int. J. Eng. Res. Technol.* **2014**, *3*(10), 1076–1079.
73. Venkata, S. R. K. G.; Mangamuri, U.; Vijayalakshmi, M.; Poda, S. Model-Based Kinetic Parameters Estimation in Batch Pullulan Fermentation Using Jaggery as Substrate. *J. Chem. Pharm. Res.* **2016**, *8*(3), 217–224.
74. Venkata, S. R. K. G.; Sambasiva, K. R. S.; Mangamuri, U. K.; Vijaya Lakahmi, M.; Poda, S. Production of Pullulan using Jaggery as Substrate by *Aureobasidium pullulans* MTCC 2195. *Curr. Trends Biotechnol. Pharm.* **2016**, *10*(2), 153–160.
75. Vijayendra, S. V. N.; Bansal, D.; Prasad, M. S.; Nand, K. Jaggery: A Novel Substrate for Pullulan Production by *Aureobasidium pullulans* CFR-77. *Process Biochem.* **2001**, *37*(4), 359–364.
76. Wu, S.; Jin, Z.; Kim, J. M.; Tong, Q.; Chen, H. Downstream Processing of Pullulan from Fermentation Broth. *Carbohydr. Polym.* **2009**, *77*(4), 750–753.
77. Wu, S.; Jin, Z.; Tong, Q.; Chen, H. Sweet Potato: A Novel Substrate for Pullulan Production by *Aureobasidium pullulans*. *Carbohydr. Polym.* **2009**, *76*(4), 645–649.

CHAPTER 8

ANTIBIOTIC RESISTANCE OF STAPHYLOCOCCUS AUREUS: A REVIEW

DIVYA LAKSHMINARAYANAN[1], JESSEN GEORGE[1,*] AND SURIYANARAYANAN SARVAJAYAKESAVALU[2]

[1]*Department of Water and Health, Faculty of Life Sciences, Jagadguru Sri Shivaathreeswara University, SS Nagar, Mysore 570015, Karnataka, India*
E-mail: divyalakshminarayanan@gmail.com

[2]*SCOPE Beijing Office, #18 Shuangqing Road, Haidian, Beijing 100085, China*
E-mail: sunsjk@gmail.com

[*]*Corresponding author.E-mail: georgejessen@gmail.com*

CONTENTS

8.1   INTRODUCTION

Over the last two decades antibiotic resistance has been given a lot of attention in the scientific community and the public. Antibiotic resistance has developed over time from resistance to single classes of antibiotics to multi-drug resistance (MDR) and extreme drug resistance.[9] *Staphylococcus aureus* is a Gram-positive bacterium and versatile human pathogen causing skin infections ranging from mild involvement of skin and soft tissue to life-threatening systemic illness.[23,24] To treat the infections caused by penicillin-resistant *S. aureus,* the methicillin was introduced in 1959. In 1961 there were reports from the United Kingdom of *S. aureus* isolates that had acquired resistance to methicillin (methicillin-resistant *S. aureus* (MRSA))[12] and MRSA isolates were soon recovered from other European countries, an later from other countries.[8]

In general, MRSA are susceptible only to glycopeptides antibiotics such as vancomycin and investigational drugs. In recent years, MRSA isolates that have decreased susceptibility to glycopeptides and therefore are a cause of great public health concern.[8] MRSA is now a major problem worldwide because of its emerging resistance to majority of the antibiotic classes MDR.[30]

An increased rate of MRSA is being reported in both clinical and nonclinical environments. Few studies have proven wastewater as a best source for the multiplication of the MRSA. Hence there is a need to identify the discharge of the clinical strains of *S. aureus* to the environment and in turn its pathways in the environment.

This chapter has taken into consideration of available data from various studies in South India with respect to antibiotic resistance exhibited by *S. aureus* in various environments. The MRSA rates in nonclinical and clinical environment in South India are incomplete and are still poorly documented. Hence, the present study we attempt to compile the scattered information regarding the MRSA rate in South India.

8.2   METHODS AND MATERIALS

8.2.1   ENVIRONMENTAL SURVEILLANCE OF MRSA

Ongoing overuse and misuse of antibiotics are the major reasons for the increasing trend of antibiotics resistance worldwide. Antibiotic resistance is not only restricted to clinical environment because most of the antibiotics are consumed in the agricultural fields and animal farming. An animal as well as human wastes are contains antibiotics from the source of human and veterinary medicines. This may contribute potential increase of antibiotic resistance selection.

8.2.2 ANTIBIOTIC RESISTANCE GENES

The antibiotics basically target cell wall synthesis, protein synthesis; the selection pressure applied by the antibiotics that used in clinical and agricultural settings has promoted the evolution and spread of genes that confer resistance.[1] Development of antibiotic resistance is conferred by mutation and selection by medical antibiotics, resistance can occur in organism by the acquisition of a novel antibiotic resistance gene by horizontal gene transfer (HGT) has a relevant role in emergence, through conjugation, transformation, or transduction through various mobile genetic elements like plasmids, transposons, integrons and so forth.[13,29] Internal mechanisms include mutational modification of gene targets, over expression of various efflux pumps; whereas acquired resistance involves enzymatic inactivation of the drug and bypassing of the target. Practices like application of sewage sludge and manure may introduce complex mixtures of bacteria containing drug resistance genes, veterinary and medical antibiotics, and other chemicals to land, where interactions may occur with indigenous soil bacteria.[29]

8.2.3 CLINICAL AND NONCLINICAL ENVIRONMENTS IN SOUTH INDIA

In India, burden of infectious diseases is high and the antibiotics play a critical role in limiting morbidity and mortality in the country. Antibiotic resistance is vulnerable in India due to the over and inappropriate use of antibiotics.[24,28] In South India, there are many studies reported increasing prevalence of MRSA over the years.[5] MRSA is now endemic and has been increasingly reported in India.[4,6,15,22,25] The present chapter is aimed to study and correlate the MRSA resistance towards various antibiotics and to find out the prevalence of MRSA in South India.

8.3   RESULTS AND DISCUSSION

According to the data collected from various articles,[3,10,11,16,18,19,22,26,20] nearly 50 antibiotics and 10 antibiotic groups have been studied in this chapter (Table 8.1). The antibiotic resistances studied in South India have been carried out in Karnataka, Tamil Nadu and Andhra Pradesh. The overall resistance percentage of antibiotics ranges from 0.56–100%.

TABLE 8.1   Overall Resistance of the Various Antibiotic Groups in South India.

Antibiotic category	Resistance (%)
Amino glycosides	6.00–95.83
Bacteriostatic antibiotic	1.00–58.33
Beta lactam	19.00–100
Glycopeptide	0.56–100
Lincosamide	10.00–70.58
Macrolide	29.00–100
Quinolones	5.00–85.10
Rifampin	9.90–50.00
Sulfonamides	9.00–96.52
Tetracycline	13.60–83.33

The maximum value for aminoglycoside antibiotic in South India is registered for gentamycin of Andhra Pradesh (63.12%) (Fig. 8.1). All the other values for aminoglycoside antibiotics in South India are below 40% (Fig. 8.2). In South India most studies have been carried out in Karnataka

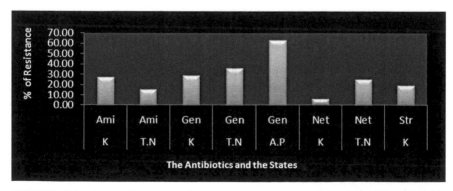

FIGURE 8.1   State wise data of percentage of resistance toward aminoglycoside antibiotics in South India.

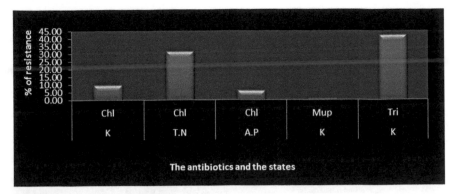

FIGURE 8.2   State wise data of percentage of resistance toward bacteriostatic antibiotics in South India.

with a resistance of trimethoprim (42.50%), mupirocin (1%), and chloramphenicol (9.78%). Tamil Nadu registered high chloramphenicol resistance (31.66%) among the South Indian States.

Beta lactams are the major class of antibiotics exhibiting MRSA resistance. Their resistance value for β-lactams ranges from 19 to 100% in South India (Fig. 8.3). In South India penicillin and oxacillin shows highest resistance compared to other antibiotics. Penicillin resistance is 100% in all states except Tamil Nadu and the oxacillin resistance is 100% in all states except Karnataka. Methicillin resistance is above 70% for all states. All other antibiotics show above 30% resistance in South India.

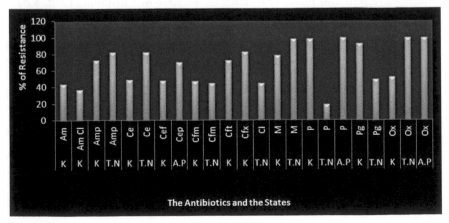

FIGURE 8.3   State wise data of percentage of resistance toward β-lactam antibiotics in South India.

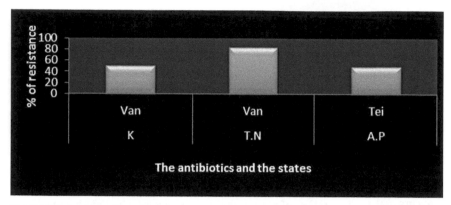

FIGURE 8.4   State wise data of percentage of resistance towards glycopeptide antibiotics in South India.

The antibiotics in glycopeptides category include vancomycin and teicoplanin. In South India vancomycin resistance was above 50% and the highest resistance was registered for Tamil Nadu (83.33%). teicoplanin resistance was reported for the state of Andhra Pradesh (48.93%) in South India (Fig. 8.4).

The lincosamide antibiotics studied in the review include lincomycin and clindamycin (Fig. 8.5). The lincomycin resistance has been registered 26.66% in Karnataka and 42.72% in Andhra Pradesh. Erythromycin is a major category of macrolide group of antibiotic registered in India. The erythromycin has been reported with a higher resistance of 69.50% in Andhra Pradesh and lower resistance of 57.75% in Karnataka (Fig. 8.6).

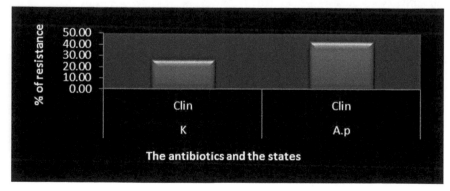

FIGURE 8.5   State wise data of percentage of resistance toward lincosamide antibiotics in South India.

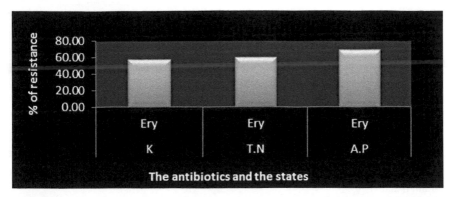

FIGURE 8.6   State wise data of percentage of resistance toward macrolide antibiotics in South India.

Quinolones include ciprofloxacin, levofloxacin, norfloxacin, spar-floxacin, ofloxacin, and nalidixic acid (Fig. 8.7). The ciprofloxacin value ranges from 40.80% in Karnataka to 64.31% in Andhra Pradesh. Levofloxacin from Andhra Pradesh (79.16%) registered the highest resistance among all quinolones.

Nalidixic acid and norfloxacin was only reported from Karnataka (42 and 48%). The ofloxacin value was higher for Tamil Nadu (75%) compared to that of Karnataka (16%). The rifampin resistance varies from 9.90% (Andhra Pradesh)—50% (Tamil Nadu) (Fig. 8.8).

The sulfonamide antibiotics studied in this review includes Co-trimoxazole and sulfamethoxazole. The sulfonamide resistance is only registered in Karnataka (42.50%). The Co-trimoxazole resistance is higher in Andhra Pradesh (76.94%) and lower in Karnataka (15.94%) (Fig. 8.9).

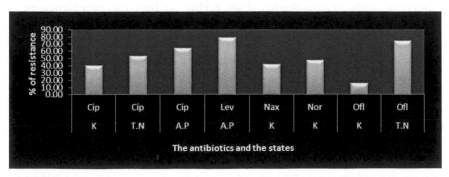

FIGURE 8.7   State wise data of percentage of resistance toward quinolones antibiotics in South India.

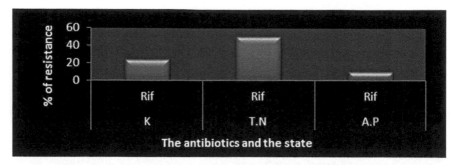

FIGURE 8.8 State wise data of percentage of resistance toward rifampin antibiotics in South India.

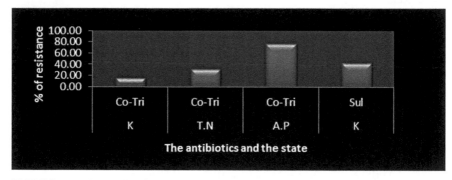

FIGURE 8.9 State wise data of percentage of resistance toward sulfonamide antibiotics in South India.

The doxycycline study is also registered in the State of Andhra Pradesh (13.60%). Tetracycline resistance ranges from 24.95% (Karnataka)—83.33% (Tamil Nadu) (Fig. 8.10). To find the prevalence of *S. aureus* and MRSA throughout India data have been collected from 40 articles contributing MRSA studies. From the articles we have derived the following results and conclusions, which are represented diagrammatically in Figure 8.11.

8.3.1 ANTIBIOTICS AND RESISTANCE GENES IN NONCLINICAL ENVIRONMENT

As the consequence of human activity natural ecosystems can suffer by the antibiotic pollution, environmental bacteria are not globally under the

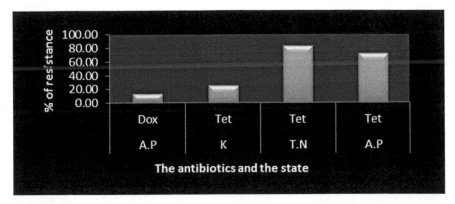

FIGURE 8.10   State wise data of percentage of resistance toward tetracycline antibiotics in South India.

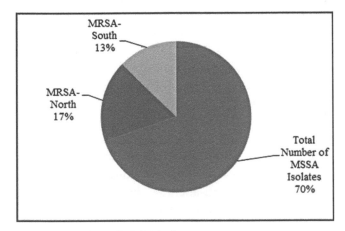

FIGURE 8.11   Total number of MRSA isolates.

strong antibiotic selective pressure suffered by human pathogens, which are challenged with antibiotics during therapy.[14] Most antibiotics used for treating infections are produced by environmental microorganisms, meaning that genes for antibiotic resistance must also have emerged in nonclinical habitats.[2,21]

Environment containing antibiotic residues exert selection pressure and contribute to the appearance of resistant bacteria. Selection of antibiotic resistance may occur in nonclinical environments such as wastewater treatment systems and agricultural environments where antibiotics may be of veterinary origin or within an environmental background where antibiotic selection is provided by bacterial antibiotic producers.[29] Compared to the surface water, higher level of antibiotic resistant bacteria and the antibiotic resistance genes are often detected in wastewater. The research in antibiotic resistance has focused on the environment only in recent years from which the antibiotics were initially extracted: soil micro-organisms and the soil ecosystem.[21]

A few studies on antibiotic resistance in nonclinical environment have been conducted India. Mostly antibiotic resistance studies from India are concentrated on coliforms. Studies conducted by Krushna et al.[17] and Tewari et al.[27] revealed the presence of antibiotic resistant *E. coli* in water systems.

In a study conducted by Dinesh et al.[7] in currency notes collected from Delhi and Dehradun suggested the presence of virulence genes in *S. aureus* isolates and these were more resistant to antibiotics. This currency may act as potent nonconventional mode for the spread of microbial pathogen.

From the above review, it is found that among the antibiotics vanco-mycin and teicoplanin are considered inferior to other category of antibiotics. Very few reports are there on the resistance *S. aureus* on vancomycin, teicoplanin, and linezolid in South India. From the above study it is concluded that the glycopeptides and oxazolidinone antibiotics should be encouraged in MRSA cases. All these antibiotics are sold in over-the-counter (OTC) without any prescription because people used to buy their own due to the time constraints and their busy life schedule. In South India majority of the drug stores provides the medicine such as pain relievers, cough remedies, anti-allergies, laxatives, antibiotics, antacids, and vitamins are sold in OTC without any prescription and this practice is increasing day by day. This leads to the emergence of human pathogens resistance.

This chapter reveals that MRSA is a major threat in South India and most of the isolates were developed multidrug resistant. Greater number of MRSA isolates was multidrug resistant as compared with the MRSA isolates. But still the vancomycin and linezolid antibiotics remain a viable option for resistant to MRSA. The present study has some limitations due to the scattered information's and the studies are not uniform in all the places. And there are very limited molecular studies that have been done so there is no clear conclusion on the antibiotic resistance genes. The study in antibiotic resistance in nonclinical environment has been more concentrated on coliforms and hence there is a need of a study of antibiotic resistance of other bacterial isolates in water, wastewater, and soil environments.

8.4   CONCLUSIONS

Methicillin-Resistant *S. aureus* is a worldwide problem but the knowledge is limited about occurrence in clinical and nonclinical environments of the developing countries. Vancomycins, linezolid are still the drug of choice for treating multidrug resistant MRSA infections. A detailed molecular study to monitor the epidemiology of MRSA in the country is highly recommended. It is also being increasingly realized that nonclinical isolates should also be tested for antibiotic resistance levels as resistance can develop by widespread use of antibiotics by farmers and veterinarians. The present study is an attempt to compile the scattered information on MRSA rate in South India and helps to predict future emergence and guide the development of strategies to counteract this resistance. The present study finding reveals that the prevalence of MRSA has increased over the years. Hence, further detailed nationwide study and continuous monitoring of MRSA is recommended for effective control of MRSA and also to prevent further selection of microbial resistance.

8.5   SUMMARY

Methicillin-resistant *S. aureus* is a causative and most frequently isolated pathogen for many nosocomial and community acquired infections. An increased rate of MRSA is being reported in both clinical and nonclinical environments. Few studies have proven wastewater as a best source for the multiplication of the MRSA. Hence there is a need to identify the discharge of the clinical strains of *S. aureus* to the environment and in turn its pathways in the environment. This chapter has taken into consideration of available data from various studies in South India with respect to antibiotic resistance exhibited by *S. aureus* in various environments. From the study it was observed that the incidence of MRSA shows a large variation, from 0.56 to 100%, in South India.

KEYWORDS

- Aminoglycoside
- bacteriostatic
- beta lactams
- glycopeptide
- lincosamide
- macrolide

- methicillin
- mRSA
- quinolones
- rifampin
- *Staphylococcus aureus*
- sulfonamides

REFERENCES

1. Alen, H. K.; Justin, D.; Helena, H. Wang.; Karen, A. C.; Julian, D.; Jo, H. Call of Wild: Antibiotic Resistance Genes in Natural Environment. *Nat. Rev.* **2010**, *8*, 251–259.

2. Alonso, A.; Sánchez, P.; Martinez, J. L. Environmental Selection of Antibiotic Resistance Genes. *Environ. Microbiol.* **2001**, *3*(1), 1–9.

3. Amruthkishan, K. U.; Sunil, K. B. Methicillin-Resistant *Staphylococcus aureus* in a Tertiary Care Hospital in North-East Karnataka: Evaluation of Antibiogram. *Curr. Res. Med. Med. Sci.* **2011**, *1*(1), 1–4.

4. Anupurba, S.; Sen, M. R.; Nath, G.; Sharma, B. M.; Gulathi, A. K.; Mohapatra, T. M. Prevalence of Methicillin-Resistant *Staphylococcus aureus* in a Tertiary Referral Hospital in Eastern Uttar Pradesh. *Indian J. Med. Microbiol.* **2003**, *21*(1), 49–51.

5. Bagga, B.; Reddy, A. K.; Garg, P. Decreased Susceptibility to Quinolones in Methicillin-Resistant *Staphylococcus aureus* Isolated from Ocular Infections at a Tertiary Eye Care Center. *Br. J. Ophthalmol.* **2010**, *94*, 1407–1408.

6. Bala, K.; Aggarwal, R.; Goel, N.; Chaudhary, U. Prevalence and Susceptibility Patterns of Methicillin-Resistant *Staphylococcus aureus* (MRSA) Colonization in a Teaching Tertiary Care Center in India. *J. Infect. Dis. Antimicrob. Agents* **2010**, *27*, 33–38.

7. Chauhan, P. B.; Desai, P. B. The Antibacterial Activity of Honey Against Methicillin-Resistant *Staphylococcus aureus* Isolated from Pus Samples. *Acta Biologica Indica* **2010**, *1*(1), 55–59.

8. D'Souza, N.; Rodrigues, C.; Mehta, A. Molecular Characterization of Methicillin-Resistant *Staphylococcus aureus* with Emergence of Epidemic Clones of Sequence Type (ST) 22 and ST 772 in Mumbai, India. *J. Clin. Microbiol.* **2010**, *48*, 1806–1811.

9. Dinesh Kumar, J.; Negi, Y. K.; Gaur, A.; Khanna, D. Detection of Virulence Genes in *Staphylococcus aureus* Isolated from Paper Currency. *Int. J. Infect. Dis.* **2009**, *13*, 450–455.

10. Enright, M.; Robinson, D.; Randle, G.; Feil, E.; Grundmann, H.; Spratt, B. The Evolutionary History of Methicillin-Resistant *Staphylococcus aureus* (MRSA). *Proc. Natl. Acad. Sci. U. S. A.* **2002**, *99*(11), 7687–7692.

11. Fiona, W. Investigating Antibiotic Resistance in NonclinicalClinical Environments. *Front. Microbiol. Antimicrob. Resist. Chemother.* **2013**, *4*(19), 1–5.

12. Habeeb, K.; Mohammad, A. Prevalence and Antibiotic Susceptibility Pattern of Methicillin-Resistant and Coagulase-Negative Staphylococci in a Tertiary Care Hospital in India. *Int. J. Med. Med. Sci.* **2010**, *2*(4), 116–120.

13. Jevons, M. P. Celbenin-Resistant *Staphylococci. Br. Med. J.* **1961**, *1*(5219), 124–125.

14. Jose, L. M. Antibiotics and Antibiotic Resistance Genes in Natural Environments. *Science* **2008**, *321*(5887) 365–367.

15. Jose, L. M. The Role of Natural Environments in the Evolution of Resistance Traits in Pathogenic Bacteria. *Proc. R. Soc. B* **2009**, *276*(1667), 2521–2530.

16. Krishnan, P. U.; Miles, K.; Shetty, N. Detection of Methicillin and Mupirocin Resistance in *Staphylococcus aureus* Isolates Using Conventional and Molecular Methods: A Descriptive Study from a Burns Unit with High Prevalence of MRSA. *J. Clin. Pathol.* **2002**, *55*, 745–748.

17. Malthi, J.; Sowmiya, M.; Margarita, S.; Madhavan, H. N.; Lily, T. K. Application of PCR Based-RFLP for Species Identification of Ocular Isolates of Methicillin-Resistant *Staphylococci* (MRS). *Indian J. Med. Res.* **2009**, *130*, 78–84.

18. Murugan, K.; Usha, M.; Malathi, P.; Saleh, A. S.; Chandrasekaran, M. Biofilm Forming Multi-Drug Resistant *Staphylococcus spp* Among Patients with Conjuctivitis. *Pol. J. Microbiol.* **2010**, *59*(4), 233–239.

19. Kumar, R.; Yadav, B. R.; Singh, R. S. Antibiotic Resistance and Pathogenicity Factors in *Staphylococcus aureus* Isolated from Mastitic Sahiwal Cattle. *J. Biosci.* **2011**, *36*(1), 175–188.

20. Sahoo, K. C.; Tamhankar, A. J.; Sahoo, S. K. Geographical Variation in Antibiotic-Resistant *Escherichia coli* Isolates from Stool, Cow-Dung and Drinking Water. *Int. J. Environ. Res. Public Health* **2012**, *9*, 746–759.

21. Joshi, S.; Ray, P.; Manchanda, V.; Bajaj, J. Methicillin -Resistant *Staphylococcus aureus* (MRSA) in India: Prevalence and Susceptibility Pattern. *Indian J. Med. Res.* **2013**, *137*, 363–369.

22. Mallick, S. K.; Basak, S. MRSA-too many Hurdles to Overcome: A Study from Central India. *Trop. Doct.* **2010**, *40*, 108–110.

23. Nadig, S.; Raju, S. R.; Arakere, G. Epidemic Methicillin-Resistant *Staphylococcus aureus* (EMRSA-15) Variants Detected in Healthy and Diseased Individuals in India. *J. Med. Microbiol.* **2010**, *59*, 815–821.

24. Varghese, G. K.; Mukhopadhya, C.; Bairy, I.; Vadana, K. E.; Varma, M. Bacterial Organisms and Antimicrobial Resistance Pattern. *J. Assoc. Physicians India* **2010**, *58*, 23–28.

25. Verma, S.; Joshi, S.; Chitnis, V.; Hemwani, N.; Chitnis, D. Growing Problem of Methicillin *Staphylococci*-Indian Scenario. *Indian J. Med. Sci.* **2000**, *54*, 535–540.

26. Shobha, K. S.; Ravikumar, T. N.; Gurumurthy, B. Y.; Subbaraju, A. Methicillin and Vanomycin Resistant Staphylococci from Cell Phones-An Emerging Threat. *Int. J. Microbiol. Res.* **2012**, *3*(3), 163–166.

27. Tewari, S.; Ramteke, P. W. Tripathi, M.; Shailendra, K.; Satyendra, K. G. Plasmid Mediated Transfer of Antibiotic Resistance and Heavy Metal Tolerance in Thermo Tolerant Waterborne Coliforms. *Afr. J. Microbiol. Res.* **2013**, *7*(2), 130–136.

28. Tambekar, D. H.; Dhanorkar, D. V.; Gulhane, S. R.; Dudhane, M. N. Prevalence and Antimicrobial Susceptibility Pattern of Methicillin-Resistant *Staphylococcus aureus* from Health Care and Community Associated Sources. *Afr. J. Infect. Dis.* **2008**, *1*(1), 52–56.

29. Vidhani, S.; Mehndiratta, P. L.; Mathur, M. D. Study of Methicillin-Resistant *Staphylococcus aureus* (MRSA) Isolates from High Risk Patients. *Indian J. Med. Microbiol.* **2001**, *19*(2), 87–90.

30. Gaze, W. H.; O'Neill, C.; Wellington, E. M. H.; Hawkey, P. M. Antibiotic Resistance in the Environment, with Particular Reference to MRSA. *Adv. Appl. Microbiol.* **2008**, *63*, 249–270.

PART III

Potential of Nano- and Bio-fertilizers in Sustainable Agriculture

CHAPTER 9

ROLE OF NANOFERTILIZERS IN SUSTAINABLE AGRICULTURE

ARTI GOEL

Amity Institute of Microbial Biotechnology, Amity University, J-3 Block, 3rd Floor, Room 307, Sector-125, Noida 201313, India

Corresponding author. E-mail: agoel2@amity.edu, Mobile: +91-8800422339

CONTENTS

9.1 INTRODUCTION

Fertilizers have an enormous potential to enhance production and quality of food, mainly after introducing high-yielding and fertilizer responsive varieties. Most of the field crops need large amounts of inorganic inputs. Besides for enhancing nutrient use efficiency and to overcome prolonged problem of eutrophication, nanofertilizers have been considered as a best alternative. Synthesis of nanofertilizers has been done for regulating the nutrients release depending on the requirements of the crops. Apart from this, nanofertilizers have proved more efficient than ordinary fertilizers. Some advantageous effects of nanofertilizers comprehend rise in nutrient use efficacy, improved yield as well as decreased level of soil pollution. The possible influence of nanofertilizers in crop development and growth improvement depends on their capacity of high absorbance and reactivity. If nanofertilizers are composed of the particles having sizes lesser than the sizes of cell wall pores (5–20 nm), then they may enter in plant cells straight through the cell wall structures which are sieve like.

Nanofertilizers basically dissolve in solution and discharge the nutrient(s) in soluble ions form. Almost equal amount of soluble nutrient ions can be absorbed by the plants from nanofertilizers and dissolved conventional fertilizers. Apart from this, amount of nanofertilizer dissolution in water/soil solution ought to be high due to its smaller particle sizes as well as higher specific surface areas. Besides various technology advancements in agriculture, applications of nanotechnology and the use of products made up of nanoparticles like nanofertilizers is still in its growing stage in agriculture. Since application of nanofertilizer not only promotes development and growth of plants and crops, but also their antioxidant activity. Therefore, nanofertilizers possess the potential of increasing production of crops and nutrition of plants.

In the present scenario, agriculture sector development is only possible by increasing of resource use efficiency through effective use of modern technologies, which will damage the production bed minimally because of limited arable lands and water resources. Among various modern technologies, nanotechnology is very important because it has the potential to transform the agricultural systems, biomedicine, environmental engineering, security and safety, water resources, energy conversion as well as various supplementary areas. Nanostructured formulations, via various mechanisms like targeted delivery as well as conditional release, might release their active ingredients more precisely as a response to environmental triggers and biological demands. Utilization of nanofertilizers leads to an enhancement in nutrients use efficacy, reduction in soil toxicity as well as decreases the possible negative effects related with over dosage of fertilizers and also reduces the regularity of the application. Hence, particularly in developing countries nanotechnology possesses a significant potential to achieve sustainable agriculture.[8]

Excessive applications of fertilizers containing nitrogen and phosphorus affect the groundwater and also lead to eutrophication in aquatic ecosystems.[12,10] According to the Royal Society, "Nanotechnologies are the design, characterization, production and application of structures, devices and systems by controlling shape and size at nanometer scale."[2] Currently, nanotechnology is gradually moved away from the experimental into the practical areas, for example: development of slow/controlled release fertilizers and conditional release of pesticides and herbicides. Practical applications of nanotechnology made it important for stimulating the expansion of environment friendly and sustainable agriculture. Certainly, nanotechnology has provided the feasibility of exploiting nanostructured or nanoscale materials in the form of fertilizer carriers or controlled-release vectors for constructing of "smart fertilizer" as new facilities for enhancing nutrient use efficiency and also reduces costs of environmental protection.[3] Thus, these findings can articulate the use of nanofertilizer, which will lead to an increased efficiency of the micro and macro elements, reduces the toxicity of the soil as well as reduces the frequency of application of conventional fertilizers.

Hence, it can be concluded that nanomaterials and nanofertilizers possess the potential to revolutionize agricultural systems.[3] Some advantages related to nanotechnological tools in fertilizers are presented in Table 9.1.

9.2 MECHANISMS OF UPTAKE OF NANOFERTILIZER BY PLANTS

Nanoparticles can directly enter into the cells of plants via sieve like cell wall structures but only with the condition when the size of the particles are smaller than the cell wall pores size (5–20 nm). Further proceeding of nanoparticles through cell membrane, interactions with cytoplasm and applications of nanoparticles carrying nutrients are also complex.[9] Though, no research has excluded it as one of the primary mechanisms that the nutrient elements could be absorbed by plant root system through dissolution of nanoparticles in water/soil solution. Nanoparticles basically dissolve in water/soil solution and discharge the nutrient(s) in the form of soluble ions, which will be extensively used by the plants as those from the dissolved conventional fertilizers. Rate of dissolution and amount of nanoparticles in water/soil solution should be higher than those of the related bulk solids due to smaller particle sizes and higher specific surface areas of the former.[7]

TABLE 9.1   Some of Advantages Related to Ransomed of Conventional Fertilizers Using Nanotechonology.

Desirable properties	Examples of nanofertilizers-enabled Technologies
Controlled-release formulations	So-called smart fertilizers might become reality through transformed formulation of conventional products using nanotechnology. The nanostructured formulation might permit fertilizers intelligently control the release speed of nutrients to match the uptake pattern of crop.
Controlled-release modes	Both release rate and release pattern of nutrients for water-soluble fertilizers might be precisely controlled through encapsulation in envelope forms of semipermeable membranes coated by resin-polymer, waxes, and sulfur.
Effective duration of nutrient release	Nanostructured formulation can extend effective duration of nutrient supply of fertilizers in to soil.
Loss rate of fertilizer nutrients	Nanostructured formulation can reduce loss rate of fertilizer nutrient supply of fertilizers in to soil by leaching and/or leaking.
Nutrient uptake efficiency	Nanostructured formulation might increase fertilizers efficiency and uptake ratio of the soil nutrients in crop production and save fertilizers resource.
Solubility and dispersion for mineral micronutrients	Nanosized formulation of minerals micronutrients may improve solubility and dispersion of insoluble nutrients in soil, reduce soil absorption and fixation, and increase the bioavailability.

9.3   TYPES OF NANOFERTILIZERS

9.3.1   MACRONUTRIENT NANOFERTILIZERS

Macronutrient nanofertilizers involve one or more elements as macronutrient elements like N, P, K, Ca, and Mg, thus being able to supply these essential macronutrients to plants. Large quantities of macronutrient fertilizers (mainly N and P fertilizers) have been used for increasing production of food, fiber, and other essential commodities. It has been reported that total macronutrient fertilizer ($N + P_2O_5 + K_2O$) consumption was 175.7 million ton (Mt) in 2011 which is projected to increase to 263 Mt in 2050.[1] Smil[11] estimated that N fertilizers have added a roughly 40% increase in per capita food production in the past 50 years, indicating an important role of these macronutrient fertilizers in production of food globally. Moreover due to the low efficiency (30–50%) and heavy application of these macronutrient fertilizers, major amounts of these nutrients (N and P) are transported into groundwater and surface bodies, which disrupted aquatic ecosystems and threatening health of human and aquatic life.[11] Hence, an urgent and practically essential research direction is to develop highly efficient and environment friendly macronutrient (N and P) nanofertilizers to substitute the conventional N and P fertilizers and to ensure the sustainable food production while protecting the environment. Therefore, macronutrient nanofertilizers development has been considered as top priority in fertilizer research.[7] Table 9.2 shows details of macronutrient-containing NPs which could potentially improve plant growth by supplying nutrients.

9.3.2   MICRONUTRIENT NANOFERTILIZERS

Micronutrients in plants comprise of manganese (Mn), iron (Fe), copper (Cu), zinc (Zn), and molybdenum (Mo). As compared with the macronutrients (P, N, and K), only trace levels of micronutrients are required for vigorous growth of plants and crops as is shown in the composition of Hoagland solution.[6] In N, P, and K fertilizers (collectively called composite fertilizers), micronutrients are often added at low rates (5 mg l^{-1}) as soluble salts for crop uptake. Micronutrients in these composite fertilizers usually provide adequate nutrients and become the source of environmental risks. Though, plant availability of the applied micronutrients may become low and micronutrient deficiency may occur in some alkaline and coarse textured soils, or those containing low soil organic matter (SOM).[5] Apparently, micronutrient nanofertilizers may enhance bioavailability of these nutrients to plants even under worst conditions. Development and application of nanofertilizers are still at initial stages, there are few if any specific researches or systemic studies on the effects and advantages of applying micronutrient nanofertilizers under field conditions. Details of micronutrient-containing NPs,[7] which could potentially improve plant growth by supplying nutrients are presented in Table 9.3.

9.4   ADVANTAGES OF NANOFERTILIZERS IN AGRICULTURE

- Nano coatings as well as technology proved to be efficient for reducing costs as well as for increasing productivity around the farm.

TABLE 9.2 Nanoparticles Reportedly Enhanced Plant Growth by Providing Macronutrient.

Nutrient provided	Nanoparticle type, particle size, concentration, (reference)	Test plants, method, and medium	Growth enhancements		
			Over control 1 (0 target nutrient)	Over control 2 (regular fertilizer with target nutrient)	Other factors
Macronutrient P	Apatite, $Ca_5(PO_4)_3OH$, 16 nm, 21.8 mg L^{-1} as P	Soybean, 5-month greenhouse test, 50% perlite and 50% peat moss, nutrient solution.	Aboveground biomass (6.5) a, belowground biomass (40), growth rate (2), yield (5.4)	Soluble $Ca(H_2PO_4)_2$, 21.8 mg L^{-1} as P, aboveground biomass (1.2), belowground biomass (1.4), growth rate (1.3), yield (1.2)	N/A
Macronutrient Ca	Calcite, $CaCO_3$, 20–80 nm, 160 mg L^{-1} as Ca	Peanut seedlings, 80-day greenhouse test, sand medium, nutrient solution	Aboveground biomass (1.2), Ca content in stems (5.2) and roots (3.7), soluble sugar(1.3) and protein (2) in shoots	$Ca(NO_3)_2$, 200 mg L^{-1} as Ca, aboveground biomass (0.96), Ca contents in stem (1.03) and roots (0.76), soluble sugar (1) and protein (1) in shoot	Highest yield achieved at a combination of 1 g L^{-1} humic acid and Ca NPs
Macronutrient Mg	Mg-NPs, 500 mg L^{-1} as Mg, others unknown	Black-eyed pea, foliar application, field experiment, soil, irrigation	1000-Seed weight (0.93), leaf Mg (1.1), stem Mg (1.2)	500 mg L^{-1} Regular Mg salts, 1000-seed weight (0.92), leaf Mg (1.2)	Highest yield achieved at a combination of 500 mg L^{-1} Fe and Mg-NPs

The enhancement value in parentheses: Biomass under NP treatment/biomass under control.

TABLE 9.3 Nanoparticles Reportedly Enhanced Plant Growth by Providing Micronutrients.

Nutrient provided	Nanoparticle type, particle size, concentration, and (reference)	Test plants, method, and medium	Growth enhancements		
			Over control 1 (no target nutrient)	Over control 2 (regular fertilizer with target nutrient)	Other factors
Micronutrient Fe	Superparamagnetic iron oxide NPs, Fe_3O_4, 18.9–20.3 nm,30, 45, and 60 mg l^{-1}	Soybean, 7-day greenhouse test perlite medium, nutrient solution.	No reported	Fe-EDTA, Chlorophyll content (1.1) a at 30 mg l^{-1} and (1) at 45mg l^{-1} but(0.8)at 60mg l^{-1}	N/A
	Fe-NPs, 0.25 and 0.5 g l^{-1}, others unknown	Black-eyed pea, foliar application, field experiment, soil, irrigation	1000-Seed weight (1.01 and 1.07), leaf Fe (2.1 and 1.3), chlorophyll content (1.05 and 1.1)	A regular Fe salt, 1000-seed weight (1.03 and 1.04), leaf Fe (2.1 and 1.5), chlorophyll content (1.05 and 1.1)	N/A
Micronutrient Mn	Metallic Mn, 20 nm, 0.05. 0.1, 0.5, and 1.0 mg l^{-1}	Mung bean plant, 15-day in growth chamber, perlite medium, nutrient solution	Root length (1.4–1.5), shoot length (1.2–1.4), dry weight (1–2), chlorophyll (1.1–1.8) and carotenoid (1.1–2) contents, photosynthesis rate (1.1–1.7)	MnSO₄, root length (1–1.6), shoot length (1.1), dry weight (1–2), chlorophyll (1–1.6) and carotenoid (1.1–1.4) contents, photosynthesis rate (1.1–1.3)	Mn-NPs did not show phytotoxicity as MnSO₄ did

TABLE 9.3 (Continued).

Nutrient provided	Nanoparticle type, particle size, concentration, and (reference)	Test plants, method, and medium	Growth enhancements		Other factors
			Over control 1 (no target nutrient)	Over control 2 (regular fertilizer with target nutrient)	
Micronutrient Zn	ZnO, 20 nm, 1–2000 mg l^{-1}	Mung bean and chickpea seedlings, 60-hour in an incubator, agar medium	Shoot height (1.3) and (1.6) biomass, root length (1.9) and biomass(1.4) for Mung bean at 20 mg l^{-1}; (1), (2.8), (1.5,) and (1.3) for chickpea at 1 mg l^{-1}	Not reported	Zn-NPs at levels higher than the optimum showed phytotoxicity
	ZnO, 10 nm, 400 and 800 mg kg^{-1} soil	Cucumber, soil mixture, a life cycle of 53-days greenhouse test		Not reported	No negative effects found
	ZnO, 20 nm,1–2000 mg l^{-1}	Rape seeds, 5-day germination, water	Root dry mass (1.1–1.6); fruit starch (1.1–1.6), glutelin (0.9–2), Zn (1.7–2.5)		Zn-NPs at levels higher than the optimum showed phytotoxicity
	metallic Zn, 35 nm, 1–2000 mg L^{-1}	Ryegrass seed, 5-day germination, water	Root elongation (1.1) at 2 mg l^{-1}		
			Root elongation (1.2) at 2 mg l^{-1}		

TABLE 9.3 *(Continued).*

Nutrient provided	Nanoparticle type, particle size, concentration, and (reference)	Test plants, method, and medium	Growth enhancements		Other factors
			Over control 1 (no target nutrient)	Over control 2 (regular fertilizer with target nutrient)	
Micro-nutrient Cu	70% CuO and 30% Cu_2O, 30 nm, 0.025, 0.25, 0.5, 1, and 5 mg l^{-1} as Cu; Metallic Cu, b 50 nm, 130 and 600 mg kg^{-1} as Cu	Elodea densa planch, 3-day incubation, water. Lettuce seeds, 15-day germination, soil.	Photosynthesis rate (1–1.4) at b 0.5 mg l^{-1}; Shoot/root ratio (1.4–1.9), total N (1.2) and organic matter (1.1) at 130 mg kg^{-1}	$CuSO_4$; photosynthesis rate (1.4–3) at b 0.5 mg L^{-1}; leaf Cu (5) at b 0.5 mg l^{-1}; Not reported	$CuSO_4$ inhibitory at all concentrations; Influences found only when Cu-NPs added in soil 15 day before seeding
Micro-nutrient Mo	Mo, 100–250 nm, 8 mg l^{-1}, others unknown	Chickpea, seed-soaked 1–2 h, a life cycle, soil, rhizo-sphere examination	Nodule number/mass (11/6.2), activity of anti-oxidant enzymes (1.9–2.6), symbiotic bacteria (1.9–3.2)	Not reported	Microbial treatment further enhanced the effects

The enhancement value in parentheses: Biomass under NP treatment/biomass under control.

- Improvement in soil aggregation, moisture retention and carbon build up can be obtained through the use of nanofertilizers.
- The yield per hectare is also much higher than conventional fertilizers, thus giving higher returns to the farmers.
- Nanofertilizers association with nanodevices leads to the release of N and P fertilizer with their uptake by crops. Hence, beneficial in avoiding undesirable nutrient losses to air, water and soil through direct internalization by crops, and avoiding the interaction of nutrients with microorganisms, water, air as well as soil.[4]

9.5 SUMMARY

Synthetic fertilizers possess a significant potential of polluting air, soil, and water. Therefore, numerous efforts have been made to minimize these problems by agricultural practices and for the formulation of improved fertilizers. Nanotechnology has opened up potential innovative applications in diverse agricultural and biotechnology field. Nanostructured formulation could release their active ingredients in responding to environmental triggers and biological demands more precisely through various mechanisms such as targeted delivery or slow/controlled release mechanisms as well as conditional release. Hence, using these mechanisms nanofertilizers can be designed and constructed probably. Utilization of nanofertilizers reduces soil toxicity and minimizes the potential negative effects associated with over dosage as well as reduces the frequency of the application. Nanofertilizer mainly delays the release of the nutrients and extends the fertilizer effect period. Apparently, there is a chance for nanotechnology to have a significant influence on energy, the economy and the environment, by improving fertilizers. Hence, nanofertilizers has been known to possess high prospective for attaining sustainable agriculture, especially in developing countries.

KEYWORDS

- Nanofertilizers
- smart fertilizers
- nanostructured formulation
- nanodevices
- antioxidant activity
- environmental triggers

REFERENCES

1. Alexandratos, N.; Bruinsma, J. *World Agriculture Towards 2030/2050: The 2012 Revision.* ESA Working Paper 12-03, FAO, Rome, 2012.
2. Chinnamuthu, C. R.; Boopathi, P. M. Nanotechnology and Agroecosystem. *Madras Agric. J.* **2009**, *96*(1/6), 17–31.
3. Cui, H. X.; Sun, C. J.; Liu, Q.; Jiang, J.; Gu, W. Applications of Nanotechnology in Agrochemical Formulation, Perspectives, Challenges and Strategies. *International conference at Nanoagri (Sao Pedro)*, 20–25 June 2010, Brazil.
4. DeRosa, M. R.; Monreal, C.; Schnitzer, M.; Walsh, R.; Sultan, Y. Nanotechnology in Fertilizers. *Nat. Nanotechnol. J.* **2010**, 5(2), 91.
5. Fageria, N. K. *Use of Nutrients in Crop Plants;* CRC Press: Boca Raton, Florida, 2009; p 430.
6. Hoagland, D. R.; Arnon, D. I. The Water-Culture Method for Growing Plants without Soil. *Calif. Agric. Exp. Stn. Circ.* 1950, *347*, 1–32.
7. Liu, R.; Lal, R. Potentials of Engineered Nanoparticles as Fertilizers for Increasing Agronomic Productions. *Sci. Total Environ.* **2015**, *514*, 131–139.
8. Naderi, M. R.; Shahraki, A. D. Nanofertilizers and their Roles in Sustainable Agriculture. *Int. J. Agric. Crop Sci.* **2013**, 5(19), 2229–2232.
9. Nair, R.; Varghese, S. H.; Nair, B. G.; Maekawa, T.; Yoshida, Y.; Kumar, D. S. Nanoparticulate Material Delivery to Plants. *Plant Sci.* **2010**, *179*(3), 154–163.
10. Shaviv, A. Advances in Controlled Release Fertilizer. *Adv. Agron.* **2000**, *71*, 1–49.
11. Smil, V. Nitrogen and Food Production: Proteins for Human Diets. *Ambio* **2002**, *31*(2), 126–131.
12. Tarafdar, J. C.; Sharma, S.; Raliya, R. Nanotechnology: Interdisciplinary Science of Applications. *Afr. J. Biotechnol.* **2013**, *12*(3), 219–226.

ROLE OF SEAWEED AND WATER HYACINTH BIOFERTILIZERS: QUALITY AND PRODUCTION OF OKRA

SAVAN D. FASARA[1], SONAL V. PANARA[1,2],
CHHAYA R. KASUNDRA[1,3], PRASHANT D. KUNJADIA[4],
GAURAV V. SANGHVI[5], BHAVESH D. KEVADIYA[6] AND
GAURAV S. DAVE[7,*]

[1]Department of Biochemistry, Saurashtra University, Rajkot, Gujarat, India, E-mail: sdfasara@gmail.com, Mobile: +91-9137668720

[2]Department of Biochemistry, Saurashtra University, Rajkot, Gujarat, India, E-mail: sonalipanara3@gmail.com

[3]Department of Biochemistry, Saurashtra University, Rajkot, Gujarat, India, E-mail:chhayachemist@gmail.com

[4]Department of Biochemistry, Saurashtra University, Rajkot, Gujarat, India B. N. Patel Institute of Paramedical and Science, Bhalej Road, Anand, Gujarat, India, E-mail: pdkunjadia@yahoo.com, Mobile: +91-9824252544

[5]Department of Pharmaceutical Sciences, Saurashtra University, Rajkot 360005, India, E-mail: sanghavi83@gmail.com

[6]University of Nebraska Medical Center, Omaha, NE, USA, E-mail: bbhaveshpatel@gmail.com, Mobile: +001-8454809317

[7]Department of Biochemistry, Saurashtra University, Rajkot 360005, Gujarat, India

*Corresponding author.E-mail: gsdspu@gmail.com, Mobile: +91-9428275894

CONTENTS

10.1   INTRODUCTION

Algae are large and diverse group of autotrophic organism that is photo-synthetic and eukaryotic. The most complex form of algae is the seaweeds. Algae are found all over the earth, both on the ground and in the water. Some algae form symbiotic relationships with fungi, marine invertebrates and sea sponges.[2] Algae are popular foods, especially in Asian cultures. Algae are good source of vitamins, minerals, and micronutrients. Algae are also high in omega-3 fatty acids.[2]

Seaweeds are used as food, feed, fodder, fertilizer, agar, alginate, carrageenan, and source of various fine chemicals.[7] In recent years, the use of natural seaweeds as fertilizer[5] has allowed for substitution in place of conventional synthetic fertilizer.[3] Seaweed extracts are marketed as liquid fertilizers and biostimulants since they contain many growth regu-lators such as cytokinins,[4,14] auxins,[15] gibberellins,[17] betaines,[1] macro-nutrients such as Ca, K, and P, and micronutrients like Fe, Cu, Zn, B, Mn, Co, and Mo,[6] necessary for the development and growth of plants. Seaweeds and seaweed extracts also enhance soil health by improving moisture holding capacity[11] and by promoting the growth of beneficial soil microbes.[6]

Water hyacinths (*Eichhornia crassipes*) are one of the most common plants seen in water gardens. This floating plant is native to South America and very sensitive to cold weather. Water hyacinths have round, glossy green leaves held on upright, fleshy stalks. Short plants usually have a large "ball" at the base of each leaf. This, along with the fleshy stem, is filled with styrofoam-like material that keeps the plant afloat. Beneath the water is a large, feathery root system that is usually black or purple. *Eichhornia crassipes* is an excellent source of biomass. One hectare of standing crop thus produces more than 70,000 m^3 of biogas. One kg of dry matter can yield 370 L of biogas, giving a heating value of 22,000 kJ/m^3 compared to pure methane (895 Btu/ft^3). The roots of *E. crassipes* naturally absorb pollutants, including lead, mercury, and strontium-90, as well as some organic compounds believed to be carcinogenic, in concentrations 10,000 times that in the surrounding water. Water hyacinths can be cultivated for waste water treatment. The plant is used as a carotene-rich table vegetable in Taiwan. Javanese sometimes cook and eat the green parts and inflorescence. The plant is also used as animal feed and organic fertilizer. According to https://en.wikipedia.org/wiki/Eichhornia_crassipes, water hyacinths can be used for paper production on a small scale."

Okra or lady finger (*Abelmoschus esculentus*) is one of the important vegetables of India and throughout the world (Appendix—A). It is grown throughout the tropical and subtropical regions and also in the warmer parts of the temperate regions. The nutritional value of 100 g of edible okra is characterized by 1.9 g protein, 0.2 g fat, 6.4 g carbohydrate, 0.7 g minerals, and 1.2 g fibers (http://www.ncpahindia.com/okra.php). Okra is a popular health food due to its high fiber, vitamin C, and folate content. Okra is also known for being high in antioxidants. Okra is also a good source of calcium and potassium. Greenish-yellow edible okra oil is pressed from okra seeds; it has a pleasant taste and odor, and is high in unsaturated fats such as oleic acid and linoleic acid. The oil content of some varieties of the seed can be quite high, about 40%. Oil yields from okra crops are also high. At 794 kg/ha, the yield was exceeded only by that of sunflower oil. The Okra oil is suitable for use as a biofuel. Okra has a good potential as a foreign exchange crop and accounts for 60% of the export of fresh vegetables in India. According to the report of Indian Horticulture Database 2013, it is cultivated in 0.5308 million ha area with productivity of 12 metric ton/ha. Gujarat is the fourth largest productive state of India. The crop is also used in paper industry as well as for the extraction of fiber.[16]

The present research study was planned to evaluate the biochemical and food quality of okra (*A. esculentus*) grown under fertilizer treatment of seaweed (brown and red seaweeds) and water hyacinth (*E. crassipes*) combination.

10.2   MATERIALS AND METHODS

10.2.1   PREPARATION OF SEAWEED SOLID FERTILIZER

The specimens of brown seaweed (*Sargassum wightii*) and red seaweed (*Galaxaura oblongata*) were collected from Mangrol coast of Gujarat, India. Whole adult plants were collected early in the morning and washed in the field with seawater initially to remove macroscopic epiphytes, sediment, and organic matter. Algae were packed in plastic bags and kept on ice until returned to the laboratory. In the laboratory, samples were gently brushed under running sea water, rinsed with distilled water, and were shade dried for seven days followed by oven drying at 40°C. Then the dry material was hand crushed and made into a coarse powder using a grinder. This powder was used as a *biofertilizer*.

10.2.2   PREPARATION OF WATER HYACINTH SOLID FERTILIZER

The specimen of water hyacinth was collected from river of Bedi village (Rajkot). The leaves of water hyacinth were collected early in the morning, and were packed in plastic bags and kept on ice until returned to the laboratory. In the laboratory, sample were gently brushed under running tap water, rinsed with distilled water, and were shade dried for seven days followed by oven drying at 40°C. Then material was hand crushed and made into a coarse powder using a grinder. This powder was used as *biofertilizer*.

10.2.3   FIELD TRIALS

Field trial was concluded at Garden of *Saurashtra University*. The experimental area was ploughed thoroughly two times followed by a final ploughing accompanied by sowing okra seeds along with furrows at an interval of one foot and leveled off. These experimental trials were conducted in rows. For each experiment, six plants per row were taken. Plants were irrigated every fifteen days. Different combinations of seaweed and *Hyacinth* biofertilizers were made. *Hyacinth, Sargassum,* and *Galaxaura* solid fertilizers were applied in different combinations (Table 10.1).

TABLE 10.1   Different Combinations of Fertilizer.

Group No.	Sets of fertilizer combination
1	Control
2	*Hyacinth* autoclaved
3	*Hyacinth*
4	*Sargassum* autoclaved
5	*Sargassum*
6	*Galaxaura* autoclaved
7	*Galaxaura*
8	*Hyacinth* + *Sargassum* autoclaved (1:1)
9	*Hyacinth* + *Sargassum* (1:1)
10	*Hyacinth* + *Galaxaura* autoclaved (1:1)
11	*Hyacinth* + *Galaxaura* (1:1)
12	*Sargassum* + *Galaxaura* autoclaved (1:1)
13	*Sargassum* + *Galaxaura* (1:1)
14	*Hyacinth* + *Sargassum* +*Galaxaura* autoclaved (1:1 :1)
15	*Hyacinth* + *Sargassum* +*Galaxaura* (1:1 :1)

10.2.4   BIOCHEMICAL MEASUREMENTS

Reducing sugar concentration and protein concentration were measured by DNSA[10] and Lowry[8] method, respectively. Shoot length, number of leaves, number of fruits, and length of okra fruits were measured at different time intervals.

10.3   RESULTS AND DISCUSSION

The Figure 10.1 indicates that carbohydrate percentages in fruits were increased and the groups 5, 7, 9, 10, 11, and 15 exhibited around 300 mg% of total carbohydrates in fruits.

This reflects *Hyacinth* and *Galaxaura* combination without autoclaved contains nutrients which increase the carbohydrate content in okra fruits. These combinations of biofertilizers are beneficial for the development of high quality vegetables like okra in agricultural practice. As the different types of seaweeds contain various micronutrients, extract of

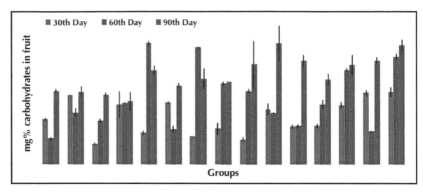

FIGURE 10.1   Reducing sugar content of fruit.

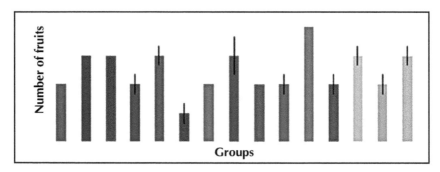

FIGURE 10.2   Number of fruits in a group.

these seaweeds and *Hyacinth* helps in the nourishment of soil, ultimately it is reflected in growth and development of a crop.

Number of fruits were increased in groups 5, 8, 11,13, and 15 (Fig. 10.2). Maximum number of fruits were observed in group 11. This agrees with the positive effect of group 11 biofertilizer combination on the increased carbohydrate content in okra fruits. *Hyacinth* and *Galaxaura* without autoclaved combination is responsible for the increased number of fruits in plants.

Fruit length was maximum in group 11, which furhter confirms beneficial and fruit quility improving capacity of *Hyacinth* and *Galaxura* without autoclaved combination (Fig. 10.3). These results can further be evaluated for the properties of okra crop under the above combinations on pilot scale field trial in different parts of India.

Fruit protein content was evaluated in different groups treated with different combination of seaweed and *Hyacinth*. Surprisingly, group 11 showed

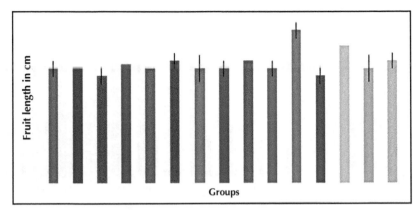

FIGURE 10.3   Fruit length in a group.

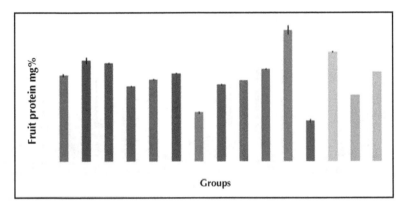

FIGURE 10.4   Protein content of fruit.

maximum protien content, which further supports quality improvement in fruits treated with the *Hyacinth* and *Galaxaura* without autoclaved combination (Fig. 10.4). Beneficial effect of water *Hyacinth* and *Galaxaura* reflects its further use as a biofertilizer to improve the quality and quantity of okra. Current needs of food quality and quantity improvement can be supported with these experiments to improve food crop with biofertilizer treatment.

Shoot length of plant was shown to increase as days progressed, maximum growth was observed in terms of shoot length in group 8, treated with *Hyacinth:Sargassum* (1:1) autoclaved at 75th Day (Fig. 10.5). In general, authors found that Sargassum without autoclaved exhibited

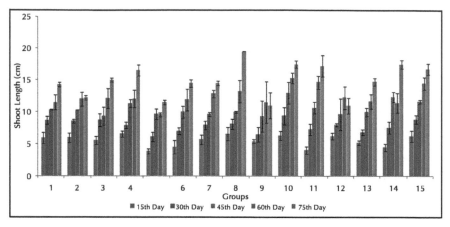

FIGURE 10.5   Shoot length at different time interval.

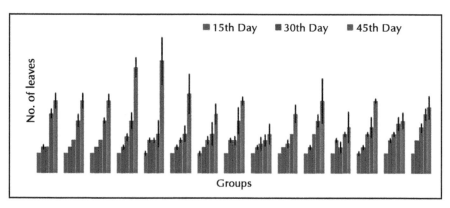

FIGURE 10.6   Number of leaves at different time interval.

negative effect on shoot length compared to autoclaved sample. This observation supports the compounds of seaweed contains various plant growth promoting nutrients like potassium, calcium, manganese, magnesium, iron, zinc[12,13] as well as plant growth regulators that is indole acetic acid, abscisic acid (ABA), gibberellin, indole butyric acid.[9] Upon autoclaving, these regulators degraded and could not give its effect.

Similar results were observed for the number of leaves of plant, maximum number was observed in group 5, treated with *Sargassum* without autoclaved at 75th Day (Fig. 10.6). Moreover, *Sargassum* with autoclaved extract showed similar high number in leaves, which indicates positive effect of *Sargassum* on leaf development and production.

10.4 CONCLUSIONS

The changes of different physiological characteristics and the selected parameters can get affected by algae fertilizers and *Hyacinth*. Different combinations of algae and *Hyacinth* have shown significant changes in shoot length, number of leaves, and number of fruits. Biochemical constitution of shoot and fruit showed significant change. Studied combinations of algae and *Hyacinth*, with or without autoclave, exhibited significant impact on plant growth, fruits quality, and quantity in comparison to control group. *Hyacinth:Galaxaura* showed significant increase in shoot length of plants and increase in number of fruits and fruit length. Combination of all three showed increased protein concentration in every cycle and number of leaves was also increased in *Hyacinth:Sargassum:Galaxaura* and it also showed maximum carbohydrate concentration. Present study suggests further experiments to formulate various combinations of seaweed and *Hyacinth* to improve lady finger quality and quantity in different crop fields of India.

10.5 SUMMARY

On the basis of this study, authors prepared different sets of fertilizer like *Hyacinth, Sargassum, Galaxaura* and mixture of all these in different sets with autoclave or without autoclaved. Therefore, 15 groups were formed. Authors observed different parameters on different period. For shoot length, five cycles of 15 days were observed. The highest value of length was seen in *Hyacinth:Sargassum* autoclaved. Number of leaves was highest in *Sargassum* treated group. Highest concentration of carbohydrate in fruits was found in *Hyacinth:Galaxaura*, compared to highest concentration of protein in *Galaxaura* autoclaved treated group. Highest length of fruits for cycle at 75 days was found in *Hyacinth:Galaxaura*. Number of fruits at 75th day was highest in *Hyacinth:Galaxaura* autoclaved. These observations were made at the end of the experiments and it was concluded to use different combinations of seaweed along with water *Hyacinth* for okra crop quality and quantity improvement. In future, it is recommended to test the different formulation of seaweed and water *Hyacinth* in form of customized tablets to replace the chemical biofertilizers. Long term use of natural fertilizer will help in revival of soil fertility and quality, hence quality and nutrient value of crop.

KEYWORDS

- *Abelmoschus esculentus*
- biofertilizer
- *Galaxura*
- *Hyacinth*
- lady finger
- plant hormones
- protein
- reducing sugar
- seaweed

REFERENCES

1. Blunden, G.; Morse, P. F.; Mathe, I.; Hohmann, J.; Critchleye, A. T.; Morrell, S. Betaine Yields from Marine Algal Species Utilized in the Preparation of Seaweed Extracts used in Agriculture. *Nat. Prod. Commun.* **2010**, *5*, 581–585.
2. Brown, B. J.; Preston, J. F. L-Guluronan-Specific Alginate Lyase from a Marine Bacterium Associated with *Sargassum*. *Carbohydr. Res.* **1991**, *211,* 91–102.
3. Crouch, I. J.; Van Staden, J. Commercial Seaweed Products as Biostimulants in Horticulture. *J. Home Consum. Hortic.* **1994**, *1*, 19–76.
4. Durand, M.; Beaumatin, P.; Bulman, B.; Bernalier, A.; Grivet, J. P., Serezat, M.; Gramet, G.; Lahaye, M. Fermentation of Green Alga Sea- Lettuce (*Ulva sp*) and Metabolism of its Sulphate by Human Colonic Microbiota in a SemiContinuous Culture System. *Reprod. Nutr. Dev.* **1997**, *37*, 267–283.
5. Hong, D. D.; Hien, H. M.; Son, P. N. Effect of Seaweed Extracts (swe) on Germination of *Trigonella foenum-graecum* Seeds. *Appl. Phycol.* **1997**, *19*, 817.
6. Khan, M. N.; Yoon, S. J.; Choi, J. S.; Park, N. G.; Lee, H. H.; Cho, J. Y.; Hong, Y. K. Anti-Edema Effects of Brown Seaweed (*Undaria pinnatifida*) Extract on Phorbol 12-myristate 13-acetate-induced Mouse Ear Inflammation. *Am. J. Chin. Med.* **2009**, *37*, 373–381.
7. Kumar, G.; Sahoo, D. Effect of Seaweed Liquid Extract on Growth and Yield of *Triticum aestivum* var. Pusa Gold. *Appl. Phycol.* **2011**, *23*, 251–255.
8. Lowry, O. H.; Rosebrough, N. J.; Farr, A. L.; Randall, R. J. Protein Measurement with the Folin Phenol Reagent. *J. Biol. Chem.* 1951, *193*, 265–275.
9. Mabeau, S.; Fleurence, J. Seaweed in Food Products: Biochemical and Nutritional Aspects. *Trends Food Sci. Technol.* **1993**, *4*(4), 103–107. https://doi.org/10.1016/0924-2244(93)90091-N
10. Miller, G. L. Use of Dinitrosalicylic Acid Reagent for Determination of Reducing Sugar. *Anal. Chem.* **1959**, *31*, 426–428.
11. Moore, A. Blooming Prospects? Humans have Eaten Seaweed for Millennia; now Microalgae are to be Served up in a Variety of Novel Health Supplements, Medicaments and Preparations. *EMBO Rep.* **2001**, *2*, 462–464.
12. Pena-Rodriguez, A.; Mawhinney, T. P.; Ricque-Marie, D.; Cruz-Suarez, L. E. Chemical Composition of Cultivated Seaweed *Ulva clathrata* (Roth) C. Agardh. *Food Chem.* **2011**, *129*, 491–498.

13. Roosta, H. R.; Mohsenian, Y. Effects of Foliar Spray of Different Fe Sources on Pepper (*Capsicum annum* L.) Plants in Aquaponic System. *Sci. Hortic.* **2012**, *146*, 182–191.

14. Stirk, W. A.; Staden, J. V. Screening of Some South African Seaweed for Cytokinin-Like Activity. *S. Afr. J. Bot.* **1997**, *63*, 161–170.

15. Strik, W. A.; Novak, O.; Hradecka, V.; Pencik, A.; Rolcik, J.; Strnad, M.; Staden, J. V. Endogenous Cytokinins, Auxins and Abscisic Acid in *Ulva fasciata* (Chlorophyta) and *Dictyota humifusa* (*Phaeophyta*): Towards Understanding their Biosynthesis and Homoeostasis. *Eur. J. Phycol.* **2009**, *44*, 231–240.

16. Tiwari, R. K.; Mistry, N. C.; Singh, B.; Gandhi, C. P. *Indian Horticulture Database—2013*. National Horticulture Board, Government of India, New Delhi; 2014.

17. Wildgoose, P. B.; Blunden, P. B.; Jewers, K. Seasonal Variations in Gibberellin Activity of Some Species of *Fucaceae* and Laminariaceae. *Bot. Mar.* **1978**, *21*, 63–65.

CHAPTER 11

ROLE OF SEAWEED AND WATER HYACINTH BIOFERTILIZERS: QUALITY AND PRODUCTION OF COWPEA (*VIGNA UNGUICULATA*)

BHUMIKA P. CHADAMIYA[1], HETAL J. CHADAMIYA[1,2],
VAISHALI D. PATEL[1,3], GAURAV V. SANGHVI[1,4],
PRASHANT D. KUNJADIA[5], DEVENDRA VAISHNAV[4,6],
GAURAV S. DAVE[1,*]

[1]*Department of Biochemistry, Saurashtra University, Rajkot 360005, Gujarat, India, E-mail: bhumikachadamiya32@gmail.com*

[2]*Department of Biochemistry, Saurashtra University, Rajkot 360005, Gujarat, India, E-mail: hetalchadmiya@gmail.com*

[3]*Department of Biochemistry, Saurashtra University, Rajkot 360005, Gujarat, India, E-mail: vaishalichadmiya@gmail.com*

[4]*Department of Pharmaceutical Sciences, Saurashtra University, Rajkot 360005, India, E-mail: sanghavi83@gmail.com*

[5]*B. N. Patel Institute of Paramedical and Science, Bhalej Road, Anand, Gujarat, India, E-mail: pdkunjadia@yahoo.com*

[6]*Department of Pharmaceutical Sciences, Saurashtra University, Rajkot 360005, India, E-mail: devvaishnav@gmail.com*

[1]*Department of Biochemistry, Saurashtra University, Rajkot 360005, Gujarat, India*

Corresponding author. E-mail: gsdspu@gmail.com

CONTENTS

11.1   INTRODUCTION

Marine algae are used in agricultural and horticultural crops and many beneficial effects, in terms of enhancement of yield and quality, have been documented.[26] The effect of many components that may work synergistically at different concentrations is the advantageous effect of seaweed extract, but mechanism of effect is to be evaluated in detail.[11]. In recent years, the use of seaweed extracts have gained popularity due to their potential use in organic and sustainable agriculture,[29] especially in rain-fed crops, as a means to avoid excessive fertilizer applications and to improve mineral absorption.

Sargassum is one of the marine macroalgae genera belonging to the class *Phaeophyceae*; it is widely distributed in tropical and temperate oceans and is reported to be used as animal feed, food ingredients, and fertilizer. Several preclinical studies on *Sargassum* species revealed numerous physiological and biological activities such as antioxidant, antitumor, antiangiogenic, anti-inflammatory, anticoagulant, anti-vasculogenic.[12] *Sargassum wightii* contains significant amount of flavonoids in support of its antioxidant activity;[18] they also possess sulphated polysaccharides, substances that are responsible for wide pharmacological actions like free radical scavenging,[23, 29] antioxidant,[32] antifungal,[6] anti-inflammatory,[8] and hepatoprotective potential.[5] It has been reported that agonic acid extracted from brown algae possess hypocholesterolemic effects[14] and exert antihypertensive potential.[22] Apart from beneficial effects, seaweeds are reported for delayed germinating effect on chickpea seed.[30]

Cultural and morphological studies showed that *Galaxaura oblongata* (Ellis et. Solander) Lamouroux has a triphasic life history with conspicuous gametophytes and small filamentous tetrasporophytes.[17] *Galaxaura oblongata* extract is found to be antioedemic in tissue plasminogen activator (TPA) induced model in mice as well as demonstrated anti-inflammatory effects on carrageenan-induced mouse paw oedema.[24] *Galaxaura oblongata* has been studied for metal remover with mean biosorption efficiency of 84%, which constitute a promising, efficient, cheap, and biodegradable sorbent biomaterial for lowering the heavy metal pollution in the environment.[13] Moreover, *G. oblongata* oil has been studied for biodiesel production in previous reports,[1] and this will be pioneer work for the evaluation of *G. oblongata* for its application as a fertilizer.

One of the fastest growing plants known as water hyacinth reproduces primarily by way of runners or stolons, which eventually form daughter plants. Each plant can produce thousands of seeds each year, and these seeds can remain viable for more than 28 years, and some water hyacinths were found to grow up to 2–5 m per day at some sites in Southeast Asia. The roots of *Eichhornia crassipes* naturally absorb pollutants, including lead, mercury, and strontium-90, as well as some organic compounds believed to be carcinogenic, in concentrations 10,000 times more that in the surrounding water. Water hyacinths can be cultivated for waste water treatment. Water hyacinth is reported for its efficiency to remove nitrogen and potassium from water.[4] "The roots of water hyacinth were found to remove particulate matter and nitrogen in a natural shallow eutrophicated wetland. The plant is also used as animal feed and organic fertilizer although there is controversy stemming from the high alkaline pH value of the fertilizer," (https://en.wikipedia.org/wiki/Eichhornia_crassipes).

Cowpea (*Vigna unguiculata* L.) is a good protein source and one of the most ancient human food sources probably since Neolithic times.[20] It is an important multipurpose grain legume extensively cultivated in arid and semiarid tropics. It is an important source of nutrients and provides high quality, inexpensive protein to diets based on the cereal grains and starchy foods.[28] Cowpea is a good source of food, forage, fodder, vegetable, and certain snacks.[21] Moreover, it has been reported about its ability to fix atmospheric nitrogen in soil at the rate of 56 kg/ha through symbiotic bacteria under favorable conditions.[2,9,10] In India, the capita/day availability of pulses had decreased from 69 g during 1960s to 35 g today, as against the Food And Agriculture Organization/World Health Organization (FAO/

WHO)'s current recommendation of 80 g per day.[3] Worldwide, cowpeas are cultivated in approximately 8 million hectares. Area under cowpea in India is 3.9 million hectares with a production of 2.21 million tons with the national productivity of 683 kg/ha.[28]

The present research study evaluates the biochemical and food quality of cowpea (*Abelmoschus esculentusa*) grown under fertilizer treatment of seaweed (brown and red seaweeds) and water hyacinth (*E. crassipes*) combination.

11.2 MATERIALS AND METHODS

11.2.1 PREPARATION OF SEAWEED SOLID FERTILIZER

The specimens of brown seaweed (*S. wightii*) and red seaweed (*G. oblongata*) were collected from Mangrol coast of Gujarat, India. Whole adult plants were collected early in the morning and washed in the field with seawater initially to remove macroscopic epiphytes, sediment, and organic matter. Algae were packed in plastic bags and kept on ice until returned to the laboratory. In the laboratory, samples were gently brushed under running sea water, rinsed with distilled water; and were shade dried for 7 days followed by oven drying at 40°C. Then the dry material was hand crushed and made into a coarse powder using a grinder. This powder was used as a *biofertilizer*.

11.2.2 PREPARATION OF WATER HYACINTH SOLID FERTILIZER

The specimen of water hyacinth was collected from river of *Bedi* village (Rajkot). The leaves of water hyacinth were collected early in the morning, and were packed in plastic bags and kept on ice until returned to the laboratory. In the laboratory, samples were gently brushed under running tap water, rinsed with distilled water, and were shade dried for seven days followed by oven drying at 40°C. Then material was hand crushed and made into a coarse powder using a grinder. This powder was used as *biofertilizer.*

11.2.3   FIELD TRIALS

Field trial was concluded at garden of *Saurashatra University*. The experimental area was ploughed thoroughly two times followed by a final ploughing accompanied by sowing cowpea seeds along with furrows at an interval of one foot and leveled off. These experimental trials were conducted in rows. For each experiment, six plants per row were taken. Plants were irrigated every fifteen days. Different combinations of seaweed and *Hyacinth* biofertilizers were made. *Hyacinth, Sargassum,* and *Galaxura* solid fertilizers were applied in different combinations (Table 11.1).

TABLE 11.1   Different Combinations of Fertilizer.

Group No.	Sets of fertilizer combination
1	Control
2	*Hyacinth* autoclaved
3	*Hyacinth*
4	*Sargassum* autoclaved
5	*Sargassum*
6	*Galaxura* autoclaved
7	*Galaxura*
8	*Hyacinth* + *Sargassum* autoclaved (1:1)
9	*Hyacinth* + *Sargassum* (1:1)
10	*Hyacinth* + *Galaxura* autoclaved (1:1)
11	*Hyacinth* + *Galaxura* (1:1)
12	*Sargassum* + *Galaxura* autoclaved (1:1)
13	*Sargassum* + *Galaxura* (1:1)
14	*Hyacinth* + *Sargassum* +*Galaxura* autoclaved (1:1:1)
15	*Hyacinth* + *Sargassum* +*Galaxura* (1:1:1)

11.2.4   BIOCHEMICAL MEASUREMENTS

Reducing sugar concentration, protein concentration was measured by DNSA[19] and Lowry[15] methods, respectively. Shoot length, number leaves, number of fruits, and length of cowpea fruits were measured at different time intervals.

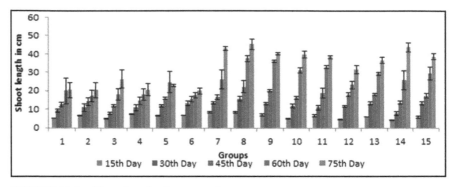

FIGURE 11.1   Shoot length of cowpea plant.

11.3   RESULTS AND DISCUSSION

Shoot length of plants were studied at interval of 15 days (Fig. 11.1). Maximum shoot length was observed through the course of study in group supplied with autoclaved extract of water hyacinth and *Sargassum*. This reflects *Hyacinth*, and *Sargassum* autoclaved combination contains nutrients and rich content of precursors for hormone to increase the shoot length in cowpea plants. It also suggests the advantageous combination to grow cowpea plants with immediate effect on shoot length. As the different types of seaweeds contain various micronutrients, extract of these seaweeds and *Hyacinth* helps in the nourishment of soil, ultimately it is reflected in growth and development of a crop. Moreover, authors found comparatively high shoot length in plants grown with autoclaved combination containing *Sargassum* as a one of the components. This observation supports that the compounds of seaweed contain various plant growth promoting nutrients like potassium, calcium, manganese, magnesium, iron, zinc[25,27] as well as plant growth regulators, that is, indole acetic acid, abscisic acid, gibberellin, indole butyric acid.[16]

The number of leaves were maximum in group eight, treated with *Hyacinth*:*Sargassum* (1:1) with autoclaved at 75th Day (Fig. 11.2). Moreover, *Sargassum* and *Hyacinth* combination with autoclaved extract showed high number in leaves, which indicates active components present in the extract are responsible for leaf development and production. Seaweeds and seaweed products are known to enhance plant chlorophyll content,[7] for example, application of a lower concentration of seaweed extract to soil or on foliage produce leaves with higher chlorophyll content. This increase in chlorophyll content was a result of reduction in chlorophyll degradation, which might be caused in part by betaines in the seaweed extract.[31]

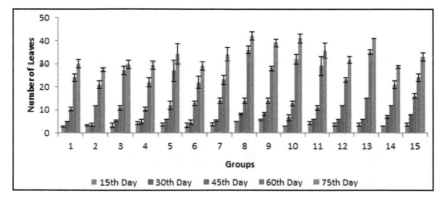

FIGURE 11.2   Number of leaves in a group.

As shown in Figure 11.3, reducing sugar content was found to be increased in groups 14 and 15 with maximum value in comparison to other studied groups. Results reflect the effect of *Sargassum* and *Galaxura* on sugar content of fruit (beans), apart from *Galaxura* and *Sargassum*, *Hyacinth* autoclaved extract showed measurable difference in comparison to control group. Increase in reducing sugar content in the abovementioned groups is resulted because of structural components of seaweeds and its hormonal precursor molecules.

Fruit protein content was evaluated in different groups treated with different combination of seaweed and *Hyacinth*. Surprisingly, group six showed maximum protien content, which further supports quality improvement

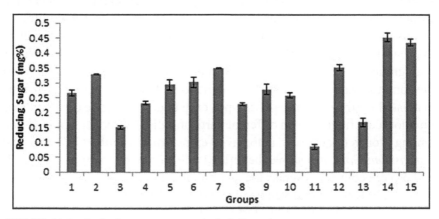

FIGURE 11.3   Reducing sugar content in fruit (beans).

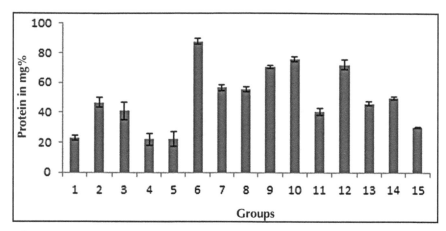

FIGURE 11.4   Protein content in fruit (beans).

in fruits treated with *Galaxura* autoclaved (Fig. 11.4). Beneficial effect of *Galaxura* reflects its further use as a biofertilizer to improve the quality and quantity of cowpea beans. These results are in contradiction with results of reducing sugar, which clearly indicates role of *Galaxura* and *Sargassum* in increasing protein and reducing sugar level of cowpea, respectively. Current needs of food quality and quantity improvement can be supported with these experiments to improve food crop with biofertilizer treatment.

Number of fruit pods were calculated in each group of plants (Fig. 11.5). Authors found high fruit pods with good quality in groups 6, 8, 11, 13 in comparison to control group. Maximum number of pods were observed in group eight, treated with *Hyacinth + Sargassum* autoclaved (1:1) extract. Root system playing a major role in transportation of water and nutrients from soil to whole plant. Authors found maximum root length (Fig. 11.6) in group 10 and groups 12,13, and 14 exhibited nonsignificant differences with group 10. Root length was maximum in a group where *Galaxura* was the common seaweed in studied combinations. These results can further be evaluated for the properties of cowpea crop under the above combinations on pilot scale field trials in different parts of India. Moreover, compartison among groups treated with autoclaved extract of *Hyacinth, Galaxura,* and *Sargassum*; *Galaxura* shown beneficial effect on pod structure (Fig. 11.7) and protein content, whereas the combination of *Hyacinth* and *Sargassum* exhibited effect on total yeild and sugar content. Further experiment will be executed to formulate the fertilizer with optimized ratio of *Hyacinth, Galaxura,* and *Sargassum* for high yeild of cowpea with high content of protein, as cowpea is mainly a rich source of protein.

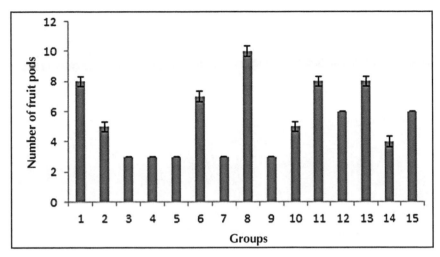

FIGURE 11.5 Number of fruit pods.

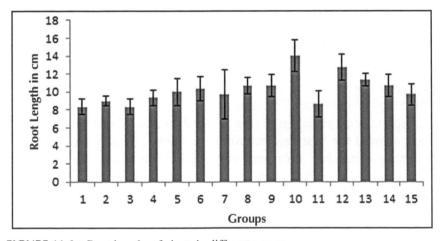

FIGURE 11.6 Root lengths of plants in different groups.

11.4 CONCLUSION

Present study evaluated different combinations of seaweed and *Hyacinth* as a biofertilizer for cultivation of cowpea. Authors found remarkable changes in shoot length, number of leaves and fruit pod, and root length. Protein and reducing sugar concentration of fruit (beans) was found in difference among different groups of fertilizer combination. Observable

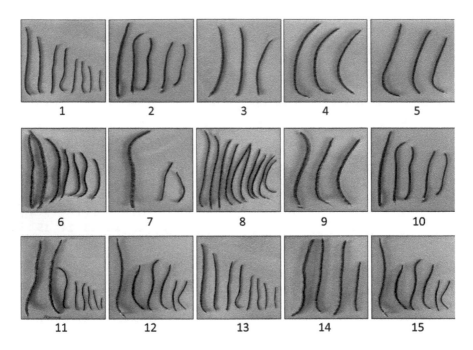

FIGURE 11.7   Fruit pods of cowpea of different groups.

difference was exhibited among different groups in compared to control group of plants for studied parameters.

Especially combination of *Hyacinth* and mixture of seaweed with auto-clave showed the best results. *Galaxura* showed significant increase in fruit protein, number leaves and root length. With best of our knowledge we report *Galaxura* as a fertilizer in combination with water hyacinth for the crop cowpea. Seaweed and water hyacinth have important chemical constituent's property, which acts as a growth promoter and quality improver. Accordingly, it could be said that *Hyacinth, Sargassum,* and *Galaxura* can be used as a biofertilizer. Present study suggests further experiments to formulate various combinations of seaweed and *Hyacinth* to improve cowpea quality and quantity in different crop fields of India.

11.5   SUMMARY

Authors prepared different sets of fertilizers like *Hyacinth, Sargassum, Galaxura*, and mixture of all these in different sets with autoclave or without autoclaving, and total 15 groups were formed including control without any treatment of fertilizer. Different parameters were measured at interval of 15–75 day-time period. For shoot length, five cycles of 15 days were observed. The highest value of length was seen in *Hyacinth: Sargassum* autoclaved. Number of leaves was highest in *Hyacinth: Sargassum* treated group. Highest concentration of carbohydrate in fruits was found in autoclaved *Hyacinth: Sargassum:Galaxura*, compared to highest concentration of protein in *Galaxura* autoclaved treated group. The number of fruit pods was highest in *Hyacinth: Sargassum* autoclaved, but it was not significantly different than the group supplied with *Galaxura*. These observations were made at the end of experiments and it was concluded to use different combinations of seaweed along with water hyacinth for cowpea crop quality and quantity improvement. In future, it is recommended to test different formulation of seaweed and water hyacinth in the form of customized tablets to replace the chemical biofertilizers. Long term use of natural fertilizer will help in revival of soil fertility and quality, hence increase in the quality and quantity of different crops across India.

KEYWORDS

- *Vigna unguiculata*
- biofertilizer
- *Galaxura*
- *Hyacinth*
- cowpea
- protein
- *Sargassum*
- seaweed

REFERENCES

1. Afify, A. E. M. M. R.; Shalaby, E. A.; Shanab, S. M. M. Enhancement of Biodiesel Production from Different Species of Algae. *Grasas Aceites.* **2010**, *61,* 416–422.
2. Ahlawat, I. P. S.; Shivkumar, B. G. Kharif Pulses. In *Text Book of Field Crops Production;* Prasad, R., Ed.; Indian Council of Agriculture Research: New Delhi, India, **2005**, volume 1, p. 248–319.

3. Ali, M.; Gupta, S. Carrying Capacity of Indian Agriculture: Pulse Crops. *Curr. Sci.* **2012**, *102*(6), 874–881.

4. Ansari, A. A.; Gill, S. S.; Khan, F. A.; Naeem, A. Phytoremediation Systems for the Recovery of Nutrients from Eutrophic Waters. In *Eutrophication: Causes, Consequences and Control;* Ansari, A. A., Gill, S. S., Eds.; Springer: Netherlands, 2013; p. 17–27.

5. Anthony, J.; Kalaiselvam, N.; Ganapathy, A.; Veena-Coothan, K.; Sreenivasan, P. P.; Palaninathan, V. Role of Sulphated Polysaccharides from *S. wightii* in Cyclosporine A-induced Oxidative Liver Injury in Rats. *Pharmacol.* **2008**, *8,* 1–9.

6. Aruna, P.; Mansuya, P.; Sekaran, S. Pharmacognostical and Antifungal Activity of Selected Seaweeds from Gulf of Mannar Region. *Sci. Technol.* **2010**, *2*(11), 5–9.

7. Blunden, G.; Jenkins, T.; Liu, Y. W. Enhanced Leaf Chlorophyll Levels in Plants Treated with Seaweed Extract. *J. Appl. Phycol.* **1996**, *8*(6), 535–543.

8. Dar, A.; Baig, H. S.; Saifullah, S. M. Effect of Seasonal Variation on the Anti-Inflammatory Activity of *Sargassum wightii* Growing on the N. Arabian Sea Coast of Pakistan. *Exp. Mar. Biol. Ecol.* **2007**, *351,* 1–9.

9. Eke, O.; Ikeorgu, J. E. G.; Okorocha, E. O. A. Comparative Evaluation of Five Legume Species for Soil Fertility Improvement, Weed Suppression and Component Crop Yields in Cassava/Legume Intercrops. *Afr. J. Roots Tuber Crops.* **1999**, *3,* 17–54.

10. Fatokum, C. A.; Taarawale, S. S.; Singh, B. B.; Korimawa, P. M.; Tamo, M. Proceedings of the World Cowpea Conference III held at IITA Ibadan, Nigeria, 2000.

11. Fornes, F.; Sánchez-Perales, M.; Guadiola, J. L. Effect of a Seaweed Extract on the Productivity of 'de Nules' Clementine Mandarin and Navelina Orange. *Bot. Mar.* **2002**, *45,* 486–489.

12. Gamal-Eldeen, A.; Ahmed, E. M. In vitro Cancer Chemopreventive Properties of Polysaccharide Extract from the Brown Algae, *Sargassum latifolium. Food Chem. Toxicol.* **2009**, *47*(13), 78–84.

13. Ibrahim, W. M. Biosorption of Heavy Metal Ions from Aqueous Solution by Red Macroalgae. *J. Hazard. Mater.* **2011**, *192*(3), 1827–1835.

14. Kim, I. H.; Lee, H. Antimicrobial Activities Against Methicillin Resistant *Staphylococcus aureus* from Macroalgae. *Ind. Eng. Chem.* **2008**, *14*(5), 68–72.

15. Lowry, O. H.; Rosebrough, N. J.; Farr, A. L.; Randall, R. J. Protein Measurement with the Folin Phenol Reagent. *J. Biol. Chem.* **1951**, *193*(1), 265–275.

16. Mabeau, S.; Fleurence, J. Seaweed in Food Products: Biochemical and Nutritional Aspects. *Trends Food Sci. Technol.* **1993**, *4*(4), 103–107.

17. Magruder, W. H. Reproduction and Life History of the Red Algae *Galaxaura oblongata* (Nemaliales, Galaxauraceae). *J. Phycol.* **1984**, *20*(3), 402–409.

18. Meenakshi, S.; Manicka, G. D.; Tamilmozhi, S. Total Flavanoid and in vitro Antioxidant Activity of Two Seaweeds of Rameshwaram Coast. *Global J. Pharmacol.* **2009**, *3,* 59–62.

19. Miller, G. L. Use of Dinitrosalicylic Acid Reagent for Determination of Reducing Sugar. *Anal. Chem.* **1959**, *31*(3), 426–428.

20. Ng, N. Q.; Marechal, R. Cowpea Taxonomy, Origin and Germplasm. In *Cowpea Research Production and Utilization;* Singh, R. S., Rachie, K. O., Eds.; John Wiley and Sons: New York, **2005**; pp 11–21.

21. Nirmal, R.; Kalloo, G.; Kumar, R. Diet Versatility in Cowpea (*Vigna unguiculata*) Genotypes. *Indian J. Agric. Sci.* **2001**, *71*, 598–601.

22. Nishide, E.; Uchida, H. Effects of Ulva Powder on the Ingestion and Excreation of Cholesterol in Rats. In *Proceedings of the 17th International Seaweed Symposium;* Chapman, A. R. O., Anderson, R. J., Vreeland, V. J., Davison, I. R., Eds.; Oxford University Press: Oxford, **2003**; pp 165–168.

23. Park, P. J.; Heo, S. J.; Park, E. J. Reactive Oxygen Scavenging Effect of Enzymatic Extracts from *Sargassum thunbergii*. *Agric. Food Chem.* **2005**, *53*, 66–72.

24. Payá, M.; Ferrándiz, M. L.; Sanz, M. J.; Bustos, G.; Blasco, R.; Rios, J. L.; Alcaraz, M. J. Study of the Antioedema Activity of Some Seaweed and Sponge Extracts from the Mediterranean Coast in Mice. *Phytother. Res.* **1993**, *7*(2), 159–162.

25. Peña-Rodríguez, A.; Mawhinney, T. P.; Ricque-Marie, D.; Cruz-Suárez, L. E. Chemical Composition of Cultivated Seaweed *Ulva clathrata* (Roth) C. Agardh. *Food Chem.* **2011**, *129*(2), 491–498.

26. Pramanick, B.; Brahmachari, K.; Ghosh, A. Efficacy of *Kappaphycus* and *Gracilaria* Sap on Growth and Yield Improvement of Sesame in New Alluvial Soil. *J. Crop Weed.* **2014**, *10*, 77–81.

27. Roosta, H. R.; Mohsenian, Y. Effects of Foliar Spray of Different Fe Sources on Pepper (*Capsicum annum* L.) Plants in Aquaponic System. *Sci. Hortic.* **2012**, *146*, 182–191.

28. Singh, A. K.; Bhatt, B. P.; Sundaram, P. K.; Kumar, S.; Bahrati, R. C.; Chandra, N.; Rai, M. Study of Site Specific Nutrients Management of Cowpea Seed Production and Their Effect on Soil Nutrient Status. *J. Agric. Sci.* **2012**, *4*(10), 191–198.

29. Thirumaran, G.; Arumugam, M.; Arumugam, R.; Anantharaman, P. Effect of Seaweed Liquid Fertilizer on Growth and Pigment Concentration of *Abelmoschus esculentus* (l) medikus. *Am.-Eurasian J. Agron.* **2009**, *2*, 57–66.

30. Viththalpara, R. D.; Saiyad, A. R.; Jani, A. J.; Vara, D. R.; Kunjadia, P. D.; Sanghvi, G. V.; Vaishnav, D.; Dave, G. S. Seaweed Pretreatment of Chickpea (*Cicer arietinum*) Enhances Post Harvest Preservation by Reducing Germination. *J. Bioprocess. Biotech.* **2015**, *5*, 257–259.

31. Whapham, C. A.; Blunden, G.; Jenkins, T.; Hankins, S. D. Significance of Betaines in the Increased Chlorophyll Content of Plants Treated with Seaweed Extract. *J. Appl. Phycol.* **1993**, *5*(2), 231–234.

32. Yuan, Y. V.; Walsh, N. A. Antioxidant and Anti-Proliferative Activities of Extracts from a Variety of Edible Seaweeds. *Food Chem. Toxicol.* **2006**, *44*(7), 1144–1150.

PART IV
Emerging Focus Areas in Biological Systems

CHAPTER 12

FISHPOND WASTEWATER: THE POTENTIAL IRRIGATION SOURCE

LALA I. P. RAY[1,*], B. C. MAL[2,3], S. MOULICK[4] AND
P. K. PANIGRAHI[5,6]

[1]School of Natural Resource Management, College of Postgraduate Studies (Central Agricultural University, Imphal), Umiam, Barapani, Meghalaya 793103, India

[2]Chhattisgarh Swami Vivekananda Technical University, Bhilai, Chhattisgarh, India

[3]Department of Agricultural and Food Engineering, Indian Institute of Technology, Kharagpur, West Bengal 721302, India,
E-mail: bmal@agfe.iitkgp.ernet.in

[4]Department of Civil Engineering, Kalinga Institute of Industrial Technology (KIIT) University, Bhubaneswar, Odisha, India,
E-mail: sanjib_moulick72@yahoo.co.uk

[5]Department of Agricultural and Food Engineering, Indian Institute of Technology, Kharagpur, West Bengal, 721302, India

[6]Government of Odisha, India,
E-mail: panigrahi.prasanta@gmail.com.

*Corresponding author. E-mail: lalaipray@rediffmail.com

CONTENTS

12 1   INTRODUCTION

Reuse of freshwater is being advocated to overcome its exploitation in the agricultural sector. Because of which several potential water sources of irrigation (municipal water, brackish water, industrial wastewater, wastewater from agricultural and allied processed industries, and wastewater from aqua cultural firms) have emerged.[8,13,14,16,17,32,34,36,38] Irrigation water from these sources with some treatment has been used suitably in irrigating agricultural crops. The reuse of water for irrigation is often viewed as a positive means of recycling, the advantage being a constant, reliable source and reduction in the amount of water extracted from the environment.[3,37] The practice of wastewater reuse for landscape irrigation in Saudi Arabia was a success story.[1] The use of wastewater can save up to 50% application of inorganic nitrogen fertilizer when it contains 40 mg of N L^{-1}.[14] The feasibility study was conducted in Brazil by using fishpond effluent to irrigate cherry tomatoes, grown with different types of organic fertilizers.[7] Higher productivity was observed in the effluent treatments. Researchers have reported that water reclamation, recycling, and reuse address the challenges of water scarcity by resolving water resource issues, creating new sources of high-quality water supplies in an integrated way.[19]

This chapter discusses vegetable-based remunerative cropping integrated with a semi-intensive aquaculture system. The polluted water exchanged from the aquaculture fishponds was used to irrigate tomato during the winter season. The efficacy of exchanged water from fishponds stocked with the high densities of three species of Indian Major Carps (IMC) as an irrigation source was monitored for three consecutive growing seasons.

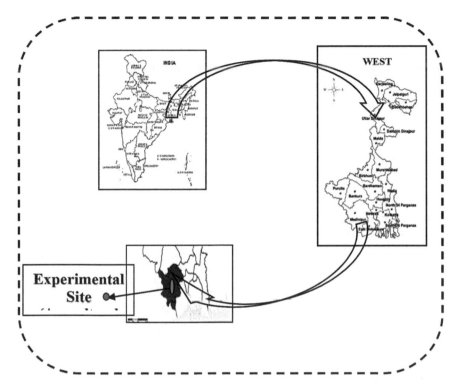

FIGURE 12.1   Location map of the experimental site.

12.2   MATERIALS AND METHODS

Field experiments were conducted at the experimental farm of the Department of Agricultural and Food Engineering, Indian Institute of Technology, Kharagpur, West Bengal in eastern India for three consecutive years (2006, 2007, and 2008). The site is located at latitude of 22° 19' North and longitude of 87° 19' East with an altitude of 48 m above the mean sea level with an average annual rainfall of 1200 mm. The experimental site is shown in Figure 12.1. The average soil type of this region is light textured, acidic lateritic with pH ranging from 4.0 to 6.8. Soil at the experimental site is lateritic with sandy loam texture and very low fertility. The physical and chemical properties of the soil at the site are presented in Tables 12.1 and 12.2.

Dugout ponds with a depth of 1.5 m and average water spreading area of 150 m^2 were constructed. The ponds were covered with suitable polythene sheets (blue colored silpaulin with 250 gage thickness). Fishponds were

TABLE 12.1   Physical Properties of Soil at the Experimental Site.

Soil depth (cm)	Particle size distribution (%)			Bulk density (g cm^{-3})	FC (mm cm^{-1})	WP (mm cm^{-1})	K_s (cm hr^{-1})
	Clay	Silt	Sand				
15	14.5	26.2	59.3	1.61	2.0	0.9	0.487
30	21.2	19.4	59.4	1.56	2.2	0.9	0.375
45	27.8	20.1	52.1	1.59	2.2	1.1	0.278
60	28.2	19.2	52.6	1.63	2.4	1.2	0.162
90	29.6	24.8	45.6	1.69	2.6	1.6	0.107

TABLE 12.2   Chemical Properties of Soil at the Experimental Site.

Parameters	Values
pH (1:2.5: soil: water)	5.2
Electrical conductivity (1:1: soil: water)	0.56 dS m^{-1} at 25°C
Cation exchange capacity	6.00 meq per 100 g soil
Organic carbon	0.28%
Available nitrogen	0.025%
Available phosphorus	0.004%
Available potassium	0.015%
Total nitrogen	0.035%
Total phosphorus	0.045%
Total Potassium	0.420%

stocked with IMC with three different stocking densities (SD: 2.0, 3.5, and 5.0 numbers per square meter of water spread area), with a stocking ratio of 4:3:3 for *Catla, Rohu,* and *Mrigala,* respectively. Depending on the fish stocking density and the supply of enriched feed, the fishpond water gets polluted with time. Fishpond water needs to be exchanged to be used as nutrient-rich irrigation water. This exchanged water was used as a source of irrigation for tomato crop in this study.

Thirty six plots (6 × 5 m size each) were prepared adjacent to the fishponds along with three control plots. The plots were separated by 60 cm bunds. Field study was taken up with tomato (*Lycopersicum esculentum* L.) cultivar MHTM-256 (Suparna). The seedlings were raised in the nursery inside a polyhouse. Three weeks old seedlings were planted

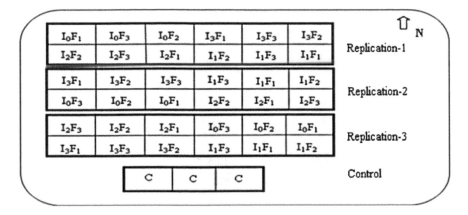

FIGURE 12.2 Schematic layout of the experimental plot: I = Main treatment, source of irrigation; F = Sub treatment, fertilizer and C = control plot.

in the experimental plots with 0.75 m row to row and 0.6 m plant to plant spacings. Irrigation was based on the volume of water available from the fishponds. Split-plot experimental design was followed with the irrigation as main treatment and suboptimal doses of fertilizer as sub-treatment.

Four different sources of irrigation water (I_0 with direct tube-well water, I_1 from fishpond with stocking density of $5.0/m^2$, I_2 from fishpond with stocking density of $3.5/m^2$, and I_3 from fishpond with stocking density of $2.0/m^2$) constituted the main treatment, three reduced doses of nitrogen fertilizers constituted the sub-treatments, and three replications were followed. Recommended full dose of fertilizer for the experiment was 80–40–40 kg ha^{-1} (NPK). However, for nitrogen application, three sub-treatments (90% N (F_1), 80% N (F_2), and 70% N (F_3)) were followed. All recommended doses of fertilizers except nitrogen were applied as a basal dose. Nitrogen fertilizer was applied in all cases as 20% of the treatment amount as a basal dose, and rest in two equal splits during the crop growth period. The schematic layout of the field experiment is shown in Figure 12.2.

The volume of water supplied to a given plot was known by measuring the discharge obtained from the pump and the time of application. Weather parameters (temperature, solar radiation, wind speed, rainfall, and evaporation during the crop growing seasons) are presented in Table 12.3.

The total of 2.8, 148.9, and 0 mm of rainfall was received during each winter growing seasons for tomato crop. The details of experimental

TABLE 12.3 Rainfall, Temperature, Solar Radiation, and Wind Speed During Crop Growing Seasons.

Parameter	Experiment-1 (2005–2006)	Experiment-2 (2006–2007)	Experiment-3 (2007–2008)
Rainfall (mm)			
R_{max}	1.02	38.3	0
Total	2.8	148.9	0
Temperature (°C)			
T_{max}	38.5	35.5	33.4
T_{min}	9.2	9.0	8.6
Mean	25.1	24.6	23.54
Solar radiation ($KW\ m^{-2}\ h^{-1}$)			
average	0.18	0.21	0.15
Wind speed ($m\ s^{-1}$) maximum	6.48	7.26	7.44
mean	0.36	0.28	0.26

layout and the site are shown in Figure 12.3. The yield obtained from the control plot with 100% recommended dose of fertilizer and tube-well water was compared with the yield from the treatment plots. The nutrient supplementing potential of the fishpond water was also studied in a framework of proper experimental design.

12.3 RESULTS AND DISCUSSION

12.3.1 WATER QUALITY

Average values of various water quality parameters from two different sources, *viz.*, tube-well water and fishpond wastewater along with their permissible limits are listed in Table 12.4. The average values of the water quality parameters show wide variations between the tube-well water and fishpond wastewater. The values for ammoniacal N (NH_3-N), orthophosphate (PO_4-P), and nitrite-N (NO_2-N) were almost zero for tube-well water. The physicochemical characteristics of water in fishponds are one of the deciding factors in optimizing the conditions for fish productivity in small

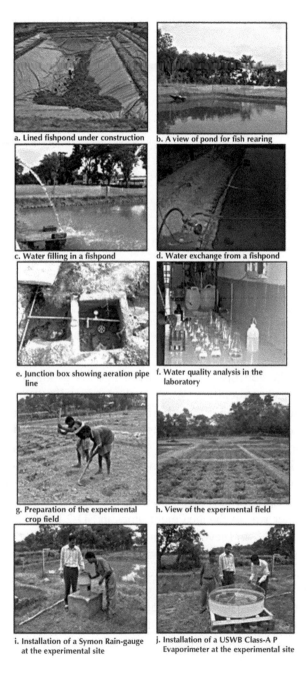

a. Lined fishpond under construction   b. A view of pond for fish rearing

c. Water filling in a fishpond   d. Water exchange from a fishpond

e. Junction box showing aeration pipe line   f. Water quality analysis in the laboratory

g. Preparation of the experimental crop field   h. View of the experimental field

i. Installation of a Symon Rain-gauge at the experimental site   j. Installation of a USWB Class-A P Evaporimeter at the experimental site

FIGURE 12.3   Details of field layout and site.

TABLE 12.4   Mean Values of Water Quality Parameters of Tube-well Water and Fishpond Wastewater and Their Ideal Values.

Parameter	Tube-well water	Fishpond wastewater	Ideal value (range)[#]
Temperature (°C)	28.64 ± 4.26	26.55 ± 4.85	25–32
pH	6.65 ± 0.87	7.24 ± 0.56	6.7–8.5
DO (mg L^{-1})	2.2 ± 1.35	5.66 ± 1.05	5–10
TSS (mg L^{-1})	74.65 ± 18.66	88.65 ± 22.5	30–200
NO$_3^-$ -N (mg L^{-1})	0.021 ± 0.004	0.55 ± 0.42	0.1–3.00
Ammonia nitrogen (NH$_3$-N) (mg L^{-1})	–	< 0.1	
unionized		< 1.0	0–0.1
Ionized			0–1.0
NO$_2^-$ -N (mg L^{-1})	–	0.07 ± 0.05	0–0.5
Total -N (mg L^{-1})	–	2.12 ± 0.74	0.05–4.5
PO$_4^{3-}$-P (mg L^{-1})	–	0.12 ± 0.06	0.05–0.4

[#]The ideal values of water quality parameters (physical and chemical) for aquaculture practices in freshwater, prescribed by Central Institute of Freshwater Aquaculture (CIFA)

fishponds. The fertility status of the fishponds is known to be directly related to the water quality.[22] Water exchange has direct influence over the water quality of the pond, growth of fish, and economy of the fish culture.

Repeated water exchange during the later stage of fish growth could reduce the total ammonia nitrogen (TAN) concentration and other nitrogenous parameters in the fishpond water for all the three stocking densities and for the whole time period of the study. More frequent water exchange was needed for the higher stocking density ponds than the lower density ponds. For example, fishpond water was exchanged eight times in SD-2.0 fishpond compared to 13 times in SD-3.5 and seventeen times in SD-5.0 during 2005–2006. Similar requirements were noticed in the remaining two years of the study.

Boyd et al.[6] suggested that water exchange is an effective measure in improving the water quality in small fishponds. Out of 11 months of the culture period, no water exchange was required during the initial three months of culture (i.e., June–August). It may be due to low biomass of fish in the culture ponds and dilution effect of pond water due to heavy rainfall during these months.[25,27,28,30] From the month of September onwards, monthly water exchange ranging from 10 to 40% was needed based on the degree of pollution of the fishpond water. The total amount of water exchange during the culture period varied from 80 to 170%.

It may be noted that about 2.08 M m^3 of exchanged water is available from fishpond of 1 ha area with a stocking density of 5 per meter square. For semi-intensive IMC culture, with stocking density ranging from 2.0 to 5.0 per meter square, about 1.13–2.08 M m^3 of water is required for exchange in a year. In a recirculatory aquaculture system, the values of TAN or other parameters are easily controlled by filtering the water through a suitable filtration system. In the case of intensive pond culture system, water exchange is considered as a better option for controlling TAN and other parameters. From the present study, it is estimated that large volume of water should be exchanged for maintaining the TAN values within its permissible range. Disposal of the huge volume of polluted water from an intensive fish farm to the adjacent environment is a dangerous concern.[9,21,28,29,33,41,9] On the other hand, the scarcity of freshwater is found as a limitation of water exchange.[4] The introduction of integrated aquaculture-cum-irrigation (IAI) has been recommended as a solution to the problem by many researchers. In this chapter, the water was used twice, first for aquaculture and then for irrigation. The water with high concentration of different inorganic nutrients is considered to be polluted water for fish culture. However, it is enriched with different nutrients for agricultural crop production.

12.3.2 SOIL MOISTURE DYNAMICS

The variation of soil moisture content during the cropping season in tomato crop is shown in Figure 12.4. The variation was less from January 15–29 and from February 5 to harvest date in 2006–2007 due to the rainfall received. But large variations in moisture level between irrigation treatments were observed during 2005–2006 and 2007–2008 throughout the cropping seasons as there was no rainfall. In 2005–2006, the maximum variation (about 4%) was found in case of I_1 during the latter part of flowering because of less number of water exchanges from the SD of 2.0 for fishpond. The soil moisture variation in case of I_0 was not conspicuous during all the seasons due to application of water at the time of crop need directly from the tube-well. The variation in the treatments was narrowed down in the later stage of the crop growth due to frequent water exchange resulting in availability of more irrigation water. The least variation in soil moisture content with I_3 treatment was observed due to frequent availability of exchanged water from SD-5.0 fishpond.

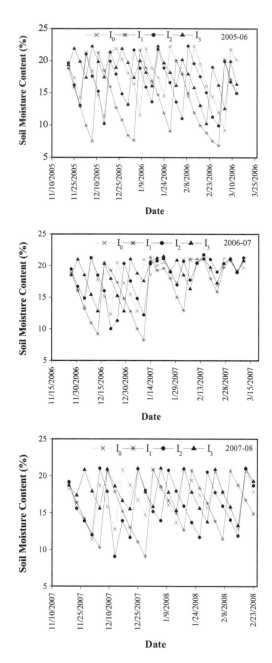

FIGURE 12.4 Variation of soil moisture in different treatments during the winter season for Tomato.

TABLE 12.5 Number of Irrigation for Tomato Crop in Different Crop Stages and Years.

Crop seasons	Irrigation sources	Number of Irrigation in Different stages of Crop					Total number of irrigation
		Initial establishment	Vegetative	Flowering	Fruit initiation/Fruiting	Fruit harvesting	
2005–2006	I_0	2	1	2	1	—	6
	I_1	2	1	1	1	—	5
	I_2	2	2	2	1	—	7
	I_3	2	2	2	2	2	10
2006–2007	I_0	1	1	2	1	—	5
	I_1	1	1	1	1	—	4
	I_2	2	1	2	1	—	6
	I_3	2	2	2	1	1	8
2007–2008	I_0	2	1	2	1	—	6
	I_1	2	1	1	1	—	5
	I_2	2	1	2	1	—	6
	I_3	2	2	2	1	1	8

I_0: Irrigation with tube-well water, I_1: Irrigation from SD-2.0, I_2: Irrigation from S.D.-3.5, I_3: Irrigation from S.D.-5.0

12.3.3   NUMBER OF IRRIGATIONS

Irrigation was provided to the crop based on the availability of water from fishpond due to exchange. The stage of application with the numbers of irrigation was monitored regularly and the data are presented in Table 12.5. In 2005–2006, very less amount of rainfall (2.8 mm) was received during the latter part of the crop when it was about to be harvested. During 2007–2008, the crop received no rainfall whereas in 2006–2007 the rainfall received was 149.3 mm during the crop growth. Since the rainfall was received during harvesting time, there was rotting of tomato in the field resulting in low yield during that year.

The treatment I_3 received the highest number of irrigation (10 in 2005–2006 and 8 both in 2006–2007 and 2007–2008). The number of irrigation provided under I_1 treatment was the least (4–5) due to availability of less water from fishpond water exchange. Irrigation from the treatment I_0 was provided by tube-well water based on the moisture depletion pattern of the soil in the field. The total number of irrigation provided under I_0 varied from 5 to 6. As the water of the fishpond (SD 5) was polluted early, more number of irrigation was applied under I_3 and at late fruiting stage of the crop this excess irrigation water was not utilized properly by the crop. Irrigation from source I_2 was almost uniform in different crop growth stages.

12.3.4   NUTRIENT RECOVERY FROM FISHPOND

The amount of nutrient recovered from the fishpond was estimated from the nutrient loads of exchanged wastewater. It is found that as the number of water exchange was more in high stocking density fishpond (I_3), the amount of recovered nutrient was also more with I_3 as compared to the two other SDs. The amount of nutrient recovered from different fishponds is presented in Table 12.6. It may be noted that the nutrient recovery (N) is higher in case of SD of 5.0 fishpond (33.27 kg ha^{-1}) compared to the other two treatments. The total nutrient recovered from the fishponds through water exchange was not fully utilized during the cropping season due to mismatch of crop stage with the water exchange calendar. However, about 65–75% of the total nutrients were utilized by the crop. The recovery of phosphate from the exchanged water of fishpond was estimated to be about 70%. Higher recovery (1.70 kg ha^{-1}) occurred with SD of 5.0 in fishpond as compared to the other two treatments.

TABLE 12.6 Nutrient Recovery (kg ha^{-1}) from Fishpond.

Fishpond	Nitrogen (recovered)				Nitrogen (utilized)			
	2005–2006	2006–2007	2007–2008	Mean ± SD	2005–2006	2006–2007	2007–2008	Mean ± SD
SD-2.0	12.45	12.09	15.54	13.36 ± 1.55	6.70	6.49	9.75	7.65 ± 1.49
SD-3.5	26.33	15.06	22.01	21.13 ± 4.64	11.98	10.82	13.95	12.25 ± 1.29
SD-5.0	37.40	27.34	35.07	33.27 ± 4.30	21.5	16.39	21.55	19.81 ± 2.42
Phosphate (recovered)					*Phosphate (utilized)*			
SD-2.0	0.79	0.40	0.92	0.70 ± 0.22	0.43	0.4	0.92	0.58 ± 0.24
SD-3.5	2.01	1.15	2.33	1.83 ± 0.50	1.00	0.89	1.40	1.10 ± 0.22
SD-5.0	2.75	1.88	3.60	2.74 ± 0.70	1.57	1.24	2.29	1.70 ± 0.44

12.3.5   CROP PERFORMANCE

The application of suboptimal and full doses of fertilizer favored the growth and yield of tomato. The crop yield was found to be statistically different under different treatments. An increase in yield was also observed with the increase in fertilizer dose. The control plot yield of tomato with reduced doses of N fertilizer was found to be statistically at par with the SD-5.0 treatment.

There was no significant effect of N levels on fruit yield. The yield was low during 2006–2007 due to unseasonal high rainfall at the maturity stage of the crop; many fruits were damaged and rotted. The maximum yield (66.85–70.19 t ha^{-1}) was recorded with I_3 and the minimum (53.07–61.29 t ha^{-1}) was from I_1. The mean data indicate that I_3 produced the highest yield of 68.27 t ha^{-1} which was 5.6–20.8% more than the other treatments. I_1 gave the lowest yield of 56.53 t ha^{-1}. It is concluded from the mean data that F_1 produced the highest yield of 63.03 t ha^{-1} and F_3 gave the lowest yield of 62.69 t ha^{-1}. The interaction effect of irrigation and N was not significant. However, the maximum yield of 69.3 t ha^{-1} was recorded with I_3F_2 followed by I_3F_3 (67.8 t ha^{-1}) and I_3F_1 (67.7 t ha^{-1}).

The average yield obtained from the control plot was 70.48 t ha^{-1}. The comparison of yield between I_3 and control plot shows that both the yields are at par. Therefore, it can be inferred that even with the reduced dose of fertilizer application, the yield of tomato can be at par with that of 100% fertilizer dose use if the crop is irrigated with fishpond wastewater (Table 12.7; Figs. 12.5(a, b)).

The yield of tomato was the highest with irrigation from SD-5.0 treatment in fishpond and application of suboptimal fertilizer dose. This highest yield was almost at par with the result obtained from control plot with 100% of the recommended doses of fertilizer. There was a significant variation of the yield of tomato irrigated with tube-well water and with the wastewater from SD-2.0 fishpond. It may be attributed to the less number of irrigation that could be possible from SD-2.0 fishpond wastewater and without any scientific scheduling. Ray et al.[24] reported a tomato yield of 64.5 t ha^{-1} irrigated with tube-well water and it increased to 95.8 t ha^{-1} due to irrigation with fishpond effluent. Similar findings were also reported by Castro et al.[7] Pinto[23] observed an increase in tomato productivity in the range of 19.5–21.8% when fertilizer application was changed from conventional method to fertigation. This is supported with the findings by some other

TABLE 12.7 Effect of Irrigation Source and Nitrogen Level on Fruit Yield (t ha^{-1}) of Tomato.

Treatment	2005–2006	2006–2007	2007–2008	Mean
Irrigation sources				
I_0	63.25	56.82	65.53	61.87
I_1	55.22	53.07	61.29	56.53
I_2	67.39	59.98	66.52	64.63
I_3	67.77	66.85	70.19	68.27
SEm (±)	1.57	1.13	1.80	0.88
CD (0.05)	5.44	3.92	NS	2.62
Nitrogen levels				
F_1	64.32	58.89	65.87	63.03
F_2	62.88	59.67	65.70	62.75
F_3	63.02	58.98	66.08	62.69
SEm (±)	1.76	1.70	1.86	1.03
CD (0.05)	NS	NS	NS	NS
Interaction (I × F)				
SEm (±)	3.28	3.01	3.53	1.89
CD (0.05)	NS	NS	NS	NS
Control Yield				
Control plot	70.52	68.26	69.67	69.48

I_0: Irrigation with tube-well water; I_2: irrigation from SD-3.5; I_1: irrigation from SD-2.0; I_3: irrigation from SD-5.0; F_1: 90% of recommended N; F_2: 80% of recommended N; F_3: 70% of recommended N.

studies done on the same.[10,15,35,39] Research conducted with aqua effluent irrigation claimed to have reduced the recommended fertilizer by almost 50% in a field trial in Saudi Arabia with wheat as the trial crop. However, in this literature there is no mention of the type of fish reared and stocking details.[13] On the contrary, aqua effluent irrigation needs to be provided along with the recommended doses of fertilizer, as it contains the least amount of nutrients.[40] Similar findings were obtained in the present field investigation with IMC stocked fishpond wastewater carrying low nutrient value.[40] Castro et al.[7] also reported that there was nonsignificant interaction between the types of irrigation and fertilizers on fruit mean weight. Only types of irrigation had significant effect, as the plants irrigated with well water had higher fruit mean weight than the plants irrigated with fish

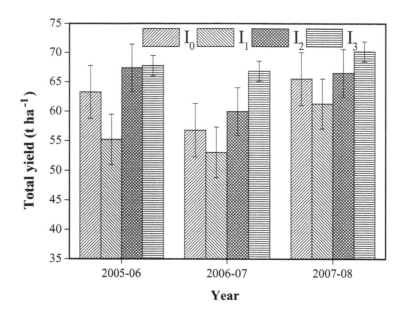

FIGURE 12.5a   Effect of source of irrigation on fruit yield.

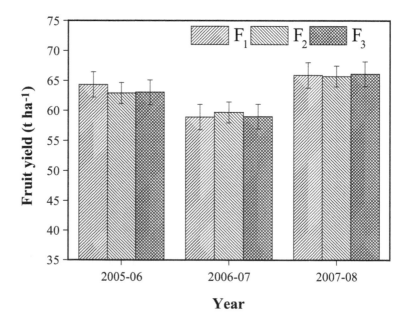

FIGURE 12.5b   Effect of Nitrogen level on fruit yield.

effluent. The increase in tomato fruit yield with fertilizer intervention was reported by several researchers. But the research work was done mostly on drip fertigation. There was an increase in tomato yield with an increase in N level as reported by several investigators.[2,5,10–12,18,20,26,31,32,35]

12.4 CONCLUSIONS

It may be concluded that the fishpond wastewater is one of the potential sources to supplement irrigation water to the crop as well as bridge the gap of chemical fertilizer requirement, especially nitrogen, to a marginal amount. Farmers owning irrigation source, fishpond and crop land adjacent to each other can supply the water to the pond instead of direct crop irrigation and the pond wastewater to the crop land for irrigation to increase the stocking density of fish, supplement the nitrogen fertilizer to the crop, and thereby, increase the income and avoid environmental pollution.

12.5 SUMMARY

Wastewater from fishponds cultured at three stocking densities (SD) of IMC was evaluated to find out its efficacy as irrigation source with tomato as a test crop. A field trial was conducted in sandy loam soil of Indian Institute of Technology (IIT) Kharagpur, West Bengal, India during 2006–2008. Water quality parameters such as temperature, pH, dissolved oxygen, nitrite, nitrate, TAN, orthophosphate, and total suspended solid for fishpond wastewater were monitored on every alternate date for the entire growth period of IMC. Water exchange was performed before the fishpond water attained critical level of TAN value as it is harmful for the pond ecosystem and fish growth. The exchanged water was used for irrigation. The maximum numbers of water exchange for fishponds for a culture period of 300 days were 10, 13, and 17 for SD-2.0, 3.5 and 5.0, respectively. The highest and lowest yields for tomato were 68.27 t ha^{-1} and of 61.87 t ha^{-1} with irrigation from SD-5.0 and tube-well water, respectively. There was a recovery of inorganic nitrogen to the tune of 13.36–33.27 kg ha^{-1} and phosphate to the tune of 0.70–2.74 kg ha^{-1} from the fishpond wastewater. The ratio of pond area to crop area for an integrated agri–aquaculture system was estimated to be 35:65 for SD-2.0; 30:70 for SD-3.5 and 22:78 for SD-5.0.

12.6   ACKNOWLEDGEMENT

Research funds available from the Indian Council of Agricultural Research (ICAR), New Delhi, for conducting the field research at Indian Institute of

- Aqua effluent
- fishpond
- irrigation scheduling
- lined pond
- mrigala
- olericulture
- pond aeration
- ratio of pond area to crop area
- silpaulin
- total ammonia nitrogen
- total soluble solid
- water exchange
- water toxicity

Technology, Kharagpur, West Bengal, India are thankfully acknowledged by the authors.

KEYWORDS

REFERENCES

1. Al-Ama, M. S.; Nakhla, G. F. Waste Water Reuse in Jubail, Saudi Arabia. *Water Res.* **1995**, *29*(6), 1579–1584.
2. Al-Mutaz, I. S. Treated Wastewaters as a Growing Water Resource for Agriculture Use. *Desalination* **1989**, *73*, 27–36.
3. Al-Shammiri, M.; Al-Saffar, A.; Bohamad, S.; Ahmed, M. Waste Water Quality and Reuse in Irrigation in Kuwait Using Micro-Filtration Technology in Treatment. *Desalination* **2005**, *185*, 213–225.
4. Avnimelech, Y.; Kochba, M. Evaluation of Nitrogen Uptake and Excretion by Tilapia in Biofloc Tanks, Using [15]N Tracing. *Aquaculture* **2009**, *287*(1), 163–168.
5. Bafna, A. M.; Daftardar, S. Y.; Khade, K. K.; Patel, V. V.; Dhotre, R. S. Utilization of Nitrogen and Water by Tomato Under Drip Irrigation System. *J. Water Manage.* **1993**, *1*(1), 1–5.
6. Boyd, C. E.; Pillai, U. K. *Water Quality Management in Aquaculture*; Special Publication No. 22. Central Marine Fisheries Research Institute: Cochin, India, 1984; p 76.

7. Castro, R. S.; Borges Azevedo, C. M. S.; Bezerra-Neto, F. Increasing Cherry Tomato Yield Using Fish Effluent as Irrigation Water in Northeast Brazil. *Sci. Hortic.* **2006**, *110*, 44–50.

8. David, D. J.; Williams, C. H. Effects of Cultivation on the Availability of Metals Accumulated in Agricultural and Sewage-Treated Soils. *Prog. Water Technol.* **1979**, *11*, 257–264.

9. FAO (Food and Agriculture Organization). *FAO Statistics Database 2006.* http:// faostat. fao.org/ (accessed June 10, 2008).

10. Hebbar, S. S.; Ramachandrappa, B. K.; Nanjappa, H. V.; Prabhakar, M. Studies on NPK Drip Fertigation in Field Grown Tomato (*Lycopersicon esculentum* Mill.). *Eur. J. Agron.* **2004**, *21*, 117–127.

11. Herrera, F.; Castillo, J. E.; Chica, A. F.; Lopez Bellido, L. Use of Municipal Solid Waste Compost (MSWC) as a Growing Medium in the Nursery Production of Tomato Plants. *Bioresour. Technol.* **2008**, *99*(2), 287–296.

12. Hussain, Z. Problems of Irrigated Agriculture in Al-Hassa, Saudi Arabia. *Agric. Water Manage.* 1982, *5*(4), 359–374.

13. Hussain, G.; Al-Jaloud, A. A. Effect of Irrigation and Nitrogen on Water Use Efficiency of Wheat in Saudi Arabia. *Agric. Water Manage.* **1995**, *27*, 143–153.

14. Hussain, G.; Al-Saati, A. Wastewater Quality and its Reuse in Agriculture in Saudi Arabia. *Desalination* 1999, *123*, 241–251.

15. Ismail, S. M.; Ozawa, K.; Khondaker, N. A. Influence of Single and Multiple Water Application Timings on Yield and Water Use Efficiency in Tomato (var. First power). *Agric. Water Manage.* **2008**, *95*, 116–122.

16. Leeper, G. W. *Managing the Heavy Metals on the Land*; Marcel Dekker: New York, 1978, p 233.

17. Li, T. Y.; Baozhong, P.; Houquin, H.; Korong, J.; Monggae, C.; Disheitane, K. *Experimental Irrigation Project at Glen Valley Water Care Works: Final report.* Sino-Botswana Government, **2001**, p 100.

18. Mahajan, G.; Singh, K. G. Response of Greenhouse Tomato to Irrigation and Fertigation. *Agric. Water Manage.* **2006**, *84*, 202–206.

19. Miller, G. W. Integrated Concepts in Water Reuse: Managing Global Water Needs. *Desalination* **2006**, *187*, 65–75.

20. Mofoke, A. L. E.; Adewumi, J. K.; Babatunde, F. E.; Mudiare, O. J.; Ramalan, A. A. Yield of Tomato Grown Under Continuous-Flow Drip Irrigation in Bauchi State of Nigeria. *Agric. Water Manage.* **2006**, *84*, 166–172.

21. Naylor, R. L.; Goldburg, R. J.; Primavera, J. H.; Kautsky, N.; Beveridge, M. C. M.; Clay, J.; Folke, C.; Lubchenco, J.; Mooney, H.; Troell, M. Effect of Aquaculture on World Fish Supplies. *Nature* **2000**, *405*, 1017–1024.

22. Ntengwe, F. W.; Edema, M. O. Physico-Chemical and Microbiological Characteristics of Water for Fsh Production Using Small Ponds. *Phys. Chem. Earth.* **2008**, *33*, 701–707.

23. Pinto, J. M. Doses and Period of Application of Irrigation on Tomato Crop (Portuguese). *Hortic. Bras.* **1997**, *15*(1), 15–18.

24. Prinsloo, J. F.; Schoonbee, J. H. Investigations into the Feasibility of a Duck-Fish-Vegetable Integrated Agriculture: Aquaculture System for Developing Areas in South Africa. *Water Sci.* **1987**, *13*(2), 109–118.

25. Ray, L. I. P.; Bag, N.; Mal, B. C.; Das, B. S. Feasibility of Irrigation Options with Aquaculture Waste Water. *Proceedings of 2nd International Conference on Hydrology and Watershed Management,* volume 2, **2006**; pp 873–885.

26. Ray, L. I. P.; Panigrahi, P. K.; Moulick, S.; Bag N.; Mal, B. C.; Das, B. S. Multiple Usage of Fresh Water in Aquaculture and Olericulture—A Case Study. *National Workshop on sustainability of Aquacultural Industry,* IIT Kharagpur, September 28–29, **2007**, pp 88–95.

27. Ray, L. I. P.; Panigrahi, P. K.; Mal, B. C. *Adequacy of Aquaculture Effluent as an Irrigation Source.* XLII (42nd) ISAE Annual Convention and Symposium February at CIAE: Bhopal, 2008a; pp 1–3, paper #SWE-28.

28. Ray, L. I. P.; Moulick, S.; Mal, B. C.; Panigrahi, P. K. Effect of Intensification on Water Quality and Performance of Aqua-Effluent as an Irrigation Source. *Proceedings of 2nd World Aqua Congress,* Nov 26–28, **2008b**, *2,* 402–410.

29. Ray, L. I. P.; Moulick, S.; Mal, B. C.; Panigrahi, P. K. Effect water quality on irrigation. *Proceedings of 2nd World Aqua Congress,* November 26–28, **2008c**, *2,* 397–402.

30. Ray, L. I. P.; Mal, B. C.; Das, B. S. Temporal Variation of Water Quality Parameters in Intensively imc Cultured Lined Pond. In: *e-Proceedings of International Conference on Food Security and Environmental Sustainability (FSES-2009),* December 17–19, 2009, pp 1–10.

31. Ray, L. I. P.; Panigrahi, P. K.; Moulick, S.; Mal, B. C.; Das, B. S.; Bag, N. Aquaculture Waste Water- an Irrigation Source. *Int. J. Sci. Nat. (IJSN).* **2010**, *1*(2), 148–155.

32. Ray, L. I. P. Techno-Economic Feasibility of Integrated Agri-Aquaculture System. Unpublished PhD Thesis, Submitted to Indian Institute of Technology, Kharagpur-721302, West Bengal, 2014, p 211.

33. Read, P.; Fernandez, T. Management of Environmental Impacts of Marine Aquaculture in Europe. *Aquaculture.* **2003**, *226*(1), 139–163.

34. Shuval, H. Water Pollution Control in Semi-Arid and Arid Xones. *Water Res.* **1967**, *2,* 297–308.

35. Singandhupe, R. B.; Rao, G. G. S. N.; Patil, N. G.; Brahmanand, P. S. Fertigation Studies and Irrigation Scheduling in Drip Irrigation System in Tomato Crop (*Lycopersicon esculentum* L.). *Eur. J. Agron.* **2003**, *19,* 327–340.

36. Sopper, W. Disposal of Municipal Wastewater Through Forest Irrigation. *Environ. Pollut.* 1971, *1,* 263–284.

37. Toze, S. Reuse of Effluent Water-Benefits and Risks. *Agric. Water Manage.* **2006**, *80,* 147–159.

38. USAEPA (United States Environmental Protection Agency). *Process Design Manual for Land Treatment of Municipal Wastewater.* EPA 625/1-81-013, 1981, p 254.

39. Wang, D.; Kang, Y.; Wan, S. Effect of Soil Matric Potential on Tomato Yield and Water Use Under Drip Irrigation Condition. *Agric. Water Manage.* **2007**, *87,* 180–186.

40. Wood, C. W.; Meso, M. B.; Karanja, N.; Veverica K. L. Use of Pond Effluent for Irrigation in an Integrated Crop/Aquaculture System. *Ninth Work Plan, Effluents and Pollution Research (9ER1) Final Report,* 2005, p 112.

41. Yokoyama, H. Environmental Quality Criteria for Fish Farms in Japan. *Aquaculture* **2003**, *226*(1), 45–56.

CHAPTER 13

URBAN WASTEWATER TREATMENT SYSTEMS: ASSESSMENT OF REMOVAL EFFICIENCY BASED ON MICROBIAL PATHOGENS: A CASE STUDY IN MYSORE, INDIA

JESSEN GEORGE[1*], DIVYA LAKSHMINARAYANAN[2], SEVERENI ASHILI[3] AND SURIYANARAYANAN SARVAJAYAKESAVALU[4]

[1]Department of Water and Health, Faculty of Life Sciences, Jagadguru Sri Shivaathreeswara University, SS Nagar, Mysore, 570015, Karnataka, India

[2]Department of Water and Health, Faculty of Life Sciences, Jagadguru Sri Shivaathreeswara University, SS Nagar, Mysore, 570015, Karnataka, India, E-mail: divyalakshminarayanan@gmail.com

[3]Hifikepunye Pohamba Campus, University of Namibia, Ongwediva, Namibia, E-mail: severeni.ashili@gmail.com

[4]SCOPE Beijing Office, #18 Shuangqing Road, Haidian District, Beijing, 100085, China, E-mail: sunsjk@gmail.com

[*]Corresponding author.E-mail: georgejessen@gmail.com

CONTENTS

13.1   INTRODUCTION

Urban wastewater contains numerous potentially pathogenic microorganisms and a high content of organic matter. Therefore, it poses number of risks for public health.[6] There are many pathogenic microorganisms which will always be present in partially treated or untreated wastewater and sewage sludge[1,7] It is mainly associated with enteric diseases in humans through consuming wastewater irrigated food crops and vegetables.[5,11,14] Wastewater and sewage sludge reuse is an essential method of water management in many regions of the developing world.[4]

Urban area's population in the world is increasing daily and the situation has become intense in low income or developing countries, where it is expected that an additional 2.1 billion people will be living in cities by 2030.[18] As a result, these cities produce gallons of wastewater which sometimes is collected, treated, and be used directly or indirectly without beneficial use. Information on wastewater collection, treatment, and use is scattered, and needs comprehensive review. However, recent efforts from global organizations such as FAO, IWMI through AquaStat[19] provide a more updated review on this subject.

Irrigation with partially treated or untreated wastewater and sewage sludge may cause negative impacts on human health and risk of infection. Reuse of partially treated wastewater irrigation is a common practice in India and developing regions of the world. In India, the reuse of raw or partially treated wastewater mainly happens due to high treatment costs.[12] An estimated 80% of wastewater generated especially from China and India is mainly used for *irrigation purposes*. The cross-city comparisons emphasized the contrast between developing and developed countries, and the capacity for collection and treatment in developing countries is limited and the same goes for treatment.[13] This then implies that the large portion of untreated municipal wastewater ends up in natural water bodies either directly or indirectly.

In India, more than 80% of wastewater being generated is directly discharged into natural water bodies without any treatment due to lack of infrastructure and resources for treatment.[2,12,20] Both untreated and treated wastewaters in Mysore, are used for *irrigation purpose* due to scarcity of freshwater resources.[15]

This chapter is an attempt to assess the microbial pathogens reduction in wastewater treatment plants in Mysore city, India, and to determine the microbial risk associated with reuse of partially treated wastewater.

13.2   MATERIALS AND METHOD

13.2.1   CHARACTERISTICS OF SEWAGE TREATMENT PLANTS (STPs) IN MYSORE

Mysore watershed is divided into five drainage districts, namely, A, B, C, D, and E. The city has been provided with three wastewater treatment plants (Fig. 13.1): Rayankere sewage treatment plant (RSTP); Vidyaran-yapuram sewage treatment plant (VSTP); and Kesare sewage treatment plant (KSTP).[1] The present inflow to the STPs is about 145 millions of liters per day (MLD) with the facultative aerated lagoons at the treatment process. The untreated and treated UW is largely used for irrigation.

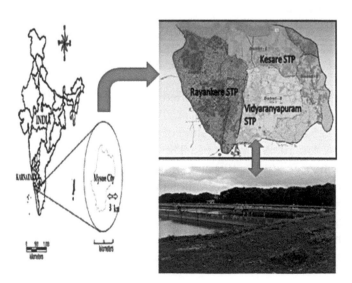

FIGURE 13.1   Sewage treatment plants (STPs) in Mysore.

13.2.2   SAMPLING AND ANALYSIS

The raw and treated wastewater samples were collected from the above-mentioned sewage treatment plants in Mysore and were preserved at 4°C during transportation to laboratory. They were immediately analyzed for total coliform count (TCC), fecal coliform count (FCC), pH, COD, and BOD. All analyses were carried out following the standard methods.[17]

Total coliforms and fecal coliforms were enumerated using the most probable number method (MPN).[16,10] In the presumptive test for coliforms, three 10-mL, three 1-mL, and three 0.1-mL volumes of the appropriate dilution of the water sample were inoculated in nine fermentation tubes with a Durham vial in MacConkey broth. The inoculated tubes were incubated for 48 h at 37°C, and those presenting gas and acid were confirmed in EMB agar (Himedia, Mumbai) at 37°C for TC and in MacConkey broth with a Durham vial at 44°C for 24 h for fecal coliform (FC).[8,9,16,17]

13.3   RESULTS AND DISCUSSION

The physicochemical and microbiological characteristics of raw and treated wastewater were carried out in each treatment plant. Tables 13.1–13.3 indicate the raw sewage characteristics of the STPs that were investigated during this study.

13.3.1   STATUS OF RAYANAKERE STPs

The pH of the raw sewage was varied from 7.2 to 7.5 and BOD values for raw sewage ranged from 264 to 280 mg/L. The COD values of raw sewage were 273–295.5 mg/L. The TCC values of raw sewage were observed in the range of $2.2–2.4 \times 10^8$ MPN/100mL. The FCC values were observed maximum during monsoon (4.5×10^6) and minimum post-monsoon (4.1×10^6).

The average seasonal variations of physicochemical and microbiological parameters are mentioned in Table. 13.1. The treatment efficiency in the terms of microbiological parameters is shown in Figure 13.2.

The pH of the treated effluent was found to be higher compared to raw effluent ranging from 8.1 to 8.2. The BOD and COD values for treated

TABLE 13.1 Average Seasonal Variation of Physicochemical and Microbiological Parameters in Rayanakere Sewage Treatment Plant (RSTP).

Parameters	Pre-monsoon		Monsoon		Post-monsoon	
	Raw	Treated	Raw	Treated	Raw	Treated
pH	7.4	8.2	7.2	8.2	7.5	8.1
BOD	272.1	22.9	264.4	25.7	280	23
COD	290.8	138.2	273	123.5	295.5	138.4
TCC	2.2×10^8	1.1×10^8	2.4×10^8	1.6×10^8	2.4×10^8	1.6×10^8
FC	4.2×10^6	2.1×10^6	4.5×10^6	2.2×10^6	4.1×10^6	1.9×10^6

TABLE 13.2 Average Seasonal Variation of Physicochemical and Microbiological Parameters in Vidyaranyapuram Sewage Treatment Plant (VSTP).

Parameters	Pre-monsoon		Monsoon		Post-monsoon	
	Raw	Treated	Raw	Treated	Raw	Treated
pH	7.3	8.0	7.3	8.1	7.2	8.2
BOD	266.5	18.7	262.6	26.1	261.1	22.1
COD	295.9	151.3	270.7	119	301.3	143.9
TCC	2.4×10^8	1.2×10^8	2.4×10^8	1.8×10^8	2.2×10^8	1.4×10^8
FC	4.4×10^6	2.2×10^6	4.6×10^6	2.4×10^6	4.2×10^6	2.1×10

TABLE 13.3 Average Seasonal Variation of Physicochemical and Microbiological Parameters in Kesare Sewage Treatment Plant (STP).

Parameters	Pre-monsoon		Monsoon		Post monsoon	
	Raw	Treated	Raw	Treated	Raw	Treated
pH	7.3	8.0	8.0	8.0	8.3	8.2
BOD	275.8	21.6	280.3	16.8	269.5	22.9
COD	300.0	146.6	311.1	166.1	344.4	215.2
TCC	2.3×10^8	1.1×10^8	2.4×10^8	1.9×10^8	2.2×10^8	1.5×10^8
FC	3.9×10^6	1.8×10^6	4.2×10^6	2.0×10^6	4.1×10^6	1.9×10^6

sewage were in the range of 22.9–25.7 mg/L, 123.5–138.4 mg/L, respectively. The TCC and FCC values for treated sewage were observed in the range of $1.1–1.6 \times 10^8$ MPN/100 mL, $1.9–2.2 \times 10^6$ MPN/100 mL, respectively (Table 13.1).

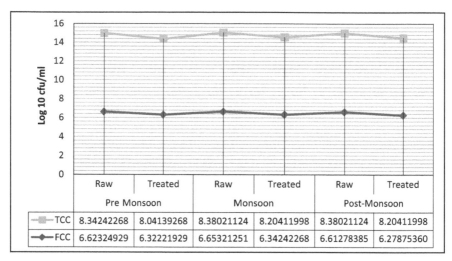

The following data table is part of the figure:

	Pre Monsoon		Monsoon		Post-Monsoon	
	Raw	Treated	Raw	Treated	Raw	Treated
TCC	8.34242268	8.04139268	8.38021124	8.20411998	8.38021124	8.20411998
FCC	6.62324929	6.32221929	6.65321251	6.34242268	6.61278385	6.27875360

FIGURE 13.2   Treatment efficiency for microbiological parameters in RSTP.

13.3.2   STATUS OF VIDYARANYAPURAM STPs

The pH of the raw sewage was varied from 7.2 to 7.3 and BOD values for raw sewage ranged from 269.5 to 280.3 mg/L. The COD values of raw sewage were 270.7–301.3 mg/L. The TCC values of raw sewage were observed in the range of 2.2–2.4 × 10^8 MPN/100 mL. The FCC values were observed maximum during monsoon (4.6 × 10^6) and minimum post-monsoon (4.2 × 10^6). The average seasonal variation of physicochemical and micro-biological parameters is presented in Table 13.2. The treatment efficiency in the terms of microbiological parameters are shown in Figure 13.3.

The pH of the treated effluent was found to be higher compared to raw effluent ranging from 8.0 to 8.2. The BOD and COD values for treated sewage were in the ranges of 18.7–26.1 mg/L, 119–151.3 mg/L, respectively. The TCC and FCC values for treated sewage were observed in the range of 1.2–1.8 × 10^8 MPN/100 mL, 2.1–2.4 × 10^6 MPN/100 mL, respectively.

13.3.3   STATUS OF KESARE STPs

The pH of the raw sewage varied from 7.3 to 8.3 and BOD values for raw sewage ranged from 269.5 to 280.3 mg/L. The COD values of raw sewage were 300.0–344.4 mg/L. The TCC values of raw sewage were observed in

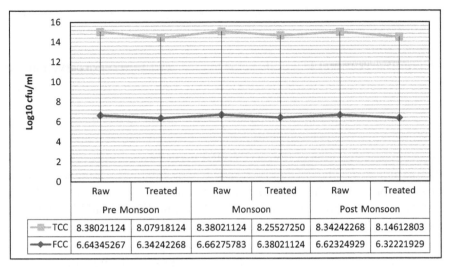

FIGURE 13.3   Treatment efficiency for microbiological parameters in VSTP.

the range of $2.2–2.4 \times 10^8$ MPN/100 mL. The FCC values were observed maximum during monsoon (4.2×10^6) and minimum during pre-monsoon (3.9×10^6).

The average seasonal variations of physicochemical and microbiological parameters are mentioned in Table 13.3. The treatment efficiency in terms of microbiological parameters is mentioned in (Fig. 13.4).

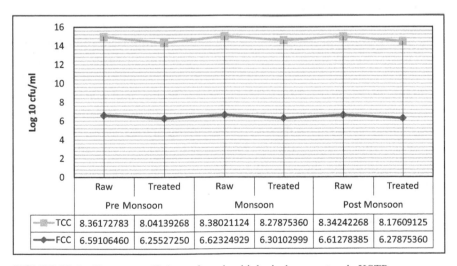

FIGURE 13.4   Treatment efficiency for microbiological parameters in KSTP.

The pH of the treated effluent found to be higher when compared to raw effluent ranging from 8.0 to 8.2. The BOD and COD values for treated sewage were in the range of 16.8–22.9 mg/L and 146–215.2 mg/L, respectively. The TCC and FCC values for treated sewage were observed in the range of $1.1–1.9 \times 10^8$ MPN/100 mL and $1.8–2.0 \times 10^6$ MPN/100 mL, respectively.

From the results in this chapter, it can be concluded that the sewage treatment plants (STPs) exhibit different physicochemical and microbiological characteristics of raw and treated wastewater. In the present study, it was observed that STPs investigated were unable to produce effluent that complies with the discharge standards in terms of TCC and FCC. As per Bureau of Indian Standards (BIS), TCC and FCC should be less than 1000 and 2500 MPN/100 mL respectively.

The higher count of microbial load during the monsoon season may be attributed due to the seasonal changes such as temperature and rainfall, which might have influenced the occurrence of microbial population in the wastewater even after the wastewater is treated. According to the research data by other investigators, reports were found on higher microbial load during the monsoon period.[3,8,9] The overall findings of the present chapter indicate that there was a potential microbial risk associated with the handling and reuse of wastewater in Mysore.

13.4 CONCLUSION

The microbiological indicators did not show an important reduction over treatment. The presence of pathogenic organisms in treated wastewater effluent is a potential public health hazard, as this water source is directly or indirectly discharged in receiving water bodies and may be used by communities for multiple purposes. The microbial risks relating to wastewater and sewage sludge depend on the pathogen load in raw wastewater; hence it requires very efficient treatment to reduce the pathogen load. This study indicates the partially treated or treated wastewater and sewage sludge will cause potential health risks to farmers who use this water for irrigation purposes.

This study also reveals that the continuous monitoring of sewage treatment plants (STPs) and sources of raw sewage need to be identified.

It requires more rigorous and effective treatment for wastewater and sewage sludge to eliminate the microbial load from such water to ensure better sanitation measures. Moreover the treatment plant needs to implement advanced and updated treatment techniques to ensure complete removal of toxic organic and inorganic pollutants as well as pathogenic microorganisms.

13.5 SUMMARY

This chapter mainly focuses on the status of urban wastewater treatment systems in Mysore, Karnataka, India. The raw and treated wastewater samples were collected from the above mentioned sewage treatment plants in Mysore and to carried out both physicochemical and microbiological characteristics such as TCC, FCC, pH, COD, and BOD. The overall results of the present study indicate that STPs investigated were unable to produce effluent in compliance with the discharge standards in terms of TCC and FCC. The microbiological indicators did not show an important reduction over treatment. The presence of pathogenic organisms in treated wastewater effluent is a potential public health hazard, as this water source is directly or indirectly discharged in receiving water bodies and may be used by communities for multiple purposes. The microbial risks relating to wastewater and sewage sludge depend on the pathogen load in raw wastewater; hence it requires very efficient treatment to reduce the pathogen load. On the whole this study indicates the partially treated or treated wastewater and sewage sludge will cause potential health risk to farmers who use this water.

13.6 ACKNOWLEDGEMENTS

The authors are thankful to Department of Water and Health, JSS University, for their constant support throughout this work. We acknowledge Indian council of Medical Research (ICMR) (Project: 2012-14060) for financial support provided to the first author. We thankful to the sewage treatment plant authorities of Mysore for their cooperation and help during the study period.

KEYWORDS

- Fecal coliform count
- health hazard
- microbial pathogens
- removal efficiency
- sewage sludge
- total coliform count

REFERENCES

1. Divya, L.; Jessen, G.; Suriyanarayanan, S; Karthikeyan, K. Studies on Pathogenic Bacterial Strains from Selected Sewage Treatment Plants (STP's) of Mysore, Karnataka, India During Different Seasons: A Comparative Appraisal. *J. Environ. Res. Dev.* **2014**, *9*(1), 24–30.
2. Divya, L.; Jessen, G.; Midhun, G.; Magesh, S. B.; Suriyanarayanan, S. Impacts of Treated Sewage Effluent on Seed Germination and Vigor Index of Monocots and Dicot Seeds. *Russ. Agric. Sci.* **2015**, *41*(4), 252–257.
3. García-Armisen, T.; Inceoglu, N. K.; Ouattara, A. A.; Verbanck, M. A.; Brion, N.; Pierre, S. Seasonal Variations and Resilience of Bacterial Communities in Sewage Polluted Urban River. *PLoS ONE* **2014**, *9*(3), e92579.
4. Gerba, C. P.; Rose, J. B. International Guidelines for Water Recycling: Microbiological Considerations. *Water Sci. Technol.: Water Supply* **2003**, *3*(4), 311–316.
5. Hamilton, A. J.; Stagnitti, F.; Premier, R.; Boland, A. M. Quantitative Microbial Risk Assessment Models for Consumption of Raw Vegetables Irrigated with Reclaimed Water. *Appl. Environ. Microbiol.* **2010**, *72*(5), 3284–3290.
6. Howard, I.; Espigares, E.; Lardelli, P.; Martin, J. L.; Espigares, M. Evaluation of Microbiological and Physicochemical Indicators for Wastewater Treatment. *Environ. Toxicol.* **2004**, *19*(3), 241–249.
7. Jessen, G; Divya, L.; Suriyanarayanan, S. A Review of Quantitative Microbial Risk Assessment in the Management of *Escherichia coli* 0157:H7 via Drinking Water. *J. Environ. Res. Develop.* **2013**, *8*(1), 60–68.
8. Jessen, G.; Divya, L.; Magesh, S. B.; Suriyanarayanan, S. An Assessment of Removal Efficiency for the Bacterial Pathogens in Mysore Urban Water Treatment System, Karnataka, India: A Case Study. *Desalin. Water Treat.* **2015a**, *57*(23), 10886–10893.
9. Jessen, G.; Divya, L.; Sajith, K. S.; Severeni, A.; Suriyanarayanan, S. Evaluation of Bacterial Pathogens in Surface Water of Cauvery River, Near Mysore, Karnataka, South India. *Bull. Environ. Sci. Res.* **2016**, *5*(1), 14–18.

10. Jessen, G; Wei, A.; Dev, J.; Dongqing, Z.; Min, Y.; Suriyanarayanan, S. Quantitative Microbial Risk Assessment to Estimate Health Risk in Urban Drinking Water Systems of Mysore, Karnataka, India. *Exposure Health* **2015b**, *7*(3), 331–338.

11. Mara, D. D.; Sleig, P. A.; Blumenthal, U. J.; Carr, R. M. Health Risks in Wastewater Irrigation: Comparing Estimates from Quantitative Microbial Risk Analyses and Epidemiological Studies. *J. Water Health* **2007**, *5*(1), 39–50.

12. Mekala, G. D.; Davidson, B.; Samad, M.; Boland, A. M. *Wastewater Reuse and Recycling Systems: A Perspective into India and Australia*. International Water Management Institute: Colombo, Sri Lanka, 2008; p 35.

13. Raschid-Sally, L.; Jayakody, P. *Drivers and Characteristics of Wastewater Agriculture in Developing Countries: Results from a Global Assessment*; IWMI Research Report 127; International Water Management Institute: Colombo, 2008.

14. 15. Razak, S.; Arve, H.; Philip, A.; Pay, D.; Petter, D.; Jenssen, G.; Thor-Axel, S. Quantification of the Health Risk Associated with Wastewater Reuse in Accra, Ghana: A Contribution Toward Local Guidelines. *J. Water Health* **2008**, *6*(4), 461–471.

15. Shakunthala, B.; Shivanna, S.; Doddaiah, S. Urban Wastewater Characteristic and its Management in Urban Areas—A Case Study of Mysore City, Karnataka, India. *J. Water Res. Prot.* **2010**, *2*, 717–726.

16. Sravani, M.; Divya, L.; Jessen, G.; Suriyanarayanan, S. Assessment of Physicochemical and Microbiological Characteristics of Water Samples in Suttur Village, Nanjangud Taluk, Mysore, Karnataka. *Int. Res. J. Environ. Sci.* **2014**, *3*(7), 1–4.

17. *Standard Methods for the Examination of Water and Wastewater*. 20th Edition; American Public Health Association/American Water Works Association/Water Environment Federation, Washington DC, USA, 1998.

18. United Nations. *World Population Prospects: the 2012 Revision*. United Nations, Department of Economic and Social Affairs, Population Division: New York, 2012.

19. Wichelns, D.; Pay, D.; Manzoor, Q., Eds. *Wastewater: Economic Asset in an Urbanizing World. Springer: USA,* E-book, 2015; Chapter 1; pp 3–14.

20. Winrock International India. *Urban Wastewater: Livelihoods, Health and Environmental Impacts in India*; Research report submitted to International Water Management Institute, 2007.

CHAPTER 14

PHYTOCHEMICAL SCREENING, FORMULATION AND EVALUATION OF POLYHERBAL OINTMENT CONTAINING MIMOSA PUDICA AND SAMADERA INDICA

P. H. RAJASREE, JESSEN GEORGE[1,*] AND B. KRISHNA PRASAD[2]

[1]*Department of Pharmaceutics, JSS College of Pharmacy, Mysore, 570015, Karnataka, India*

[2]*Bioclinica Safety and Regulatory Solutions, Metagalli, 570016, Mysore, India, E-mail: krishnaprasad8710@gmail.com*

[*]*Corresponding author. E-mail: rajasreeph@gmail.com*

CONTENTS

14.1   INTRODUCTION

Medicinal plants are important agents for treating various diseases.[1] Various medicinal plants have been used for years in daily life to treat diseases all over the world.[9] Herbal medicine is making a dramatic comeback and an increasing number of patients are visiting alternative medicine clinics.[7,8] Today, there is widespread interest in drugs derived from medicinal plants. This interest primarily stems from the belief that green medicine is safe and dependable, compared with costly synthetic drugs that may have adverse effects.[8] There are many natural antimicrobials that can be derived from medicinal plants.[3,5] Use of plants and traditional practices will continue to play a significant role in the sociocultural life of village communities.[2,4] The plants of *Mimosa pudica* (Family—*Fabaceae*) known as "sensitive plant, sleepy plant, Dormilones, or shy plant" and *Samadera indica* (Family-Simaroubaceae) are well known for its medicinal properties (Fig. 14.1).

This research study investigates possible antimicrobial activity of the formulated ointment made from the methanolic extracts of *M. pudica* and *S. indica* plants. Ointments were evaluated for its physicochemical property and antibacterial activity.

14.2   MATERIALS AND METHODS

14.2.1   COLLECTION OF PLANTS AND EXTRACTION

Plants of *M. pudica* and *S. indica* were collected from Wayanad district of Kerala, South India. The plant parts were washed under running tap

FIGURE 14.1   Plants of *Mimosa pudica* (left) and *Samadera indica* (right).

water, and the leaves were cut into small pieces and shade dried at 300°C, 50 ± 5% relative humidity for 15 days. The powdered leaves were passed through sieve no. 22/8 to get a coarse powder.[8] The powder was stored in air tight container for further use.

The powder of *M. pudica* and *S. indica* were subjected to soxheletation. The amount of dried powder taken was 500 g and it was defatted using petroleum ether and the marc obtained for each plant was extracted using methanol in the Soxhlet apparatus for 24 h. After this procedure, the solvent was distilled out and the concentrated residues were analyzed by various chemical tests.

14.2.2   PHYTOCHEMICAL ANALYSIS

After extraction, the methanolic extract was subjected to various phytochemical screenings as per the standard procedure.[6]

14.2.3   FORMULATION OF EMULSIFYING OINTMENT BASE

Required quantities of emulsifying wax, liquid paraffin, and white soft paraffin were weighed and melted. To this, adequate quantities of methanolic extract of the plants were added. By using fusion method, herbal ointment was formulated.[8]

14.2.4   PRELIMINARY AND PHYSICOCHEMICAL EVALUATIONS

14.2.4.1   COLOR AND ODOR

The color and odor of the prepared ointment was studied by visual examination and feeling the ointment.

14.2.4.2   DETERMINATION OF PH

The pH value of a solution was determined by Mettler Toledo digital pH meter.

14.2.4.3 SPREADABILITY

Spreadability of the formulation was determined by an apparatus that has been designed by Multimer. An excess of the formulated polyherbal ointment was placed on the ground plate of the apparatus and sandwiched between the plate and another glass plate having the dimension of fixed ground plate and provided with the hook. Excess of the ointment was scraped off from the edges and top plate was then subjected to a force of 80 g. The time (in seconds) required by the top plate to cover a distance of 10 cm was noted using a string attached to the hook. A shorter interval indicates better spreadability. Spreadability is measured by the following formula:

$$S = (M \times L)/T \qquad (1.1)$$

where, M = weight tide to upper slide; L = length of glass slide; and T = Time.

14.2.4.4 EXTRUDABILITY

The polyherbal ointment was filled in collapsible tubes. The extrudability of the different ointment formulations was determined in terms of weight in grams required to extrude a 0.5 cm ribbon of ointment in 10 s.

14.2.5 MICROBIOLOGICAL STUDIES

The antibacterial activity of different concentrations of polyherbal ointment was evaluated against the strain of *Staphylococcus aureus*. Nutrient agar and Mueller-Hinton agar (MH) media were used for bacterial culture and incubated at temperature $37 \pm 2°C$ for 48 h. After 24 h, the growth was measured. Determined MIC was at the lowest test concentration needed to ensure that the culture did not grow over 10% of the relative cell density.

14.2.6 DIFFUSION STUDIES

The formulated ointment was accurately weighed and placed in the donor part of the Franz diffusion cell and a semi permeable cellulose membrane with 1000 MW cut off. The diffusion apparatus was filled with phosphate buffer, pH 6.8. The receptor phase was stirred thoroughly. Intervals of 20, 40, 60, 80, 100, and 150 min were chosen for sampling time, and samples were analyzed for drug content spectrophotometrically at 258 nm.[9]

14.2.7 STABILITY STUDIES

The most satisfactory formulation was sealed in a glass vial and kept at 4 ± 2°C and 25 ± 2°C at RH of 65 ± 5% and 37 ± 5% RH for 2 months. At the end of 1 and 2 months, the samples were analyzed for the drug content and in vitro diffusion study.

14.3 RESULTS AND DISCUSSION

Literature review revealed that the selected two herbs of *M. pudica* and *S. indica* have antibacterial activity. In the present study, polyherbal ointments were prepared by fusion method using emulsifying ointment as the base. Extraction and the phytochemical screening was done using methanol as the solvent. Phytochemical screening confirmed the presence of various phytoconstituents like carbohydrate, glycosides, flavonoids, and tannins. The formulation was evaluated for various physicochemical studies, such as: color and odor, pH, spreadability, extrudability, diffusion studies, microbiological study, and stability analysis. The results of the formulations are mentioned in Tables 14.1–14.4.

Methanolic extracts of *M. pudica* and *S. indica* were selected for formulating polyherbal ointment. Phytochemical analyses confirmed the presence of carbohydrates, flavonoids, tannins, and glycosides. Using emulsifying ointment base, polyherbal ointment was prepared by fusion method. The formulation was evaluated for physicochemical studies such as color and odor, pH, spreadability, extrudability, diffusion studies, microbiological study, and stability analysis. The physicochemical parameters are within the acceptable range. The antimicrobial activity of the prepared ointment was compared with betadine (marketed formulation) using *S. aureus*. The formulations F2 and F3 showed greater activity on *S. aureus* compared to betadine. The release studies confirmed that at 133

TABLE 14.1   Composition of the Formulated Ointment.

Ingredients	F1 (2%)	F2 (4%)	F6 (6%)
Mimosa pudica Methanolic extract	2 g	4 g	6 g
Samadera indica Methanolic extract	2 g	4 g	6 g
Emulsifying ointment	q.s to 100 gm	q.s to 100 gm	q.s to 100 g

TABLE 14.2   Phytochemical Analyses of Extracts of *Mimosa pudica* and *Samadera indica*.

Constituents	Name of test	Methanolic extract of *Mimosa pudica*	Methanolic extract of *Samadera indica*
Carbohydrates and reducing sugars	Molisch's test	+	+
	Fehling's test	+	−
	Benedict's test	+	+
	Barfoed's test	−	−
Proteins	Millon's test	−	−
	Biuret test	−	−
	Xanthoprotein test	−	−
	Legal's test	−	−
	Keller-Killiani test	−	−
Glycosides	Borntrager's test	−	+
	Modified Borntrager's test	+	+
Flavonoids	Shinoda test	+	+
	Lead acetate	+	+
	Sodium hydroxide	+	+
Tannins	5 % Fecl₃	−	−
	Lead acetate solution	−	+
	Bromine water	−	+
	Dilute iodine solution	−	+

TABLE 14.3 Physicochemical Evaluation of the Formulations.

Physicochemical parameters	F1 (2%)	F2 (4%)	F3 (6%)
Color	Dark green	Dark green	Dark green
Odor	Characteristic	Characteristic	Characteristic
pH	6.00	6.20	6.88
Spreadability (sec)	12	14	16
Extrudability (g)	160	165	177
Diffusion study (cm)	0.22	0.46	0.78
Stability (4, 25 and 37°C)	Stable	Stable	Stable

TABLE 14.4 Microbiological Studies

Ointments	Zone diameter in cm (*Staphylococcus aureus*)
F1 (2%)	0.78
F2 (4%)	3.1
F3 (6%)	3.5
Standard	1.5

min, 95% of the drug was released. The stability study confirmed that the formulation was stable.

14.4 CONCLUSION

The formulated polyherbal formulation is a potent and an effective antiseptic ointment to treat *S. aureus* infections. The results concluded that the prepared formulation containing methanolic extracts of *M. pudica* and *S. indica* is an effective antiseptic ointment with acceptable characteristics. Hence this chapter concludes that an efficient antiseptic ointment with antimicrobial activities can be formulated from the methanolic plant extracts of methanolic extracts of *M. pudica* and *S. indica*, which can also be used for wound healing and various skin infections. Further research may be possible in future in the areas of in vivo studies and wound healing.

14.5 SUMMARY

The present study was designed to formulate and to evaluate polyherbal ointment with antiseptic activity. Ointments were formulated using methanolic extracts of *M. pudica and S. indica* and were evaluated for its physicochemical property and antibacterial activity. Phytochemical screening confirmed the presence of various phytoconstituents: like carbohydrate, glycosides, flavonoids, and tannins. The formulation was evaluated for various physicochemical studies such as color and odor, pH, spreadability, extrudability, diffusion studies, microbiological study, and stability analysis. The antibacterial activity of the prepared ointment was compared with betadine (marketed formulation) using *S. aureus*. Compared to betadine, formulations F2 and F3 showed greater activity on *S. aureus*. The release studies confirmed that at 140 min, 90% of the drug was released. The stability study confirmed that the formulation was stable. The results concluded that the prepared formulation containing *M. pudica and S. indica* is an effective antiseptic ointment with acceptable characteristics.

KEYWORDS

- **Antibacterial activity**
- **betadine**
- **extrudability**
- **herbal medicine**

- **methanolic extracts**
- *Mimosa pudica*
- *Samadera indica*
- *Staphylococcus aureus*

REFERENCES

1. Abayomi, S.; Eyitope, O.; Adedeji, O. The Role and Place of Medicinal Plants in the Strategies for Disease Prevention. *Afr. J. Tradit. Complement Altern. Med.* **2013**, *10*(5), 210–229.
2. Chhetri, H. P.; Yogol, N. S.; Sherchan, J.; Anupa, K. C.; Mansoor, S. Formulation and Evaluation of Antimicrobial Herbal Ointment. *Kathmandu University J. Sci. Eng. Technol.* **2010**, *6*(1), 102-107.
3. Gidwani, B; Alaspure, R. N.; Duragkar, N. J.; Singh, V.; Rao, S. P. Shukla, S. S. Evaluation of a Novel Herbal Formulation in the Treatment of Eczema with *Psoralea corylifolia. Iran J. Dermatol.* **2010**, *13*, 122–127.

4. Joshi, K.; Joshi, A. R. Ethnobotanical Plants Used for Dental and Oral Healthcare in the Kali Gandaki and Bagmati Watersheds, Nepal. *Ethnobotanical Leaflets* **2006**, *10,* 174–178.

5. Kokate, C. K. *Practical Pharmacognosy.* 4th ed.; Vallabh Prakashan: Delhi, 1997; pp 107–111.

6. Parmar, R. B.; Baria, A. H.; Faldu, S. D.; Tank, H. M.; Parekh, D. H. Design and Evaluation of Poly-Herbal Formulation in Semisolid Dosage form for its Antibacterial Activity. *J. Pharm. Res.* **2009**, *2*(6), 1095–1097.

7. Rajasree, P. H.; George, J.; Gritta, S.; Gowda, D. V. Formulation, Phytochemical Screening and Physicochemical Evaluation of an Antiseptic Ointment Containing *Azadiracta indica* and *Chromolena odorata. Int. J. Pharm. Biol. Sci.* **2015**, *5*(4), 114–118.

8. Rajasree, P. H.; Vishwanad, V.; Cherian, M.; Eldhose, J.; Singh, R. Formulation and Evaluation of Antiseptic Polyherbal Ointment. *Int. J. Pharm. Biol. Sci.* **2012**, *3*(10), 2021–2031.

9. Swarupa, R. G.; Gunda, R. K.; Mansa, Y.; Khushi Vardhan, C. H.; Kumar, D. V.; Venkateswarlu, G. Preparation and Evaluation of Antimicrobial and Antioxidant Activity of Polyherbal Ointment. *Int. J. Curr Trends Pharm. Res.* **2015**, *3*(4), 997–1003.

CHAPTER 15

BIOINFORMATIC ADVANCEMENTS IN POST TRANSLATIONAL MODIFICATIONS (PTMs): AN EXPERIMENTAL APPROACH

SAURABHH JAIN[1], SURBHI PANWAR[2], ASHWANI KUMAR[3,*] AND TEJPAL DHEWA[4]

[1]*Department of Applied Sciences, Seth Jai Parkash Mukandlal Innovative Engineering and Technology Institute (JMIETI), Chota Bans, Radaur, 135133, Yamuna Nagar, Haryana, India, E-mail: Saurabh.Jain83@gmail.com*

[2]*Department of Genetics and Plant Breeding, Chaudhary Charan Singh University, Meerut, 200005, Uttar Pradesh, India, E-mail: surbhipanwar11086@gmail.com*

[3]*Department of Nutrition Biology, School of Interdisciplinary and Applied Life Sciences, Central University of Haryana, Mahendergarh, Haryana, India*

[4]*Department of Nutrition Biology, School of Interdisciplinary and Applied Life Sciences, Central University of Haryana, Jant Pali, Mahendergarh, 123029, Haryana, India, E-mail: tejpaldhewa@gmail.com, tejpal_dhewa07@rediffmail.com*

[*]*Corresponding author. E-mail: ashwanindri@gmail.com*

CONTENTS

15.1   INTRODUCTION

Posttranslational modification (PTM) is a biochemical mechanism in which amino acid residues in a protein are covalently modified.[4] It is crucial for regulating conformational changes, activities, functions of proteins, and is part of most of the cellular processes. Therefore, the identification of protein PTMs is the foundation for understanding cellular and molecular mechanisms. Bioinformatics tools can generate rapid, accurate and valuable results for PTM prediction.

This chapter presents advances in PTMs.

15.2   ProP 1.0 SERVER

ProP 1.0 server predicts arginine and lysine propeptide cleavage sites in eukaryotic protein sequences using an ensemble of neural networks. Furin-specific prediction is the default. General proprotein convertase (PC) can also be predicted. This server is combined with the SignalP server predicting the presence and location of signal peptide cleavage sites.[3]

15.2.1   PROCEDURE

15.2.1.1   STEP I. SPECIFY THE INPUT SEQUENCES

All the input sequences must be in one-letter amino acid code. The allowed alphabets are given below:

A C D E F G H I K L M N P Q R S T V W Y and **X** (unknown)

All the other symbols will be converted to **X** before processing. The sequences can be submitted in the following two ways:

- Paste a single sequence of amino acid or a number of sequences in FASTA format into the upper window of the main server page.
- Select a FASTA file on local disk, either by typing the file name into the lower window or by browsing the disk.

Both ways can be employed at the same time: all the specified sequences will be processed. However, there may not be more than 2000 sequences and 200,000 amino acids in total in one submission. The sequences may not be longer than 4000 amino acids. Figure 15.1 depicts the file format of submitted sequence in ProP 1.0 server.

15.2.1.2   STEP II. CUSTOMIZE THE RUN

- By default the server produces graphical output of the predictions. However, one can suppress that by un-checking the button labeled "Generate graphics."
- Prediction of the presence and location of signal peptide cleavage sites by the SignalP server is included by default. One can suppress that by un-checking the button labeled "Include the four individual neural networks alongside the average signal peptide prediction."
- Check the button labeled "Verbose output" to display the scores produced by score. The default is to show the average score only.
- The server performs Furin-specific pro-peptide cleavage site prediction by default. Check the button labeled "General PC prediction" to perform that prediction instead.

15.2.1.3   STEP III. SUBMIT THE JOB

Press the "Submit" button. The job status will be shown in the browser window and regularly updated until it ends. The job status can be acquired through email by entering email address. The e-mail will contain the URL of the results that will remain on the server for 24 h.

15.2.1.4   STEP IV. OUTPUT FORMAT

The output format will be produced under the following heads:

- **Header line:** Citing the name and length of the sequence.

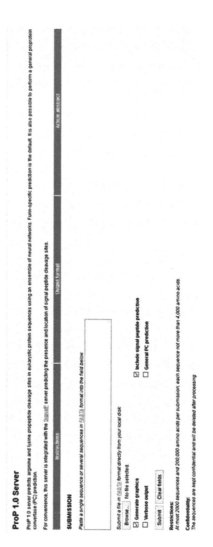

FIGURE 15.1   File format of submitted sequence in ProP 1.0 Server.

- **Sequence:** In one-letter code, as it was interpreted by the server.
- **Annotation overview:** Displaying the predictions residue by residue, as. (dot)' for neutral place-holder, 's' predicted to be within a signal peptide and 'P' predicted to be followed by a pro-peptide cleavage site
- **Prediction score table:** Citing the position, context and score for each arginine (R) and lysine (K) residue in the sequence. If the score is >0.5, the residue is predicted to be followed by pro-peptide

cleavage site; the higher is the score, more is the confidence of prediction.

- **Graph:** In GIF, showing the predictions. For each arginine (R) and lysine (K), the prediction score is plotted against the position in the sequence. If a signal peptide cleavage site has been predicted, it is also shown in the graph. Figure 15.2 shows format of output of cleavage site predictions.

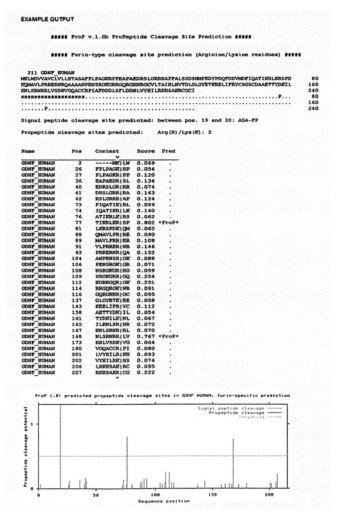

FIGURE 15.2   Output format of cleavage site predictions for a submitted sequence.

15.3   NetPicoRNA 1.0 SERVER

The NetPicoRNA 1.0 server yields neural network predictions of cleavage sites of picornaviral proteases.[1,2]

15.3.1   STEP I. SPECIFY THE INPUT SEQUENCES

All the input sequences should be specifying as per usage instructions under section 15.2.1. The sequences shorter than 9 or 15 amino acids (depending on the selected prediction options) or longer than 4000 amino acids will be ignored. Figure 15.3 shows the file format of submitted sequence in NetPicoRNA 1.0 server.

15.3.2   STEP II. CUSTOMIZE THE RUN

Select the below mentioned prediction type button (Table 15.1).

Press the button labeled "Verbose output" to consist of the above descriptions in the output.

15.3.3   STEP III. SUBMIT THE JOB

Press the Submit button. The status of job will be displayed as described earlier.

15.3.4   STEP IV. OUTPUT FORMAT

The output format will be produced under the following heads.

- **Header line:** Citing the name of the sequence.
- **Sequence:** In one-letter code, as interpreted by the server.
- **List** of selected prediction alternatives.
- **For every prediction alternative**, the output comprises predictions for all the input residues matching the pattern for that specific

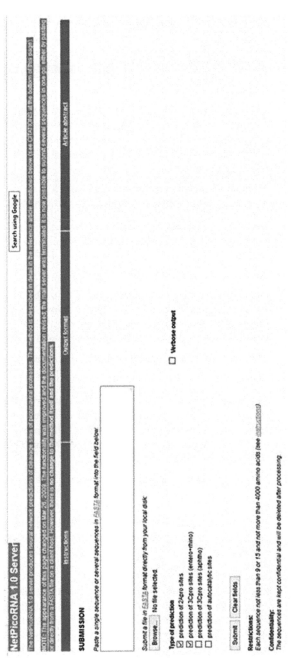

FIGURE 15.3   File format of submitted sequence in NetPicoRNA 1.0.

TABLE 15.1   Prediction Type Buttons for NetPicoRNA 1.0 Server.

Type	Limitations
2Apro sites	This network was trained on cleavage sites in entero- and rhinovirus only. This prediction will only reflect potential sites matching the following sequence:
	xp3 – xp2 – xp1 * G – xp2′ – xp3′ that is glycine (G) must be present at position P1′. The 2Apro algorithm has an input window of 15 residues and observes the central residue of this window. Therefore, the first and last seven residues of the input sequence will not be processed.
	Minimally required sequence would be of 15 aa."xxxxxxxxGxxxxxx" (G required, x is any amino acid)
3Cpro sites (entero + rhino)	It was trained on cleavage sites in entero- and rhinovirus only. This prediction will only reflect potential sites matching the following sequence:
	xp3 – xp2 – Q/E * xp1′ – xp2′ – xp3′
	that is glutamine (Q) or glutamic acid (E) must be present at position P1. The 3Cpro algorithm has an input window of nine residues and observes the central residue of this window. Therefore, the first and last four residues of the submitted sequence will not be processed.
	Minimally required sequence would be of nine aa.
	"xxxx(Q/E)xxxx" (Q or E required)
3Cpro sites (aphtho)	The network was trained on data from aphthovirus (FMDV). This algorithm was trained to look at all probable input windows of nine residues and observes the central residue of each such window. Therefore, the first and last four residues of the submitted sequence will not be processed.
	Minimally required sequence would be of nine aa.
	"xxxxxxxxx"
Autocatalytic sites	The network was trained on data from all picornaviruses. This prediction will only consider potential sites matching the following sequence:
	xp3 – L – xp1 * xp1′ – xp2′ – xp3′ – xp4′ – E
	that is leucine (L) must be present at position P2 and glutamic acid (E) must be present at position P5′. The 'auto' algorithm has an input window of 15 residues and observes the central residue of this window. Therefore, the first and last seven residues of the submitted sequence will not be processed.
	Minimally required sequence would be of 15 aa.
	"xxxxxxLxxxxxExx" (L and E required)

prediction. First, a sequential list of the residues is shown and then, a list sorted by the value of the cleavage score (highest scores first). Every residue has the following columns (Figure 15.4):

Residue:	The amino acid residue at position P1 of the cleavage site (P3-P2-P1*P1'-P2'-P3'), where cleavage is between P1 and P1'.
Pos:	Position of the residue in the sequence.
Clv:	Cleavage score output from the neural network in the interval 0.000–1.000. Scores above 0.500 are expected as potential cleavage sites. The higher the score, the more likely the prediction.
Surf:	Surface score output from the neural network in the interval 0.000–1.000. Scores above 0.500 are anticipated as surface exposed sites. The higher the score, the more likely the prediction.
Sequence:	A window of seven residues centered on the residue being examined.
Comment:	The word 'Potential' indicates that Clv >0.500 and the site is a potential cleavage site.

EXAMPLE

Lines containing the prediction scores may look like this:

```
Residue   Pos   Clv     Surf     Sequence                    Comment
----------------------------------------------------------------------
Q          85   0.927   0.626    QRTQGLI                     Potential
E          97   0.077   0.561    NDRERVN
```

FIGURE 15.4   Output format of score predictions for a submitted sequence.

In the output the sequences are sorted alphabetically (by sequence name). Therefore, the original order is not preserved.

15.4   NetCorona 1.0 SERVER

NetCorona predicts coronavirus 3C-like proteinase (or protease) cleavage sites using artificial neural networks on amino acid sequences. Every potential site is scored and a list is compiled along with graphical representation.

15.4.1 PROCEDURE

15.4.1.1 STEP I

Specify the input sequences as described earlier in section 15.2.1 (Fig. 15.5). There may be not more than 10 sequences in total in one submission. Sequences exceeding 10,000 amino acids will be ignored.

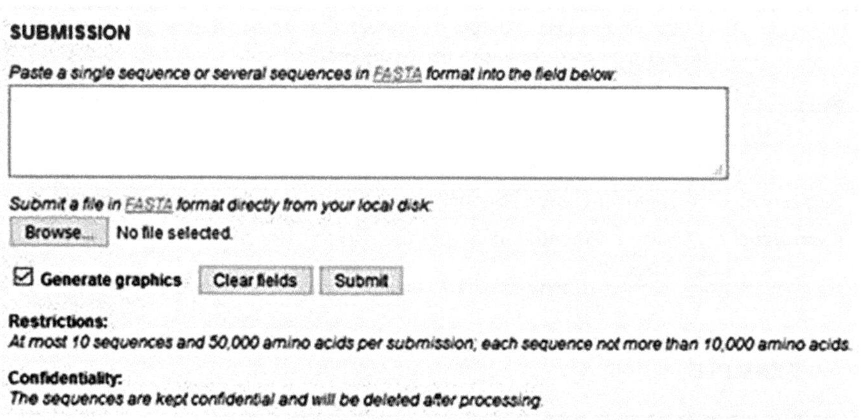

SUBMISSION

Paste a single sequence or several sequences in FASTA format into the field below:

Submit a file in FASTA format directly from your local disk:
Browse... No file selected.

☑ Generate graphics | Clear fields | Submit |

Restrictions:
At most 10 sequences and 50,000 amino acids per submission; each sequence not more than 10,000 amino acids.

Confidentiality:
The sequences are kept confidential and will be deleted after processing.

FIGURE 15.5   File format of submitted sequence in NetCororna 1.0.

15.4.1.2 STEP II. CUSTOMIZE THE RUN

The button "Generate graphics" is used to disable the graphics generated by default. If disabled only the text output is shown.

15.4.1.3 STEP III. SUBMIT THE JOB

Press the Submit button. The job status will be displayed as described previously.

15.4.1.4 STEP IV. OUTPUT FORMAT

The output with graphics enabled contains three parts (Fig. 15.6).
- The first part is an output in HOW format, which contains on the first line the number of residues and the sequence name. It will

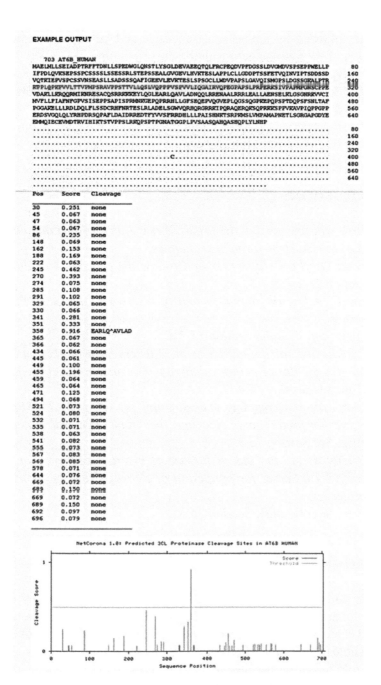

FIGURE 15.6 Output format of cleavage site predictions for a submitted sequence.

be followed by amino acid sequence and below is a representation of each amino acid with the corresponding prediction ("C" for cleavage site and "." for nothing).
* The second part is a listing of the examined residues (glutamines) and their score.
* The third part is a graphical representation of the table in the second part of the output. Multiple sequences are separated with "//".

15.5 QUESTIONS

* *How will one predict the arginine and lysine propeptide cleavage sites in eukaryotic protein sequences?*
* Hint: *Use ProP 1.0 server and follow the procedure explained in section 15.2.1.*
* *Briefly define the output format of inserted sequence in ProP 1.0 server*
* Hint: *Step IV under section 15.2.1.*
* *Explain in detail about NetCorona 1.0 Server for prediction of coronavirus 3C-like proteinase cleavage sites on amino acid sequences.*
* Hint: *See Section 15.4.*
* *Elaborate the complete experimental set up of NetPicoRNA 1.0 Server for prediction of cleavage sites of picornaviral proteases.*
* Hint: *See Section 15.3.*
* *Briefly define the output format of inserted sequence in NetPicoRNA 1.0 Server for prediction of cleavage sites of picornaviral proteases.*
* Hint: *Step IV under section 15.3.*

15.6 SUMMARY

Post translational modifications (PTMs) of protein sequences control various features of protein (activities, localization, interactions, expression and characterization). Bioinformatics centered methods produces fast and valuable information to determine the PTMs information, which plays critical role in define the mechanisms of protein. This chapter gives an overview of an experimental approach of bioinformatics tools to study the advancement in Post translational Modifications.

KEYWORDS

- **Autocatalytic sites**
- **bioinformatic tools**
- **bioinformatics**
- **FASTA file**
- **NetPicoRNA 1.0 Server**
- **neural networks**
- **post translational modifications**
- **ProP 1.0 server**

REFERENCES

1. Blom, N.; Gammeltoft, S.; Brunak, S. Sequence- and Structure-Based Prediction of Eukaryotic Protein Phosphorylation Sites. *J. Mol. Biol.* **1999**, *294*(5), 1351–1362.
2. Blom, N.; Sicheritz-Ponten, T.; Gupta, R.; Gammeltoft, S.; Brunak, S. Prediction of Post Translational Glycosylation and Phosphorylation of Proteins from the Amino Acid Sequence. *Proteomics* **2004**, *4*(6), 1633–1649.
3. Peter, D.; Søren, B.; Nikolaj, B. Prediction of Pro-Protein Convertase Cleavage Sites. *Protein Eng. Des. Sel.* **2004**, *17,* 107–112.
4. Walsh, C. T. *Post Translational Modification of Proteins*; Roberts and Company: Englewood, Colorado, 2006.

CHAPTER 16

APPROXIMATING THE QUADRATIC PROGRAMMING (QP) PROBLEM

ELIAS MUNAPO

School of Economics and Decision Sciences, North West University, Mafeking 2735, South Africa
E-mail: emunapo@gmail.com

CONTENTS

16.1   INTRODUCTION

The quadratic programming (QP) problem is the problem of minimizing a quadratic function over linear constraints. The quadratic programming has applications in areas such as public policy, transportation, engineering, finance, economics, agriculture, marketing, and resource allocation. For more on applications we refers to Gupta,[6] Horst et al.[8], and McCarl et al.[12] The quadratic problem can be classified as convex and nonconvex. There are many exact methods for solving the convex quadratic programming problem. These exact methods include simplex-based complementary pivoting, active set, interior point, gradient projection, and augmented Lagrangian methods.[1,2,4,11,13,14] Unlike the convex quadratic, the nonconvex form is very difficult to solve. Heuristics are normally used to approximate this difficult model in reasonable times.[3,10]

The chapter presents an approach for approximating the QP problem. The QP problem is approximated as a linear programming (LP) model. The optimal solution to the LP is close to the optimal solution of the original QP. The proposed approach has the strength that the QP problem can be approximated in polynomial time. The general QP model is NP hard and is nonlinear programming and is very difficult to solve.

16.2 SPECIFIC APPLICATIONS IN AGRICULTURE

Quadratic programming (QP) is used in the determination of risk in agriculture and portfolio analysis. There are several types of risks in agriculture and these include:

- Production risk
- Price and market risk
- Financial and credit risk
- Institutional risk
- Technology risk and
- Personal risk

Harry et al.[7] successfully applied a QP model to the solution of competitive equilibrium for the field crops of US agriculture. The analysis was based on nine spatially separated markets with separate demand functions for six commodities in each. The objective of the quadratic model was to maximize net profits derived from the determined demand in the various markets.

16.3 THE QUADRATIC PROGRAMMING (QP) PROBLEM

Let a QP problem be represented by Equation (16.1)

Minimize $\qquad$ $\left. f(X) = CX + \frac{1}{2} X^T DX \right\}$

Subject to $\qquad$ $AX \leq B$ $\qquad\qquad\qquad$ (16.1)

$\qquad\qquad\qquad$ $X \geq 0$

Where

$$C = (c_1, c_2, ..., c_n), \; X = (x_1, x_2, ..., x_n)^T,$$

$$D = \begin{pmatrix} d_{11} & ... & d_{1n} \\ ... & ... & ... \\ d_{n1} & ... & d_{nn} \end{pmatrix}, \quad A = \begin{pmatrix} a_{11} & ... & a_{1n} \\ ... & ... & ... \\ a_{m1} & ... & a_{mn} \end{pmatrix}$$

and

$$B = (b_1, b_2, ..., b_m)^T.$$

16.4   LINEARIZATION OF THE QUADRATIC PROBLEM (QP)

This chapter presents method of linearizing the QP problem so that the optimal solution to linear model is close to the optimal solution of the original quadratic problem. The approximated linear problem can be solved by competitive methods such as the interior point algorithms or simplex method.

16.4.1   LINEARIZING THE CONVEX QUADRATIC PROBLEM

Suppose there are no constraints. The optimal point to (Eq. 16.1) can be found in the following ways:

Approach 1: $\nabla f(X) = 0$

Approach 2: $|\nabla f(X)| = 0$

Approach 3: minimize $|\nabla f(X)|$

Approach 4: minimize ϕ such that $-\Phi \leq \nabla f(X) \leq \Phi$, where

$$\Phi = \begin{pmatrix} \phi \\ \phi \\ \cdot \\ \cdot \\ \cdot \\ \phi \end{pmatrix}, \phi \geq 0. \qquad (16.2)$$

The four approaches can give us the optimal point but approach 4 has extra and special features that may still give us the optimal point, even if constraints are added. This is shown in the following illustration.

Consider the unconstrained nonlinear model given in (Eq. 16.3).

Minimize $\qquad\qquad -8x_1 - 4x_2 + x_1^2 + x_2^2 \qquad\qquad (16.3)$

The four approaches presented above can be used to find the solution as follows.

Approach 1: $-8 + 2x_1 = 0$ and $-4 + 2x_2 = 0 \Rightarrow x_1 = 4$ and $x_2 = 2$.

Approach 2: $|-8 + 2x_1| = 0$ and $|-4 + 2x_2| = 0 \Rightarrow x_1 = 4$ and $x_2 = 2$.

Approach 3: $\min |-8 + 2x_1| = 0$ and $\min |-4 + 2x_2| = 0 \Rightarrow x_1 = 4$ and $x_2 = 2$.

Approach 4: minimize ϕ subject to: $-\phi \leq -8 + 2x_1 \leq \phi$
and $-\phi \leq -4 + 2x_2 \leq \phi \Rightarrow x_1 = 4$ and $x_2 = 2$.

16.4.2   APPROACH 4 IN GENERAL WITH NO CONSTRAINTS

Minimize $\qquad\qquad\qquad \phi$

Such that $\qquad\qquad -\Phi \leq \nabla f(X) \leq \Phi \qquad\qquad (16.4)$

$$\Phi \geq 0$$

Since $\qquad\qquad\qquad \Phi = \begin{pmatrix} \phi & \phi & \cdots & \phi \end{pmatrix}^T$

Then minimize $\qquad\qquad \phi$

Such that $\qquad\qquad -\phi \leq f_{x_1}(X) \leq \phi$

$$-\phi \le f_{x_2}(X) \le \phi \qquad (16.5)$$

$$-\phi \le f_{x_n}(X) \le \phi$$

$$\phi \ge 0.$$

This is a linear programming problem and this problem is minimized when

$$\phi = 0$$

Or when

$$f_{x_1}(X) = f_{x_2}(X) = ... = f_{x_n}(X) = 0 \qquad (16.6)$$

16.4.3   ADDITION OF LINEAR CONSTRAINTS (AX ≤ B) TO APPROACH 4

Minimize $\qquad \phi$

Such that $\qquad -\phi \le f_{x_1}(X) \le \phi$

$$-\phi \le f_{x_2}(X) \le \phi \qquad (16.7)$$

$$...$$

$$-\phi \le f_{x_n}(X) \le \phi$$

$$AX \le B$$

In this chapter, author calls the constraint $(-\phi \le f_{x_j}(X) \le \phi)$ a type 1 partial constraint. The ϕ calculated by the linear program gives the worst case of all the partial constraints. There is need also to minimize the worst case of each partial constraint.

Minimize $\qquad \phi_1 + \phi_2 + ... + \phi_n$

Such that $\qquad -\phi_1 \le f_{x_1}(X) \le \phi_1$

$$-\phi_2 \le f_{x_2}(X) \le \phi_2 \qquad (16.8)$$

$$...$$

$$-\phi_n \le f_{x_n}(X) \le \phi_n$$

$$AX \le B$$

$$0 \le \phi_1, \phi_2, ..., \phi_n \le \phi$$

In this chapter $-\phi_j \le f_{x_j}(X) \le \phi_j$ is called a type 2 partial constraint while $0 \le \phi_1, \phi_2, \ldots, \phi_n \le \phi$ are called capping constraints.

16.4.4 MULTIOBJECTIVE PROGRAMMING

The desired solution must satisfy both LPs resulting in a multiobjective linear problem. In this case there are only two objective linear functions and these can be solved in so many ways. In this paper we suggest and present only two multiobjective optimization approaches.

16.4.4.1 MULTIOBJECTIVE APPROACH 1

Minimize ϕ Minimize $\phi_1 + \phi_2 + \ldots + \phi_n$

Such that

$$-\phi \le f_{x_1}(X) \le \phi \qquad\qquad -\phi_1 \le f_{x_1}(X) \le \phi_1$$
$$-\phi \le f_{x_2}(X) \le \phi \qquad\qquad -\phi_2 \le f_{x_2}(X) \le \phi_2$$
$$-\phi \le f_{x_n}(X) \le \phi \qquad\qquad -\phi_n \le f_{x_n}(X) \le \phi_n$$
$$AX \le B \qquad\qquad\qquad\qquad\quad AX \le B$$
$$\phi \ge 0. \qquad\qquad\qquad\qquad 0 \le \phi_1, \phi_2, \ldots, \phi_n \le \phi$$

Minimize $\ell_1(\phi_1 + \phi_2 + \ldots + \phi_n) + \ell_2\phi$

Such that $-\phi \le f_{x_1}(X) \le \phi$

$-\phi \le f_{x_2}(X) \le \phi$

$\ldots$

$-\phi \le f_{x_n}(X) \le \phi$

$-\phi_1 \le f_{x_2}(X) \le \phi_1$ (16.9)

$-\phi_2 \le f_{x_2}(X) \le \phi_2$

$\ldots$

$-\phi_n \le f_{x_n}(X) \le \phi_n$

$0 \le \phi_1, \phi_2, \ldots, \phi_n \le \phi$

$AX \le B$

Where, ℓ_1 and ℓ_2 are large relative to sizes of other constants. The large constants ℓ_1 and ℓ_2 force ϕ and ϕ_j to assume small values which is what we desire and this solves the problem.

16.4.4.2   MULTIOBJECTIVE APPROACH 2

Approach 2 relies on the strategy of combining the first LP with its dual to create a system of equations whose solution minimizes ϕ. Adding the system of equations to the second LP solves our problem. Suppose the dual of (Eq. 16.7) is given by (Eq. 16.10)

Maximize τY

Such that $RY \geq \kappa$ (16.10)

                 $Y \geq 0$

where, τ, κ are constants.

Then the system of linear equations that minimizes ϕ is given by:

$$\phi = \tau Y$$
$$-\phi \leq f_{x_1}(X) \leq \phi$$
$$-\phi \leq f_{x_2}(X) \leq \phi \qquad (16.11)$$
$$\cdots$$
$$-\phi \leq f_{x_n}(X) \leq \phi$$
$$AX \leq B$$
$$RY \geq \kappa$$

With the system of linear equations that minimizes ϕ available then, the LP that solves our problem is given by (Eq. 16.12).

Minimize $\phi_1 + \phi_2 + \ldots + \phi_n$

Such that $-\phi_1 \leq f_{x_1}(X) \leq \phi_1$

                 $-\phi_2 \leq f_{x_2}(X) \leq \phi_2$

                 $\cdots$

                 $-\phi_n \leq f_{x_n}(X) \leq \phi_n$

                 $\phi = \tau Y$ (16.12)

$$-\phi \le f_{x_1}(X) \le \phi$$
$$-\phi \le f_{x_2}(X) \le \phi$$
$$\ldots$$
$$-\phi \le f_{x_n}(X) \le \phi$$
$$AX \le B$$
$$RY \ge \kappa$$
$$0 \le \phi_1, \phi_2, \ldots, \phi_n \le \phi$$

16.4.5 COMPLEXITY OF TWO LINEAR MODELS

In terms of number of constraints the linear model given in (Eq. 16.9) is better than that given in (Eq. 16.12). The linear model given in (Eq. 16.9) has $(m + 5n)$ constraints and $(2n + 1)$ variables whilst that in (Eq. 16.12) has $(6n + m + 2)$ constraints and $(4n + m + 1)$ variables.

The linear model (16.9) is a combination of (Eq. 16.7), dual of (Eq. 16.7), linear model (16.8) and the single constraint arising from equating objective function of the primal to the objective function of the dual. Common constraints are only captured once.

16.4.5.1 REARRANGING LINEAR MODEL (16.7)

Minimize ϕ

Such that $\left. \begin{array}{l} f_{x_1}(X) + \phi \ge 0 \\ f_{x_2}(X) + \phi \ge 0 \end{array} \right\}$

                $\ldots$ n constraints

                $\left. \begin{array}{l} f_{x_n}(X) + \phi \ge 0 \\ -f_{x_1}(X) + \phi \le 0 \\ -f_{x_2}(X) + \phi \le 0 \end{array} \right\}$ n constraints

                $\ldots$

                $-f_{x_n}(X) + \phi \le 0$

$AX \le B$ m constraints and this is captured once

The primal (model (16.7)) has $(n + n + m = 2n + m)$ constraints and $(n + 1)$ variables.

16.4.5.2   DUAL OF MODEL (16.7)

If the primal has $(2n + m)$ constraints and $(n + 1)$ variables then the dual has $(n + 1)$ constraints and $(2n + m)$ variables. Readers are encouraged to see Taha[15] or Winston[16] for more on duality.

16.4.5.3   EQUALITY CONSTRAINTS FROM OBJECTIVE FUNCTIONS

Primal Dual objective equality $\phi = \tau Y$ 1 constraint

16.4.5.4   REARRANGING LINEAR MODEL (16.8)

Minimize ϕ

Such that $\left. \begin{array}{l} -\phi_1 \leq f_{x_1}(X) \leq \phi_1 \\ -\phi_2 \leq f_{x_2}(X) \leq \phi_2 \end{array} \right\}$ $2n$ constraint

$$\cdots$$

$$-\phi_n \leq f_{x_n}(X) \leq \phi_n$$

$AX \leq B$ m constraints and these are already captured in (16.7)

$$\left. \begin{array}{l} \phi_1 \leq \phi \\ \phi_2 \leq \phi \end{array} \right\}$$ n constraints

$$\cdots$$

$$\phi_n \leq \phi$$

Model (16.8) has $(2n + n = 3n)$ constraints and (n) variables. In this case we exclude the common m constraints and $(n + 1)$ variables already captured from model (16.7). The linear model (16.9) is a combination of (16.7) and (16.8) and common entities are captured once.

16.4.5.5   MODEL (16.9)

Model	Constraints	Variables
Model (16.7)	$2n + m$	$n + 1$
Model (16.8)	$3n$ (excluding m already captured in (Eq. 16.8))	n (excluding X already captured in (Eq. 16.7))
Total	$5n + m$	$2n + 1$

The linear model (16.12) is a combination of (16.7), dual of (16.7), linear model (16.8), and a single constraint arising from equating the objective function of the primal to the objective function of the dual. Like the multiobjective approach 1, the common constraints are also captured once.

16.4.5.6   MODEL (16.12)

	Constraints	Variables
Model (16.7)	$2n + m$	$n + 1$
Dual of Model (16.7)	$n + 1$	$2n + m$
Primal-Dual $\phi = \tau Y$	1	0 (variables already in (Eq. 16.7))
Model (16.8)	$3n$ (excluding m already captured in (Eq. 16.8))	n (excluding X already captured in (Eq. 16.7))
Total	$6n + m + 2$	$4n + m + 1$

Using numbers of constraints and variables as a measure of problem difficulty or complexity we recommend model (16.9). It is better to work with model (16.9) since $6n + m + 2 \geq 5n + m$ and $4n + m + 1 \geq 2n + 1$.

16.4.6   REDUNDANT CONSTRAINTS

The Type 1 partial constraints appearing in both Eq. 16.9 and 16.12 are redundant.

$$-\phi \leq f_{x_1}(X) \leq \phi$$

$$-\phi \leq f_{x_2}(X) \leq \phi$$

$$\ldots$$

$$-\phi \leq f_{x_n}(X) \leq \phi$$

These constraints can be removed without necessarily changing the optimal solution as shown below. The proof is presented using Type 1, Type 2, and the capping constraints.

$$\left.\begin{array}{c} -\phi \le f_{x_1}(X) \le \phi \\ -\phi \le f_{x_2}(X) \le \phi \end{array}\right\}$$

$$\cdots \qquad\qquad\qquad \text{(Type 1 partial constraint)}$$

$$\left.\begin{array}{c} -\phi \le f_{x_n}(X) \le \phi \\ -\phi_1 \le f_{x_1}(X) \le \phi_1 \\ -\phi_2 \le f_{x_2}(X) \le \phi_2 \end{array}\right\}$$

$$\cdots \qquad\qquad\qquad \text{(Type 2 partial constraints)}$$

$$-\phi_n \le f_{x_n}(X) \le \phi_n$$

$$0 \le \phi_1, \phi_2, ..., \phi_n \le \phi \qquad \text{(Capping constraints)}$$

Suppose $X_0 = (x_1^0, x_2^0, ..., x_n^0)$ is feasible to Type 2 partial constraints, that is, $-\phi_j \le f_{x_j}(X_0) \le \phi_j$ is satisfied. Since $\phi_j \le \phi$ then $X_0 = (x_1^0, x_2^0, ..., x_n^0)$ is also feasible to the Type 1 partial constraints, that is $-\phi \le f_{x_j}(X_0) \le \phi$ is also satisfied. Thus the Type 1 partial constraints are not necessary.

Since we have selected (Eq. 16.9) as linear model of choice, discarding Type 1 partial constraints the linear model reduces to (Eq. 16.13).

Minimize $\qquad \phi$

Such that $\qquad -\phi_1 \le f_{x_1}(X) \le \phi_1$

$$-\phi_2 \le f_{x_2}(X) \le \phi_2 \qquad\qquad\qquad (16.13)$$

$$\cdots$$

$$-\phi_n \le f_{x_n}(X) \le \phi_n$$

$$0 \le \phi_1, \phi_2, ..., \phi_n \le \phi$$

$$AX \le B$$

16.4.6.1   COMPLEXITY OF THE REDUCED LINEAR MODEL

Discarding Type 1 partial constraints, reduces the number of constraints in the selected linear model from $(5n + m)$ to $(3n + m)$. There is no change in the number of variables.

16.4.7   LINEAR MODEL

Minimize $\qquad \ell_1\phi + \ell_2(\phi_1 + \phi_2 + ... + \phi_n)$

Such that $\qquad -\phi_1 \le f_{x_1}(X) \le \phi_1$

$$-\phi_2 \le f_{x_2}(X) \le \phi_2 \qquad\qquad (16.15)$$

$$-\phi_n \le f_{x_n}(X) \le \phi_n$$

$$0 \le \phi_1, \phi_2, ..., \phi_n \le \phi$$

$$AX \le B$$

Where, ℓ_1 and ℓ_2 are very large relative to all constants in the problem.

16.5   PROPOSED APPROACH

The steps for the proposed approach are summarized as follows:
    Given any convex QP

Minimize $\qquad f(X)$

Subject to $\qquad AX \le B$

$$X \ge 0$$

Step 1: Linearize QP to

Minimize $\qquad \ell_1\phi + \ell_2 I\Phi$

Subject to $\qquad -\bar{\Phi} \le \nabla f(X) \le \bar{\Phi}$

$$AX \le B$$

$$0 \le \phi_1, \phi_2, ..., \phi_n \le \phi$$

Step 2: Solve using simplex method. The optimal is the approximate solution to the original QP.

16.5.1   NUMERICAL ILLUSTRATION

Minimize $-8x_1 - 16x_2 + x_1^2 + 4x_2^2$

Subject to $x_1 + x_2 \leq 5$

$$x_1 \leq 3 \qquad\qquad (16.16)$$

$$x_1, x_2 \geq 0$$

This numerical example was taken from Jensen and Bard[9].

16.5.2   SOLUTION TO NUMERICAL EXAMPLE

Minimize $1000\phi + 500(\phi_1 + \phi_2)$

Subject to $2x_1 + \phi_1 \geq 8$

$2x_1 - \phi_1 \leq 8$

$8x_2 + \phi_2 \geq 16$

$8x_2 - \phi_2 \leq 16$ $\qquad\qquad (16.17)$

$x_1 + x_2 \leq 5$

$x_1 \leq 3$

$\phi_1 \leq \phi$

$\phi_2 \leq \phi$

$x_1, x_2, \phi, \phi_1, \phi_2 \geq 0$

Solving (16.17) using the simplex method we obtain the optimal solution given in (16.18).

$$x_1 = 3,\ x_2 = 2,\ \phi = 1,\ \phi_1 = 2,\ \phi_2 = 0,\ f(3,2) = -31 \qquad (16.18)$$

It can be noted that this solution is also optimal to the original QP given in (Eq. 16.18). This approach gives solutions that are close to the optimal solution of the original QP.

16.6   CONCLUSION

The linear model proposed in this paper has strengths. The optimal solution to the proposed linear model is close to the optimal solution of the original QP. It is worthwhile to explore the optimal solution given by the proposed linear model before thinking of other approaches. The chances of getting a solution to the QP that is near optimal quickly are high with the proposed approach. The proposed approach gives highly accurate solutions to convex quadratic problems and very good starting solutions to nonconvex quadratic problems. Competitive methods for solving the proposed linear programming model such as interior point algorithms ad simplex methods are available. Both the simplex and interior point methods are widely used and continue to compete with each other.[5]

16.7   SUMMARY

The chapter presents an approximating approach for the NP hard quadratic problem. The approach relies on the minimal change in gradient to locate the optimal solution for the quadratic problem. The proposed approximating approach is high for the convex quadratic problem and can be used to find starting solutions for the nonconvex quadratic problem. The quadratic problem has applications in agriculture.

16.8   ACKNOWLEDGEMENT

The author is grateful to the anonymous referees for constructive comments that improved this chapter.

KEYWORDS

- Convex
- interior point algorithm
- linear programming
- nonconvex
- nonlinear programming
- quadratic programming
- simplex method

REFERENCES

1. Burer, S.; Vandenbussche, D. A Finite Branch-and-Bound Algorithm for Nonconvex Quadratic Programs with Semidefinite Relaxations. *Math. Program.* Series A, **2008**, *113* (2), 259–282.

2. Burer, S.; Vandenbussche, D. Globally Solving Box-Constrained Nonconvex Quadratic Programs with Semidefinite-Based Finite Branch-and-Bound. *Comput. Optim. Appl.* **2009**, *43*(2), 181–195.

3. Cavique, L.; Luz, C, J. 'A Heuristic for the Stability Number of Graph Based on Convex Quadratic Programming and Tabu Search'. *J. Math. Sci.* **2009**, *161*(6), 944–955.

4. Freund, R. M. 'Solution Methods for Quadratic Optimization: Lecture Notes, Massachussets Institute of Technology, 2002.

5. Gondzio, J. Interior Point Methods 25 Years Later. *Eur. J. Oper. Res.* **2012**, *218*(3), 587–601.

6. Gupta, O. K. Applications of Quadratic Programming. *J. Inf. Optim. Sci.* **1995**, *16*(1), 177–194.

7. Harry, H. H; Heady, E. O.; Plessner, Y. Quadratic Programming Solution of Competitive Equilibrium for U.S. Agriculture. *Am. J. Agric. Econ.* **1965**, *50*(3), 536–555.

8. Horst, R.; Pardalos, P. M., Thoai, N. V. Introduction to Global Optimization: Nonconvex Optimization and its Applications, 2nd ed.; Kluwer Academic Publishers: Dordrecht, 2000.

9. Jensen, P. A.; Bard, J. F. *Operations Research Models and Methods;* John Wiley & Sons, Inc., New York, 2012, p. 700.

10. Li, H. C.; Wang, Y. P. An Evolutionary Algorithm with Local Search for Convex Quadratic Bilevel Programming Problems. *Appl. Math. Inf. Sci.* **2011**, *5*(2), 139–146.

11. Liu, S. T.; Wang, R. T. A Numerical Solution Methods to Interval Quadratic Programming. *Appl. Math. Comput.* **2007**, *189*(2), 1274–1281.

12. McCarl, B. A.; Moskowitz, H.; Furtan, H. Quadratic Programming Applications. *Omega* **1977**, *5*(1), 43–55.

13. Moré, J. J.; Toraldo, G. Algorithms for Bound Constrained Quadratic Programming Problems. *Numerische Mathematik* **1989**, *55*(4), 377–400.

14. Munapo, E. Minimizing Complementary Pivots in a Simplex-Based Solution Method for a Quadratic Programming Problem. *Am. J. Oper. Res.* **2012**, *2*, 308–312.

15. Taha, H. A. *Operations Research: An Introduction.* 10th ed.; Pearson, New York; 2017, p. 848.

16. Winston, W. L. *Operations Research: Applications and Algorithms.* 4th ed.; Thomson Brooks/Cole: Belmont, CA; 2003, p. 1440.

PART V

Performance of Farm Machines for Sustainable Agriculture

CHAPTER 17

CITRUS FRUIT HARVESTER

MRUDULATA DESHMUKH[1,*] AND S. K. THAKARE[2]

[1]*Department of Farm Power and Machinery, Dr. Panjabrao Deshmukh Krishi Vidyapeeth (PRDKV), Akola (MS), 444104, India*

[2]*Department of Farm Power and Machinery, Dr. Panjabrao Deshmukh Krishi Vidyapeeth, Akola (MS), 444104, India E-mail: skthakare@gmail.com*

*Corresponding author. E-mail: mrudulatad@rediffmail.com

CONTENTS

17.1   INTRODUCTION

Citrus fruits occupy an important position in India's fruit production. Citrus fruits which include sweet oranges, lemon, mandarin, lime, and so forth are primarily consumed as fresh fruits.[1,2] These fruits are processed to prepare squash, juice, marmalade and pickles. These fruits are rich source of vitamin-C and mineral salts. Mandarin orange is grown on large scale in Nagpur and Amravati districts of Vidarbha region of Maharashtra State. Area under orange crop in Vidarbha region is about 65,000 ha with annual production of about 0.425 million tons.[4,5,6] Citrus rank among top three fruits of the world with respect to area and production. In India, common citrus fruits are mandarins, sweet orange, and lime having 50, 20 and 15% of total area (3.48×10^5 ha), respectively. However these occupy only 10% of total area under all fruit crops in India with third rank after Mango and Banana. Oranges are mainly grown in Vidarbha region of the Maharashtra State. The total area under oranges is 1.48×10^5 ha with 0.93×10^5 ha as a productive area. The total production is about 8.34×10^5 M.T.[3]

The harvesting and pruning of lemon is a difficult operation due to the tallness and thorns on the branches. There are small hand tools available for harvesting and pruning. But these tools of harvesting and pruning are restricted due to tree height, unavailability of trained manual power for climbing and cost of operation, and so forth. The mechanized machines are available, but these are heavy and costly and are not suitable for small land holdings and Indian marginal famers. Harvesting and pruning of horticultural crops with the available hand tool is very difficult. The labor has to climb on the tree by carrying these hand tools, which requires skills too. The orange harvesting is generally done by manual picking. But due to the manual picking sometimes, the point of the orange where it is detached from the peduncle get opened and the fungal infection starts from that point which reduces the shelf life of the orange.

To overcome these problems of harvesting of lemon and orange, efforts are being made to develop a manual citrus fruit harvester for lemon and oranges so that the harvesting becomes easy. The details on the development of citrus harvesters are presented in this chapter.

17.2 DEVELOPMENT OF FRUIT HARVESTER

At maturity, lemon fruits fall from the tree on the ground. Due to the thorns on the tree, while falling, the fruit gets damaged and hence the quality of fruit is deteriorates resulting in the reduced shelf life. Nowadays, in the country agricultural labor is shifting toward nonagricultural jobs. Thus, the nonavailability of labor must be bridged by mechanization. During the year 2011–2012, the research study was undertaken in the Department of

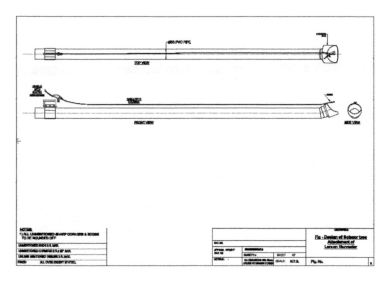

FIGURE 17.1   Views of lemon harvester.

Farm Power and Machinery, Dr. Panjabrao Deshmukh Krishi Vidyapeeth, Akola. Two harvesters were developed separately for harvesting lemon and orange (Fig. 17.1).

17.2.1   LEMON HARVESTER

It consists of main body, cutting mechanism, and collecting device. A PVC reducer of size 90 × 50 mm diameter is attached to the pipe with the inclination and is used as a main body (Fig. 17.1). A cut is given to the reducer to avoid falling of fruit on the ground. A scissor is fixed at upper side of the reducer at an angle of 62°. It consists of a pair of arm hinges at a central point providing sufficient space for cutting edge and handling portion, the scissor is operated by a clutch gear through a cable directly attached to the main pipe of diameter 5 cm and length 210 cm. A PVC pipe of 5 cm diameter and 210 cm length is provided for passing the lemon collected in the main body to the net. The lemons are finally collected in the net provided at the end of the PVC pipe. The weight of the harvester is only 1.45 kg, hence it is easy to transport and women workers can easily operate it (Fig. 17.2).

FIGURE 17.2   Overall view and harvesting with the lemon harvester.

17.2.2 ORANGE HARVESTER

All the parts of the scissor type orange fruit harvester (Fig. 17.3) are the same as that of lemon harvester, the only difference is the diameter of the collecting pipe is 9 cm and the length is 90 cm.

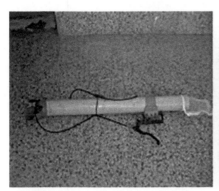

FIGURE 17.3   Harvesting of oranges with the orange harvester.

17.2.3 FIELD TRIALS

Trials of lemon harvesting were undertaken at Citrus Dieback, Central Research Station, Dr. Panjabrao Deshmukh Krishi Vidyapeeth (PDKV), Akola, and on farmer's field at Bordi, Tq. Akot, Dist Akola. Four subjects were selected for the trials and the trial was replicated thrice. The age of the subjects selected ranged from 33–39 years, and weight of each person ranged from 55–67 kg with the height of 125–158 cm. The temperature during the harvesting of lemon was 40°C and relative humidity was 20%.

17.3   RESULTS AND DISCUSSIONS

17.3.1   LEMON HARVESTER

The data on evaluation of the lemon harvester performance are given in Table 17.1. The harvesting operation was then categorized for work load on the basis of the heart rate data. The psychophysical response of subjects (Overall Discomfort Rating (ODR) and Body Parts Discomfort Scores

TABLE 17.1 Performance of the Lemon Harvester.

Subject	Replication	Heart rate (beats/min)			Energy use (kJ/min)	Output (kg/h)	Category of work load
		Resting	Working	Work pulse (ΔHR)			
Subject I	R1	67	107	40	8.293	14.8	Moderate
	R2	69	111	42	8.929	15.7	Category of work load
	R3	73	113	40	9.247	15.8	
Subject II	R1	71	110	39	8.770	15.0	
	R2	72	114	42	9.406	15.7	
	R3	74	116	42	9.724	15.9	
Subject III	R1	69	112	43	9.088	15.1	
	R2	71	111	40	8.929	15.6	
	R3	73	115	42	9.565	15.9	
Subject IV	R1	72	113	41	9.247	14.9	
	R2	74	117	42	9.883	15.6	
	R3	72	114	43	9.406	15.8	

TABLE 17.2   Psychophysical Response of the Subject During Lemon Harvesting.

Name of Subject	Replications	Overall discomfort rating, ODR		Body part discomfort score (BPDS)	
		Value	Mean	Value	Mean
Subject I	R1	4.6	4.7	21.8	23.7
	R2	4.8		24.3	
	R3	4.7		25.1	
Subject II	R1	4.9	4.9	22.7	23.4
	R2	5.2		24.5	
	R3	4.7		23.1	
Subject III	R1	4.8	4.9	24.3	24.5
	R2	5.1		25.9	
	R3	4.7		23.3	
Subject IV	R1	4.7	4.9	23.7	24.7
	R2	5.1		25.7	
	R3	5.0		24.8	

(BPDS)) was also recorded as shown in Table 17.2. The average ODR and BPDS scores were 4.8 and 24.1, respectively. Average work pulse was 41.3 beats/min (Fig. 17.4) and average energy expenditure was 9.207 kJ/min. It

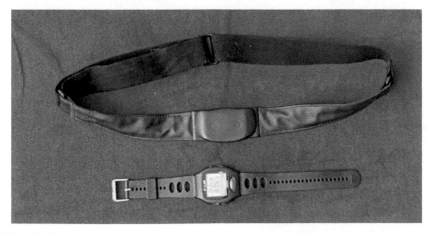

FIGURE 17.4   Polar heart rate monitor.

was concluded that the operation comes under the moderate category of work load. The average output of the lemon harvester is 15.6 kg/h.

17.3.2   ORANGE HARVESTER

The performance of the orange harvester is shown in Table 17.3. Average work pulse was 37 beats/min and average energy expenditure was 8.518 kJ/min. The average output of the orange harvester was 59 kg/h. The harvesting operation was then categorized for work load on the basis of the heart rate data and it was observed that the operation comes under the moderate category of work load (Fig. 17.5). By using this harvester, a short peduncle was left on the orange, which enhances the storage life of orange and also reduces the drudgery involved in the citrus harvesting.

The psycho-physical response is given in Table 17.4. The average ODR and BPDS scores were 4.05 and 19.7, respectively. On the basis of the heart rate data, lemon as well as orange harvesting by manual citrus harvester comes under the "moderate" category of work load.[7]

17.3.3   CATEGORIZATION OF WORK LOAD ON THE BASIS OF WORK PULSE

Very light	< 90 bpm
Light	91–105 bpm
Moderate heavy	106–120 bpm
Heavy	121–135 bpm
Very heavy	136–150 bpm
Extremely heavy	151–170 bpm

17.4   CONCLUSION

- The average output of the lemon harvester is 15.6 kg/h against the manual plucking output of 20 kg/h. The output of the orange harvester is 59 kg/h against the manual plucking output of 65 kg/h.
- Average work pulses were 41.3 beats/min. and average energy expenditure was 9.207 kJ/min.

TABLE 17.3 Performance of the Orange Harvester.

| Name of subject | Replication | Heart rate beats/min, bpm | | | Energy use kJ/min | Output kg/h | Category of work load |
		Resting	Working	Pulse			
Subject I	R1	69	105	36	7.975	57.5	Moderate category of workload
	R2	71	109	38	8.611	58.6	
	R3	73	111	38	8.929	59.3	
Subject II	R1	72	110	38	8.770	58.0	
	R2	74	111	37	8.929	59.5	
	R3	72	109	37	8.611	59.3	
Subject III	R1	67	104	37	7.816	57.7	
	R2	69	105	36	7.975	58.7	
	R3	73	110	37	8.770	59.2	
Subject IV	R1	71	108	37	8.452	57.9	
	R2	72	108	36	8.452	58.9	
	R3	74	111	37	8.929	59.3	

A   **B**

FIGURE 17.5 Psychophysical response: left (A)—subject showing score for overall discomfort; right (B)—subject showing regions for body part discomfort.

TABLE 17. 4 Psychophysical Response of the Subject During Orange Harvesting.

Name of Subject	Replications	ODR rating (ODR)		BPDS score (BPDS)	
		Value	Mean	Value	Mean
Subject I	R1	4.2	4.1	20.3	19.7
	R2	4.1		19.8	
	R3	4.0		18.9	
Subject II	R1	4.1	4.0	20.1	19.5
	R2	4.0		19.6	
	R3	3.9		18.9	
Subject III	R1	4.2	4.0	20.2	20.0
	R2	3.9		20.1	
	R3	3.9		19.8	
Subject IV	R1	4.1	4.1	19.9	19.6
	R2	4.2		19.7	
	R3	4.0		19.4	

- The average ODR and body part discomfort scores (BPDS) for lemon harvesting were 4.8 and 24.1, respectively.
- Average work pulse and energy expenditure were 37 beats/min. and 8.518 kJ/min, respectively.
- The overall discomfort rating and body part discomfort scores for the orange harvesting were 4.05 and 19.7, respectively.

- On the basis of the heart rate data, lemon as well as orange harvesting by manual citrus harvester comes under the "moderate" category of work load.

17.5 SUMMARY

At maturity, lemon fruits fall from the tree on the ground. Due to the thorns on the tree, while falling, the fruit gets damaged and hence the quality of fruit is deteriorates resulting in the reduced shelf life. Nowadays, in the country agricultural labor is shifting towards nonagricultural jobs. Thus, the nonavailability of labor is bridged by mechanization. The orange harvesting is generally done by manual picking. Sometimes by manual picking, the point of the orange where it is detached from the peduncle gets opened and the fungal infection starts from that point which reduces the shelf life of the orange. To overcome the above problems of harvesting of lemon and orange, efforts were made to develop a manual citrus fruit harvester for lemon and oranges. On the basis of the heart rate data, lemon as well as orange harvesting by manual citrus harvester comes under the "moderate" category of work load, hence it is woman friendly and will also reduce the drudery and time in harvesting of both lemon and oranges.

KEYWORDS

- **Discomfort**
- **energy**
- **harvester**
- **heart rate**
- **lemon**
- **orange**
- **oxygen consumption rate**
- **psychophysical response, resting heart rate**

REFERENCES

1. Chengappa, P. G.; Nagaraj, N. *Marketing of Major Fruits and Vegetables in and Around Bangalore;* Unpublished Report 2004–2005 by Department of Agricultural Economics; University of Agricultural Sciences: Bangalore, 2005; p 52.

2. Corlett, E. N.; Bishop, R. P. Technique for Assessing Postural Discomfort. *Ergonomics* **2006**, *19,* 175–182.
3. Indian Horticulture. Database, Government of India, 2010.
4. Indian Institute of Horticulture Research, http://www.iihr.res.in.
5. Kolhe, K. P.; Jadhav, B. B. Testing and Performance Evaluation of Tractor Mounted Hydraulic Elevator for Mango Orchard. *Am. J. Eng. Appl. Sci.* **2011**, *4*(1), 179–186.
6. Kumar, P.; Kumar, P. Demand, Supply and Trade Perspective of Vegetables and Fruits in India. *Indian J. Agric. Mark.* **2003**, *17*(3), 121–130.
7. Varghese, M. A.; Saha, P. N.; Atreya, N. A Rapid Appraisal of Occupational Work Load from a Modified Scale of Perceived Exertion. *Ergonomics* **1996**, *37,* 485–491.

CHAPTER 18

ROTARY HARVESTING MECHANISM: SORGHUM

MRUDULATA DESHMUKH[1,*]AND S. K. THAKARE[2]

[1]*Department of Farm Power and Machinery, Dr. Panjabrao Deshmukh Krishi Vidyapeeth (PRDKV), Akola (MS), 444104, India*

[2]*Department of Farm Power and Machinery, Dr. Panjabrao Deshmukh Krishi Vidyapeeth, Akola (MS), 444104, India*
E-mail: skthakare@gmail.com

[*]Corresponding author. E-mail: mrudulatad@rediffmail.com

CONTENTS

18. 1   INTRODUCTION

The effective mechanization contributes to increase the production in two major ways: first, the timeliness of operation with good quality of work and second, reduction of drudgery in operations. At present, the harvesting of sorghum is done manually with the help of sickle which involves slicing and tearing actions that result in plant structure failure due to compression, tension or shear. The total harvesting of sorghum requires two stages for cutting the plant, one at the top for separating cobs and second at the bottom for fodder which requires double labor for harvesting of sorghum.[7,10] About 25% of the total labor for grain production is required for harvesting operation alone.[1,2,6,8] Due to urbanization, there has been a continuous decline in the number of people involved in the agricultural production resulting in shortage of labor during the peak harvesting season. This condition restrains the farmer to maximize the productivity of his land and at the same time subjects his crop to losses due to untimely harvesting. This situation necessitates the introduction of a system of operation that requires minimal turnaround time in crop production in order to catch up with the next planting season and will improve harvesting operation.

This chapter presents the development of a rotary cutting mechanism for sorghum.[3,4,5,9,11]

18.2   DEVELOPMENT OF THE CUTTING MECHANISM FOR SORGHUM

The fabrication of the set up was carried out in the workshop of the department of Farm Power and Machinery, Dr. Panjabrao Deshmukh Krishi Vidyapeeth (PDKV), Akola. Initially, the main frame is fabricated and the secondary frame is mounted on the main frame. The secondary frame is fabricated to facilitate the mounting of electric motor. The main shaft made of EN 19 material is centrally mounted on the frame passing through the secondary frame. The main shaft of the mechanism will be mounted on the frame and supported with the bearings. It carries two rotating discs, one at the top for cutting cob peduncles and another at the lower side for cutting stalks. Two bearings are mounted above the rotating disc to compensate the truss developing at the shaft. In between the bearing a secondary frame attached to the main frame. Variable frequency drive (VFD) is mounted at one side on main frame. VFD was used to vary the speed of rotating disc. A torque sensor was aligned on the main shaft in between the two bearings for the measurement of cutting torque. The pulley was mounted on the shaft above the second bearing which was connected to the electric motor pulley with V belt. Two MS pipes were used to fabricate stalk

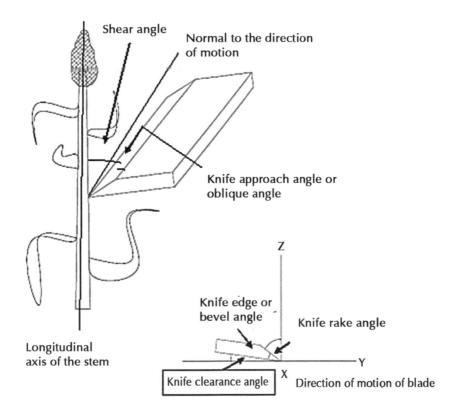

FIGURE 18.1 Terminologies of blade angles.

holder to hold stalk/cob peduncles during cutting trials. Terminologies of blade angles are shown in Figure 18.1. Various components of the rotary cutter are described as follows:

18.2.1 FRAME

The main frame made of MS angles having dimensions 1000 × 600 mm was fabricated to mount all the components of the cutting mechanism. One secondary frame is mounted on the main frame having dimensions 400 × 600 × 400 mm at a distance of 300 mm from each of the sides. The electric motor is mounted on the secondary frame.

18.2.2 MAIN SHAFT

The main shaft made of EN 19 material and size Ø55 × 458 mm length was centrally mounted on the frame passing through the secondary frame. The main shaft of the mechanism is mounted on the frame and supported with the bearings. It will carry two rotating discs, one at the top for cutting cob peduncles and another at the lower side for cutting stalks. The motion to the shaft will be provided from an electric motor with the help of belt and pulley arrangement.

18.2.3 ROTARY DISC

The disc made of mild steel having specifications Ø500 × 10 mm thick was mounted on lower (for cutting stalks) and upper end (for cutting cob peduncles) of the main shaft. The arrangement was made on the rotary disc for mounting of cutting blades at various angles. The disc rotates in the clockwise direction (Fig. 18.2).

18.2.4 CUTTING BLADES

Two types of cutting blades, that is straight plain blade and semicircular blade, were selected for the study (Fig. 18.3). The dimensions of straight plain blade were 230 × 60 × 10 mm while that of semicircular blade were Ø100 × 8 mm thick (semicircle). Both the blades are made of EN 19 material.

18.2.5 ANGLE SIM

The angle sims made of MS were fabricated for facilitating the angles of the blade. The terminologies of different blade angles are defined in the Figure 18.3.

18.2.6 BEARING

Two taper roller bearing of size Ø 40 ID with the bearing housing (size Ø 140 × 46 mm) and bearing cap (size Ø 140 × 13 mm thick) below and above the torque sensor were mounted on the shaft.

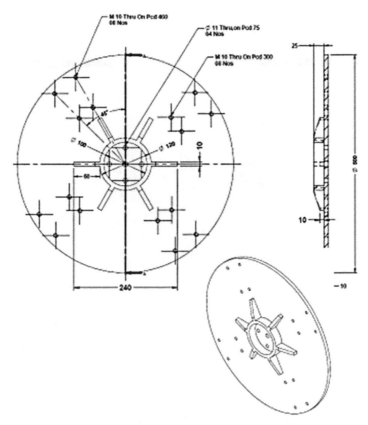

FIGURE 18.2 Sectional view of rotary cutting disc.

18.2.7   TORQUE SENSOR

Futek Sensit torque sensor (model-TRS-300) was aligned on the main shaft in between the two bearings. The torque sensor was used for measuring the cutting torque of stalks and cob peduncles at various combinations of blade angle and speed of disc.

18.2.8   PULLEY

A pulley having dimensions Ø200 and 40 mm thick was mounted on the shaft above the torque sensor. The power was transmitted from electric

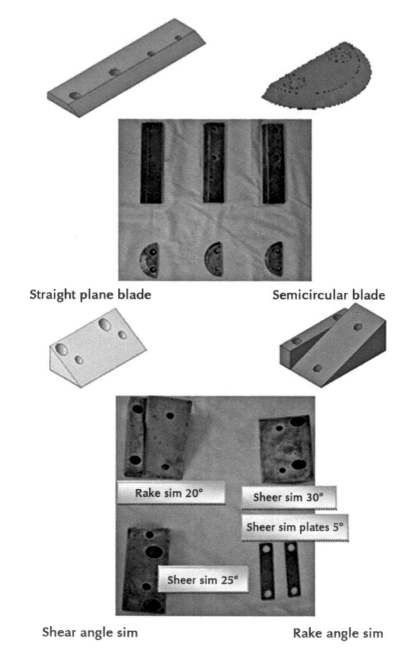

Straight plane blade   **Semicircular blade**

Rake sim 20°   Sheer sim 30°

Sheer sim plates 5°

Sheer sim 25°

Shear angle sim   **Rake angle sim**

FIGURE 18.3   Types of blades and angle sims.

motor to the shaft with the help of belt and pulley arrangement. The rpm of the shaft in the electric motor was reduced to 650 from 1440 rpm.

18.2.9   ELECTRIC MOTOR

MGM Varvel 1.5 Kw, 1440 rpm electric motor with Ø 100 and 40 mm thick pulley is used for power transmission. Electric motor was mounted on the secondary frame to facilitate the power transmission to the shaft for rotation of disc.

18.2.10   VARIABLE FREQUENCY DRIVE (VFD)

Delta make, AC drive (model VFD 015M21A) was used for speed varia-tion. The unit was mounted on the frame on one side.

18.2.11   STALK HOLDER

Two MS pipes (Ø 30 length 100 mm) were used to fabricate stalk holder to hold stalk/cob peduncles during cutting trials. These holders were welded with the circular plate and the circular plates were mounted on the front side of the frame. The two holders were spaced 20 cm apart.

18.2.12   CUTTING BLADES

Two sets of the cutting blades with different combinations of shear and rake angle sims (Figs. 18.3–18.6) are mounted on the rotating disc at the opposite side to counterbalance the vibrations developed on the disc. The stalks and cob peduncles of the sorghum varieties were held in the stalk holder and the disc was rotated by gradually increasing the speed up to 350 rpm. The observations of cutting torque in the set up were noted at 350 rpm. Then the speed was further increased up to 500 rpm and finally to 650 rpm. The cutting torque observations were then converted into specific torque and finally into force by considering the length of blade assembly.

Front view and isometric view of rotary cutting mechanism are shown in Figures 18.7 and 18.8, respectively.

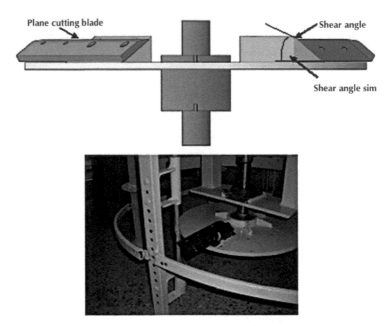

FIGURE 18.4   Shear angle (top) and image of shear angle arrangement on disc (bottom).

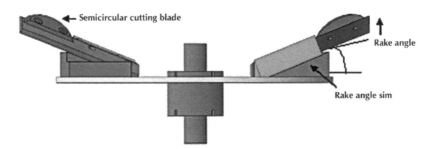

FIGURE 18.5   Positive rake angle.

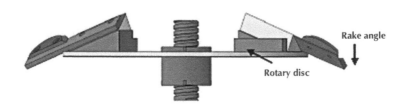

FIGURE 18.6   Negative rake angle.

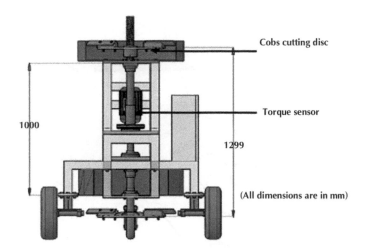

FIGURE 18.7 Front view of rotary cutting mechanism.

18.3 EVALUATION AND TESTING

Newly developed rotary sorghum harvesting mechanism was evaluated in the field for cutting stalks as well as cobs. The trials were conducted on three sorghum varieties viz. CSV-20, CSV-23 and CSH-9. The stalk-cutting efficiency with plane blade was observed maximum (92.98%) for the variety CSH-9 and minimum (90.35%) for the variety CSV-23, whereas the maximum and minimum values of cutting efficiency of cob peduncles with plane blade were 89.64 and 87.52% for the varieties CSH-9 and CSV-23, respectively.

The stalk-cutting efficiency with semicircular blade was maximum (93.80%) for the variety CSH-9 and minimum (91.33%) for the variety CSV-23, whereas the maximum and minimum values of cutting efficiency of cob peduncles with semicircular blade were 90.51 and 88.91% for the varieties CSH-9 and CSV-23, respectively.

18.4 SUMMARY

Sorghum is an important crop in the Vidarbha region of Maharashtra state. Harvesting of sorghum includes two stage cutting (reaping and nipping) hence requires double labor. The operation is time consuming and involves drudgery also. Mechanized harvesting of sorghum is a need of a day, which will reduce the drudgery and save labor input and time.

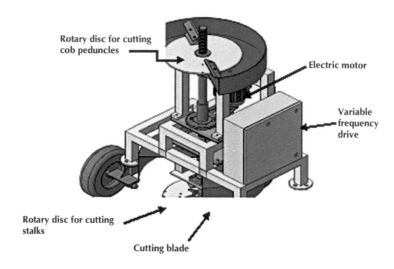

Rotary disc for cutting
cob peduncles

Electric motor

Variable
frequency
drive

Rotary disc for cutting
stalks

Cutting blade

FIGURE 18.8   Isometric view of cutting mechanism.

Hence, the present study was aimed to develop rotary cutting mechanism for harvesting sorghum. The machine consisted of different components such as: frame, central shaft, electric motor, bearings, two rotary discs, two types of blades, torque sensor, variable frequency drive, belt and pulley, plant holder and so forth. The developed cutting mechanism was tested in the field and it was observed that the said mechanism gave better performance for harvesting sorghum for both reaping and nipping operations. Hence, it can be commercialized for reducing drudgery in harvesting operation of sorghum and to overcome the labor problem in general.

KEYWORDS

- Bevel angle
- cob
- drudgery
- nipping
- peduncle
- planeblade
- rake angle

- rotary mechanism
- semicircular blade
- shear angle
- sorghum
- stalk
- stalk holder
- torque sensor

REFERENCES

1. Berge, O. I. Design and Performance Characteristics of the Flywheel Type Forage Harvester Cutter head. *Agric. Eng.* **1951,** *32*(2), 85–91.
2. Ghahraei, O.; Khoshtaghaza, M. H.; Ahmad, D. B. Design and Development of Special Cutting System for Sweet Sorghum Harvester. *J. Cent. Eur. Agric.* **2008,** *9*(3), 469–474.
3. Ghahraei, O. D.; Ahmad, A.; Khalina, H.; Othman, J. Cutting Tests of Kenaf Stems. *Trans. ASABE* **2011,** *54*(1), 51–56.
4. Kemble, L. J.; Krishnan, P.; Henning, K. J.; Tilmon, H. D. Development and Evaluation of Kenaf Harvesting Technology. *Biosyst. Eng.* **2002,** *81*(1), 49–56.
5. Lungkapin, J.; Salokhe, V. M.; Kalsirisilp, R.; Nakashima, H. Development of a Stem Cutting Unit for a Cassava Planter. *Agric. Eng. Int. CIGR E-J.* **2007,** *IX,* 1–16.
6. Metianu, A. A.; Johnson, I. M.; Sewell, A. J. A Whole Crop Harvester for the Developing World. *J. Agric. Eng. Res.* **1990,** *47,* 187–195.
7. Monroe, G. E.; Sumner, H. R. A Harvesting and Handling System for Sweet Sorghum. *Trans. ASAE* **1985,** *28,* 562–567.
8. Moontree, T.; Rittidech, S.; Bubphachot, B. Development of the Sugarcane Harvester Using a Small Engine in Northeast Thailand. *Int. J. Phys. Sci.* **2012,** *7*(44), 5910–5917.
9. Persson, S. Development of Rotary Countershear mower. *Trans. ASABE* **1993,** *36*(6), 1517–1523.
10. Rains, G. C.; Cundiff, J. S. Design and Field Testing of Whole Stalk Sweet Sorghum Harvester. *Appl. Eng. Agric.* **1993,** *9*(1), 15–20.
11. Sidahmed, M. M.; Jaber N. S. The Design and Testing of a Cutter and Feeder Mechanism for the Mechanical Harvesting of Lentils. *Biosyst. Eng.* **2004,** *88*(3), 295–304.

CHAPTER 19

PERFORMANCE OF STATIONARY POWER CUTTER: SUGARCANE SETS AND SORGHUM

S. K. THAKARE AND MRUDULATA DESHMUKH

Department of Farm Power and Machinery, Dr. Panjabrao Deshmukh Krishi Vidyapeeth, Akola (MS), 444104, India
E-mail: skthakare@gmail.com

Corresponding author. E-mail: mrudulatad@rediffmail.com

CONTENTS

19.1   INTRODUCTION

Sugarcane (*Saccharum Officinarum L.*) is the main source of sugar in Asia and Europe. It is grown primarily in the tropical and sub-tropical zones of the southern hemisphere. It is also used for chewing and extraction of juice for beverage purpose. The sugarcane cultivation and sugar industry in India plays a vital role towards socioeconomic development in the rural areas by mobilizing rural resources and generating higher income and employment opportunities. About 7.5% of the rural population, covering about 45 million sugarcane farmers, their dependents and a large number of agricultural laborers are involved in sugarcane cultivation, harvesting and ancillary activities. There are about nine states in India, where sugarcane is grown on a large extent of area with a number of varieties in India depending on the suitability of the soil (Table 19.1). India is the fourth largest exporter of sugar in the world. India is an occasional importer of sugar too, depending upon the demand and supply situation at home. During the last 10 years, India has been a net exporter of sugar.[6,7,8]

In agriculture sector, sugarcane shared is about 7% of the total value of agriculture output and occupied about 2.6% of India's gross cropped area during 2006–2007. Sugarcane provides raw material for the second largest agro-based industry after textile. About 527 working sugar factories with total installed annual sugar production capacity of about 24.2 million tons were located in the country during 2010–2011.[17]

Maharashtra and the adjoining area of Karnataka, Gujarat and Andhra Pradesh record higher sugar recoveries. Long hours of sunshine, cool nights with clear sky, and the latitudinal position of this area are highly favorable for sugar accumulation. Average recoveries of Maharashtra and Gujarat are highest in the country.

In Maharashtra, Sholapur district has highest sugar contribution of 18.62%, whereas Hingoli district has lowest percentage (1.20%) of sugar contribution (Fig. 19.1). In Vidarbha, there is higher demand of sugarcane varieties, such as: Coc-671, Co-99004, Co-85004, Co-86032, and so forth.

19.2   TYPES OF FORAGE

There are two types of forage such as sorghum (cereal forage and herbaceous (forage) mostly used for cattle feed.

TABLE 19.1   Prominent Sugar Producing States in India.

S. No	State	Sugarcane area (× 1000 ha.)	Avg. yield (Ton/ha.)	Cane crushed (105 tons)	Sugar production (× 1000 tons)	Sugar recovery (%)
A. Tropical region						
1	Maharashtra	964	81.80	802.23	9054	11.26
2	Tamilnadu	336	102.00	203.10	1846	9.09
3	Karnataka	421	89.30	337.65	3683	10.91
4	Andhra Pradesh	192	77.00	103.17	385	9.30
5	Gujarat	188	75.70	123.59	1235	9.99
B. Subtropical region						
6	U.P	2101	56.34	643.81	5887	9.14
7	Haryana	85	70.40	43.46	392	9.01
8	Punjab	70	59.60	34.33	302	8.80
9	Bihar	300	50.00	14.60	385	9.30
Total, A + B =		**4944**	**68.06**	**2398.07**	**24394**	**10.17**

19.3   USE OF FORAGE

Sorghum stalks are used primarily as silage for livestock. It is sometimes grown and harvested with soybean to improve the protein content of the silage.

Sorghum usually produces as much silage per acre as corn. The protein content of sorghum silage is similar to or slightly higher than that of corn. It is less digestible and consumption of sorghum silage by animal is also less than that of corn. Sorghum silage must be supplemented with protein, minerals, and vitamins. It is generally suggested that sorghum silage constitute not more than 50% of the forage in dairy cow feed but may be adequate alone for other categories of animals.[2] Forage legumes are mostly used for livestock feed or an essential component of most cattle and sheep feed. The farmers having 2–3 cattle can feed their animals by cutting whole stem into 2–3 pieces, which is neither recommended nor can be used efficiently.[3,4,10,15]

Presently, the farmers are using sickle or axe and manually operated wheel type forage cutter to cut the forage stalk into pieces, which is time consuming and require more human energy.[5,9] Taking above facts into consideration,

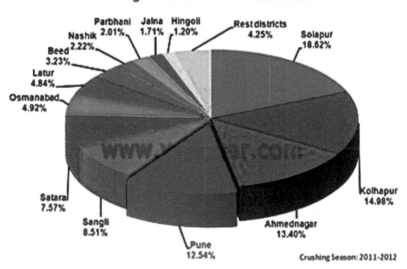

FIGURE 19.1   District wise sugar contribution percentage in total sugar production of Maharashtra.

Source: http://www.vsisugar.com/india/statistics/maharashtra_statistics.htm.

there exists a need to have a suitable power operated forage cutter.[11,12,13,14] Thought was given to develop the machine which can even be operated by one person. It should be portable and lower in cost. Pedal operated forage cutter was developed in the department of FMP at Dr. PDKV, Akola. The capacity of pedal operated forage cutter was 16–60 kg/h for making forage of dry sorghum and green maize [1] as well as around 600 sugarcane sets can be prepared by the same machine. The operation of the machine includes drudgery as the operator cannot work continuously for long time.

Thus, in view to reduce drudgery and to increase the capacity of the said machine[16,18,19] so as to be utilized by the farmers having land holding up to 5–10 acres, a power operated sugarcane set cum forage cutter was developed and evaluated to improve the working as well as capacity of newly developed sugarcane set cum forage cutter.

19.4   DESCRIPTION OF AGRICULTURAL MACHINE

A power operated mechanism was selected for operating the cutter. Specifications of the machine are given in Table 19.2. The Panjabrao Krishi Vidyapeeth (PKV) power cutter includes the following major parts namely (Fig. 19.4): stand, cutting blade, pulley, motor, belt, and foundation for pulleys and flywheel.

19.4.1   STAND

The stand to support the frame was made up from four pieces of mild steel angle of size 50 × 50 × 2 mm of 720 mm in length. The pieces were joined to the frame in such a way so that the distance between the two angles was 770 mm at the bottom for providing strong base to the cutting mechanism and 620 mm at top.

19.4.2   CUTTING BLADE

The blades were made from carbon steel material 2 mm in thickness half round in shape fixed to the handle with the help of two nuts and bolts.

TABLE 19.2 Specifications of Modified Power Operated Sugarcane Set cum Forage Cutter.

Particulars	Quantity	Materials	Size/Dimension
Belt	2	62 A	1574.8 mm
Blade	1	Carbon steel	250 × 80 × 2
Bush/bearing	2	4	P205
Eccentric rod	1	M.S.	630 × 1.5
Frame	1	M.S.	620 × 310 × 5
Handle	1	M.S.	1070 × 50 × 10
Large pulley	2	M.S.	322.5 and 355.5 mm
Motor	1		(1 hp) 1440 rpm
Shearing plate	1	M.S.	270 × 27 × 5
Small pulley	2	M.S.	55 mm
Spring	1	Steel	200
Stand	1	M.S.	770 × 620 × 72
Wheel	1	M.S.	410 (Dia.) × 2 (Thickness)
Miscellaneous	–	–	–

19.4.3 PULLEY

The larger and smaller pulleys were made of size 355.6 and 55 mm diameter, respectively. It was mounted on the shaft for transmission of power from motor to flywheel.

19.4.4 MOTOR

The single-phase 1 hp motor having 1440 rpm was used. It was placed at the bottom of the machine and used to transmit power to blade with the help of belt and pulley arrangement.

19.4.5 BELT

It was used for transmission of power from motor to pulley. The size of both belts was 62 A grade and length of belt was 1574.8 mm.

19.4.6 FOUNDATION FOR PULLEYS AND FLYWHEEL

The base support was made up of mild steel angle. The size of angle was 50 × 50 × 2 mm and 425 mm length and spacing between two angles was 160 mm

19.4.7 POWER TRAIN FOR REDUCING REVOLUTIONS (RPM) OF MOTOR

The single phase motor of 1 hp (0.746 kW) having rpm 1440 was selected/ tested to get required number of stroke of blade per minute.

19.4.8 POWER TRANSMISSION SYSTEM

A motor was provided at the base of the machine (Fig. 19.2). A power train was provided, which transmitted the power to the larger pulley with the help of a belt. The flywheel and pulley was placed on the same shaft which in turn rotated the flywheel. The flywheel transmits power to the handle through a rod which provides power to the blade.

19.4.9 CUTTING BLADE

The blades were made up from carbon steel material 2 mm in thickness half round in shape fixed to the handle with the help of two nuts and bolts.

19.4.10 CUTTING SPEED

The cutting speed of the modified power operated sugarcane set cum forage cutter was selected as 32, 34, 36, and 38 stroke/min for testing of the machine.

19.4.11 TESTING OF POWER OPERATED SUGARCANE SET CUM FORAGE CUTTER

The testing of machine was conducted on dry sorghum and green maize in the Department of Animal Husbandry and Dairy Science and Sugarcane

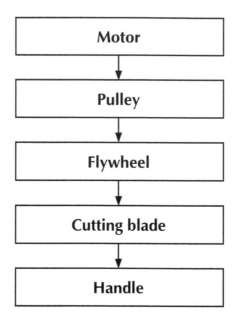

FIGURE 19.2   Power transmission system.

Research Center at Dr. PDKV, Akola. The testing included evaluation of physical properties and moisture content.

19.4.12 PHYSIOLOGICAL MEASUREMENTS

19.4.12.1 HEART RATE

The number of heart beats per minutes (pulse) was determined by the range method for the physiological workload. For this purpose, polar heart rate monitor consisting of chest belt and wrist watch was used (Fig. 19.3). This device sends the pulse data taken by polar belt from the heart region, for 5 s, to the polar clock.

$$\text{Average heart rate} = \text{average working heart rate} \\ - \text{average resting heart rate} \quad (19.1)$$

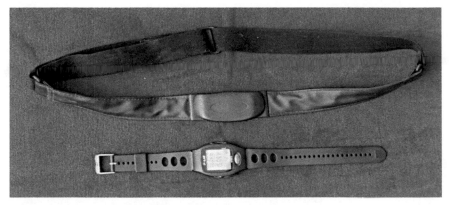

FIGURE 19.3   Polar heart rate monitor.

19.4.12.2   OXYGEN CONSUMPTION RATE

The consumption of oxygen in a given time (per minute) is termed as oxygen consumption rate and expressed in liters/minute. Oxygen consumption is the correct variable for measuring the physiological work load. But it is a difficult to measure while performing the work. Hence indirect method, that is estimation of oxygen consumption using correlation between heart rate and oxygen consumption, is used. The oxygen consumption from the heart rate data can be determined by using following equation:

$$Y = 0.0162 \, X - 1.314 \qquad (19.2)$$

Where, Y = Oxygen consumption rate, l/min; and X = Heart rate, beats/min.

19.4.13   PHYSIOLOGICAL COST OF WORK (PCW)

Total cardiac cost of work (CCW) is the total duration of activity, where total cardiac cost of work (TCCW) is TCCW = Cardiac cost of work (CCW) + Cardiac cost of recovery (CCR), where: CCW = AHR × duration of activity; AHR = average working heart rate − Average resting heart rate; and CCR = (Average recovery heart rate − Average resting heart rate) × duration of recovery.

19.5 RESULT AND DISCUSSION

The Figures 19.4–19.6 show the power operated sugarcane set cum forage cutter. In this section, authors present testing of sugarcane set cum forage cutter for dry sorghum, green maize, and sugarcane stalks (Figs. 19.7–19.9).

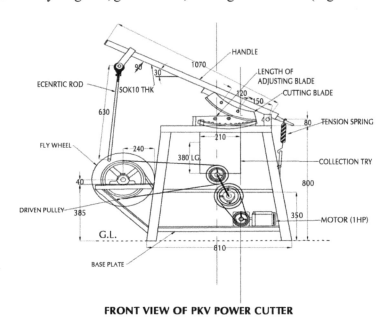

FRONT VIEW OF PKV POWER CUTTER

FIGURE 19.4　Front view of power operated sugarcane set cum forage cutter (*Panjabrao Krishi Vidyapeeth* (PKV) power cutter).

19.5.1 PHYSICAL PROPERTIES

For dry sorghum, the average stalk diameter was 29 mm and average plant height was 1560 mm. For green maize, the average diameter was 31 mm and average plant height was 1430 mm. The average stalk diameter of sugarcane was 48 mm and average plant height was 1780 mm.

19.5.2 MOISTURE CONTENT

The average moisture content was 37.23, 73.10, and 62.4% on wet basis for dry sorghum, green maize, and for sugarcane stalks, respectively.

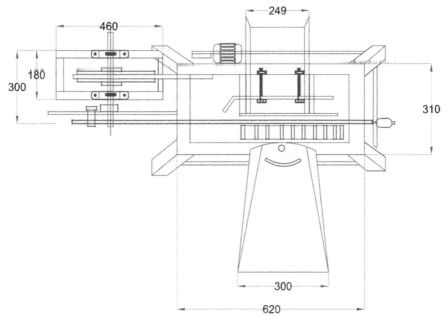

FIGURE 19.5   Top view of power operated sugarcane set cum forage cutter (PKV power cutter).

19.5.3   FIELD EVALUATION

19.5.3.1   CUTTING RATE FOR SUGARCANE SETS

19.5.3.1.1   Sugarcane Sets With Power Cutter

The experiment was conducted for 32, 34, 36, and 38 strokes/min for cutting the sugarcane stalks. The total number of sets with single and double eye buds was measured and is shown in Table 19.3.

From the Figure 19.10, it can be observed that the total number of single and double eye sets was maximum at 34 strokes/min and minimum at 38 strokes/min. From Table 19.4 and Figure 19.11, it can be observed that the damage percentage at 32 and 34 stroke/min was minimum and at par with each other. The damage percentage increased with increase in strokes from 36 to 38/min. The maximum damage was observed at 38 strokes/min. The similar trends were observed in case of double eye bud sets. The minimum was at 32 strokes/min and maximum was at 38 strokes/min.

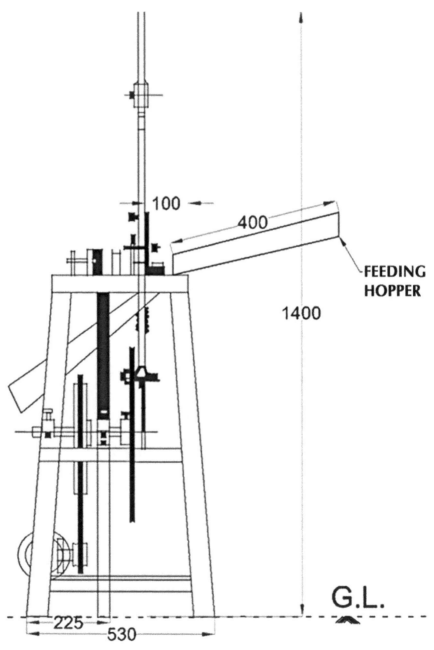

FIGURE 19.6   Side view of power operated sugarcane set cum forage cutter (PKV power cutter).

FIGURE 19.7 Left (A) PKV power cutter; right (B) cutting of sugarcane sets.

FIGURE 19.8 Left (A) single eye bud sets; right (B) double eye bud sets.

FIGURE 19.9 Left (A) cutting of dry sorghum stalks; right (B) cutting of green maize stalks.

TABLE 19.3   Cutting Rate for Sugarcane Sets with the Power Cutter.

Strokes/min	Bud sets/h		Damage, bud sets/h	
	Single eye	Double eye	Single eye	Double eye
32	1875	1868	35	32
34	1985	1977	40	41
36	1800	1777	50	57
38	1545	1530	55	60

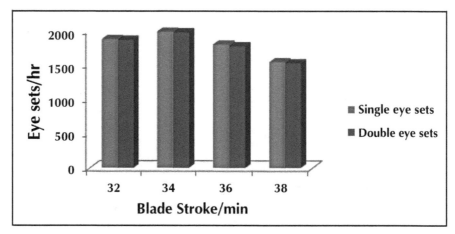

FIGURE 19.10   Eye sets per hour versus strokes per minute: single and double eye bud sets.

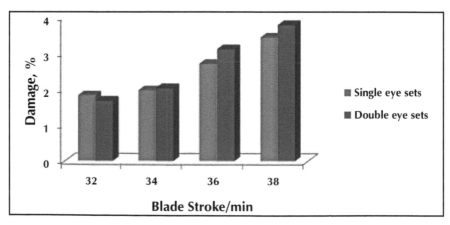

FIGURE 19.11   Damage percentage versus strokes per minute: single and double eye bud sets.

TABLE 19.4 Damage Percentage of Sugarcane Sets.

Strokes/min	Damage			
	Single eye		Double eye	
	Bud sets/h	% of Bud sets/h	Bud sets/h	% of bud sets/h
32	35	1.86	32	1.71
34	40	2.01	41	2.07
36	50	2.77	57	3.20
38	55	3.55	60	3.92

19.5.3.1.2   Sugarcane Sets with Manual Cutting

When the cutting rate and damage percentage by the manual method were compared with power cutting, it was observed that the cutting rate was almost one-fourth and damage percentage was also on higher side in manual cutting as compared to power cutter. In the case of single and double eye bud sets, damage percentage was in the range of 3.03–4.14%. The data are shown in Tables 19.5 and 19.6, and Figures 19.12 and 19.13.

19.5.3.1.3   Cutting Rate of Dry Sorghum Stalks

From Table 19.7, it was observed that the cutting rate increases from 32 to 34 stroke/min and then suddenly decreases from 36 to 38 stroke/min, due to difficulty in feeding of dry sorghum at higher stroke per minute rates. The maximum cutting rate was observed with 34 stroke/min and minimum was observed at 38 stroke/min for all cutting lengths (Fig. 19.14). Time

TABLE 19.5 Cutting Rate of Sugarcane Sets by Manual Method During 1 h.

Replication	Bud sets/h		Damage, bud sets/h	
	Single eye	Double eye	Single eye	Double eye
1	400	395	15	12
2	415	410	14	15
3	410	405	17	16
4	400	400	15	12
Average	**406.25**	**402.5**	**15.25**	**13.75**

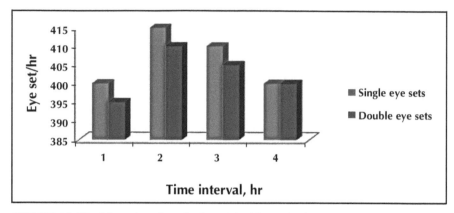

FIGURE 19.12   Manual cutting: single and double eye bud sets.

TABLE 19.6   Damage Percentage of Sugarcane Sets by Manual Cutting in 1 h.

Replication	Damage			
	Single eye		Double eye	
	Bud sets/h	% of Bud sets/h	Bud sets/h	% of Bud sets/h
1	15	3.75	12	3.03
2	14	3.37	15	3.65
3	17	4.14	16	3.95
4	15	3.75	12	3.00
Average	**15.25**	**3.75**	**13.75**	**3.40**

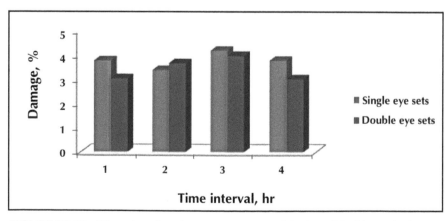

FIGURE 19.13   Manual cutting: damage percentage of single and double eye bud sets.

TABLE 19.7   Cutting Rate of Dry Sorghum Stalk at Different Stroke per min.

Length of cut mm	Stroke/min			
	32	34	36	38
	Cutting rate, kg/h			
25	25.60	30.50	26.40	20
50	53.33	60	50.50	38
75	96	107	80	65

TABLE 19.8   Time Required for Cutting 100 kg of Dry Sorghum Stalk at Different Strokes per minute.

Length of cut, mm	Strokes/min			
	32	34	36	38
	Time, h (h-min)			
25	3.90 (3.54)	3.28 (3.17)	3.78 (3.47)	5.00 (5.00)
50	1.87 (1.53)	1.66 (1.40)	1.98 (1.59)	2.63 (2.38)
75	1.04 (1.2)	0.93 (56)	1.25 (1.15)	1.53 (1.32)

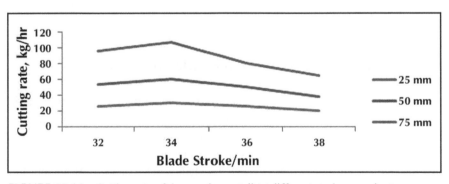

FIGURE 19.14   Cutting rate of dry sorghum stalk at different stroke per minute.

required to cut 100 kg of dry sorghum stalks was maximum at 36 to 38 stroke/min and minimum at 32 to 34 stroke/min for all the size of cutting length (Table 19.8 and Fig. 19.15).

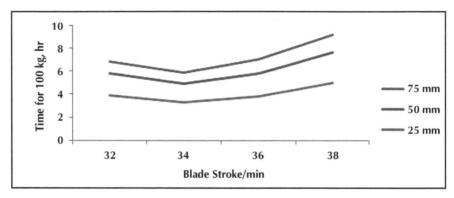

FIGURE 19.15   Time required for cutting 100 kg dry sorghum stalk at different stroke per minute.

TABLE 19.9   Cutting Rate of Green Maize Stalks at Different Strokes per minute.

Length of cut, mm	Strokes/min			
	32	34	36	38
	Cutting rate, kg/h			
25	26.51	33.05	26.80	22
50	55.04	63.08	50	40
75	98.08	110	90	70

19.5.3.1.4   Cutting Rate of Green Maize Stalks

In Table 19.9, it was observed that the cutting rate of fresh maize stalks increases from 32 to 34 stroke/min and then suddenly decreases from 36 to 38 stroke/min, due to improper feeding of green maize at higher stroke per minute rates (Fig. 19.16). The maximum cutting was observed at 34 stroke/min and minimum was observed at 38 stroke/min for all cutting lengths. Time required to cut 100 kg dry sorghum stalk was maximum at 36–38 stroke/min and minimum at 32–34 stroke/min (Table 19.10 and Fig. 19.17).

19.5.4   PHYSIOLOGICAL COST OF WORK (PCW)

The body temperature of different subjects ranged from of 97.3 to 98°F and the average body temperature was 97.8°F (Table 19.11). Resting heart

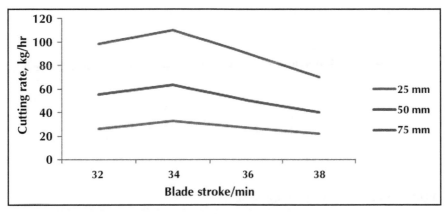

FIGURE 19.16   Cutting rate of green maize stalk at different stroke per minute.

TABLE 19.10   Time Required for Cutting 100 kg Green Maize Stalks at Different Strokes per minute.

Length of cut, mm	Stroke/min			
	32	**34**	**36**	**38**
	Time, h (h-min)			
25	3.77 (3.47)	3.02 (3.1)	3.73 (3.44)	4.54 (4.33)
50	1.81 (1.49)	1.58 (1.35)	2.00 (2.00)	2.50 (2.30)
75	1.01 (1.06)	0.90 (54)	1.11 (1.7)	1.42 (1.25)

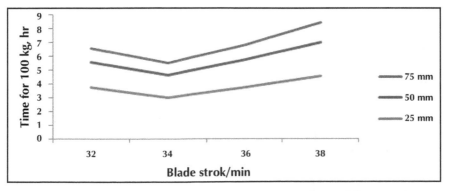

FIGURE 19.17   Time required for cutting 100 kg green maize at different strokes per minute.

TABLE 19.11 Average Measurement of Physiological Cost in Sugarcane Sets Cutting Operation.

Time interval (min)	Body temp (°F)	Heart rate (Beat/min)		Oxygen consumption rate (Liters/min.)	Work pulse rate (Beats/min)	Single eye set (nos.)	Double eye set (nos.)	Damage single eye set (nos.)	Damage double eye set(nos.)
		Resting	Working						
15	97.8	70.2	89	0.1278	18	105	104	4	4

TABLE 19.12 Scaling of Physical Work.

Category of work	Heart rate (beats/min)
Light	90
Moderate	90–110
Heavy	100–130
Severe	130

rate ranged from 70 to 71 beats/min and average resting heart rate was 70.2 beats/min. Working heart rate for five subjects ranged from 87.4 to 89.5 beats/min and average working heart rate was 89 beats/min.

Oxygen consumption rate was between 0.101–0.161 l/min and average oxygen consumption rate was 0.1278 l/min. Work pulse was between 18–19 beats/min and average work pulse was 18 beats/min for cutting operation of sugarcane sets (Table 19.12). The operation of the manual cutting was in the light—moderate weight category.

19.6   CONCLUSIONS

On the basis of the field experiments, following conclusions were drawn:

1. The power required to cutting was 0.126 kW.
2. The output of power operated cutter for 25, 50, and 75 mm length of cut was 25.6, 53.33, and 96 kg/h, 30.5, 60, and 107 kg/h, 26.4, 50.5, and 80 kg/h and 20, 38, and 65 kg/h at 32, 34, 36, and 38 stroke/min, respectively for dry sorghum stalk.

3. The output of power operated cutter for 25, 50, and 75 mm length of cut was 26.51, 55.04, and 98.08 kg/h, 33.05, 63.08, and 110.00 kg/h, 26.8, 50, and 90 kg/h and 22, 40, and 70 kg/h at 32, 34, 36, and 38 stroke/min respectively, for green maize stalk.

4. The output of power operated cutter was 1875, 1985, 1800, and 1545 for single eye sets per hour and 1868, 1977, 1777, and 1530 for double eye sets per hour at 32, 34, 36, and 38 stroke/min respectively, for sugarcane sets.

5. The output of manual cutting of sugarcane sets was 400, 415, 410 and 400 for single eye set per hour and 395, 410, 405 and 400 for double eye sets per hour.

6. The damage percentage by power operated cutter was 1.86, 2.01, 2.77 and 3.55% for single eye sets per hour; and 1.71, 2.07, 3.20 and 3.92% for double eye sets per hour at 32, 34, 36 and 38 stroke/min, respectively.

7. The damages percentage by manual cutting was 3.75, 3.37, 4.14, and 3.75% for single eye sets per hour; and 3.03, 3.65, 3.95, and 3.00% for double eye sets per hour, respectively which is almost double as compared to power operated cutter.

8. The results obtained during ergonomic evaluation of subject for manually cutting of sugarcane sets are: average heart rate of 89 beats/min; average body temperature of 97.8°F, average oxygen consumption rate of 0.1278 l/min and work pulse of 18 beats/min, respectively.

9. As per ergonomic evaluation, the manual cutting operation of sugarcane sets was in the light—moderately weight category, but the subject is able to work continuously for only 40–45 min. However in case of power operated cutter, cutting the sets can be done continuously for 2 h. After 45 min of manual cutting operation, subject required the rest for 20 min; whereas in case of power cutter, rest of 5–10 min is sufficient enough after continuously working for 2 h.

10. Hence the overall performance of the machine was satisfactory.

11. The total weight of the machine was 80 kg with motor, hence it was easy to transport.

12. The design of the machine is simple and hence local manufacturer can fabricate the machine easily and make available to the farmers.

19.7   SUMMARY

The farmer, who is having only one or two cattle, feeds them by cutting the forage stalks into two or three pieces with the help of either sickle or the axe. The cattle cannot consume it properly which leads to the wastage of about 30–40% of valuable forage. Taking this issue into consideration, the power operated forage cutter was developed which was further modified for cutting the sets of sugarcane. The hoppers are provided on the machine for ease in feeding operation and collecting the cut forage and sugarcane set directly into the bag. Sorghum stalks and green maize stalks were selected for conducting the trials on the forage cutter.

The power required to cutting was calculated to 0.23 kW. The output of power operated cutter for 25, 50, and 75 mm length of cut was found to be 30.5, 60.0, and 107.0 kg/h, respectively, for dry sorghum and 33.0, 63.0, and 110.00 kg/h, respectively, for green maize at 34 strokes/min. However in case of sugarcane set cutting, it was 1985, for single eye sets and 1977 for double eye sets per hour at 34 strokes/min, respectively, against the manual cutting of 415 for single eye sets and 410 for double eye sets per hour. The damage percentage by power operated cutter was 1.97 for single eye sets and 2.03% for double eye sets per hour at 34 strokes/min, respectively. The damage percentage by manual cutting was 3.98% for single eye sets per hour and 3.80% for double eye sets per hour, respectively, which is almost double as compare to power operated cutter.

The results on ergonomic evaluation of subject for manually cutting of sugarcane sets indicate that values were for average heart rate 89 beats/min, average body temperature 97.8°F, average oxygen consumption rate 0.1278 l/min and work pulse 18.02 beats/min, respectively. As per ergonomic evaluation, the manual cutting operation of sugarcane sets was categorized as light and moderate, but the subject is able to work continuously for only 40–45 min. However in case of power operated cutter, the sets cutting can be done continuously for 2 h. After 45 min of manual cutting operation, subject required the rest of at least 20 min whereas, in case of power cutter rest of 5–10 min is sufficient enough after continuously working of 2 h.

KEYWORDS

- Damage percentage
- double eye set
- forage
- heart rate
- oxygen consumption rate
- physiological cost of work

- power cutter
- Set
- single eye set
- sorghum
- strokes per minute
- sugarcane sets

REFERENCES

1. Aristides, C. B. The Mechanical Characteristics of Maize Stalk in Relation to the Characteristics of Cutting Blade. *J. Agric. Eng. Res.* **1974,** *19*(1), 1–12.
2. Astrand, P. O.; Rodale, K. *Text Book of Work Physiology—Physiological Bases of Exercise;* McGraw Hill Book Company: New York, 1977.
3. Belvin, F. Z.; Hansen, M. S. Analysis of Forage Harvester Design. *Agric. Eng.* **1956,** *37*(1), 21–26.
4. Chancellor, W. J. Energy Requirement for Cutting Forage. *Agric. Eng.* **1958,** *30*(10), 633–636.
5. Dange, A. R.; Thakare, S. K.; Bhaskare-Rao, I. Cutting Energy and Force as Required for Pigeon Pea Stems. *J. Agril. Tech.* **2011,** *7*(6), 1485–1493.
6. Galedar, M.; Nazari, A.; Tabatabaeefar, A.; Jafari, A. Bending and Shearing Characteristics of Alfalfa Stems. *Agric. Eng. Int.: CIGR E-J. Manuscript FP-08-001*; May 2008, *X*, 9.
7. Guzel, E.; Zeren, Y. The Theory of Free Cutting and its Application on Cotton Stalk. *Agric. Mechanization in Asia, Afr. Lat. Am.* **1990,** *21*(1), 55–56.
8. Hermitian, R.; Najafi, G.; Hosseinzadeh, B.; Tavakoli Hashjin, T.; Khoshtaghaza, M. H. Experimental and Theoretical Investigation of the Effects of Moisture Content and Internodes Position on Shearing Characteristics of Sugarcane Stems. *J. Agri. Sci. Tech.* **2012,** *14*, 963–974.
9. Liljedahl, J. B; Jackson, G. L; Degraff, R. P.; Schroedor, M. E. Measuring of Shear Energy. *Agric. Eng.* **1961,** *42*(6), 298–301.
10. Me Randal, D. M.; Nulty, P. B. Impact Cutting Behaviors of Forage Crop: Field Test. *J. Agric. Eng. Res.* **1978,** *23*, 329–338.
11. Michael, A. M.; Ojha, T. P. *Principal of Agricultural Engineering*; Jain Brothers: Jodhpur, India; 1996; p 317.
12. Miranda, C. S.; Vasconcellos, A.; Bandera, A. G. Termites in Sugarcane in Northeast Brazil; Ecological Aspects and Pest Status. *Neutron Entomol.* **2004,** *33*, 237–241.
13. Patil M.; Patil, P. D. Optimization of Blade Angle for Cutting System of Sugarcane Harvester. *Int. Indexed Refereed Res. J.* March, **2013,** *6*(42), no pages.

14. Rajput, D. S.; Bhole, N. D. Static and Dynamic Shear Properties of Paddy Stem. *The harvester (IIT –Kharagpur)*, **1973,** *15,* 17–21.
15. Richey, C. B. Discussion on Energy Requirement for Cutting Forage. *Agric. Eng.* **1958,** *39,* 636–637.
16. Singh A. K.; Singh, P. R.; Gupta, R. Mechanization of Sugarcane Harvesting in India. *J. Sugarcane Res.* **2012,** *2*(2), 9–14.
17. Taghijarah, H.; Ahmadi, H.; Ghahderijani, M. Shearing Characteristics of Sugarcane (*Saccharumofficinarum L.*) Stalks as a Function of the Rate of the Applied Force. *Aust. J. Crop Sci.* **2011,** *5*(6), 630–634.
18. Taghinezhad, J.; Alimardani, R.; Jafri, A. *Effect of Sugarcane Stalks Cutting Orientation on Required Energy for Biomass Products.* Unpublished Report by Department of Agricultural Machinery Engineering, Faculty of Agricultural Engineering and Technology University of Tehran, P.O. Box 4111, Tehran 13679–47193, Iran; 2012.
19. Visvanathan, R.; Sreenarayanan, V. V.; Swami Nathan, K. R. Effect of Knife Angle and Velocity on the Energy Required to Cut Cassava Tubers. *J. Agric. Eng. Res.* **1996,** *64,* 99–102.

CHAPTER 20

TRACTOR THREE-POINT LINKAGE SYSTEM: COMPUTER-AIDED DESIGN AND SIMULATION ANALYSIS

THANESWER PATEL[1,*], P. K. PRANAV[2], NIKHIL KUMAR[3] AND SHYAMTANU CHAUDHURI[4]

[1]*Department of Agricultural Engineering, North Eastern Regional Institute of Science and Technology (NERIST), Nirjuli (Itanagar), 791109, Arunachal Pradesh, India*

[2]*Department of Agricultural Engineering, North Eastern Regional Institute of Science and Technology (NERIST), Nirjuli (Itanagar), 791109, Arunachal Pradesh, India*
E-mail: pkjha78@gmail.com

[3]*Department of Agricultural Engineering, North Eastern Regional Institute of Science and Technology (NERIST), Nirjuli (Itanagar), 791109, Arunachal Pradesh, India*
E-mail: nikhilnerist@gmail.com

[4]*Department of Agricultural Engineering, North Eastern Regional Institute of Science and Technology (NERIST), Nirjuli (Itanagar), 791109, Arunachal Pradesh, India*
E-mail: nikhilnerist@gmail.com

Corresponding author. E-mail: thaneswer@gmail.com

CONTENTS

20.1   INTRODUCTION

Tractor is the main power source for almost all agricultural operations. The farm tractor is specifically designed to deliver a high tractive effort (or torque) at slow speeds for pulling or pushing agricultural machinery.[1] A farm tractor is also being used for various alternate operations, for example, plowing, tilling, disking, harrowing, planting, transportation and so forth. Due to the rapid increase in energy cost, power optimization has become a common objective for many engineering devices. Among the various agricultural operations, tillage is one of the highest energy demanding activities in an agricultural production system, and hence, the evaluation of tillage effort is a field of great interest.[2] The three-point linkage system is most widely used for tillage implements in the world for many years. Therefore, it is necessary to evaluate their suitability for three-point linkage (Fig. 20.1) implements for the best utilization of available usable power of tractor.

The computer-aided design (CAD) for three-dimensional solid modeling and finite element method applications is most widely accepted and used in various research fields. The use of this technology not only shortens the product design cycle, but also improves the accuracy and reliability of the product.[3,4] As a result, physical objects have widely been replaced by the computer models. The application of computer-aided design for predicting the force analysis of three-point linkages system helps to determine the relative importance of many factors affecting in the stress analysis without conducting experiments. After modeling, hitch point can be analyzed and further suggestive improvements can be incorporated to the design if needed. This leads to better and cheaper products as the CADs are simpler to analyze and easier to change design dimensions compared to empirical methods.[5] Furthermore, it would help for economic design of the three-point linkages system for the most efficient utilization of maximum available usable tractor power.

In the past, efforts have been made by many researchers to measure the force and power requirements of the tillage implements. Only a few

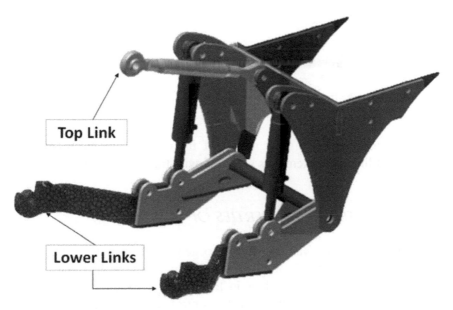

FIGURE 20.1   Tractor three-point linkage system.

researchers have studied about the design and material cost analysis of the three-point linkages system. Keeping the abovementioned facts in view, the study was undertaken, and the efforts were made to study the existing design and material requirements for lower links of the three-point linkage system with the objectives to analyze von Mises stress of the existing lower link using CATIA software and to design modified lower link dimensions based on the maximal stress developed.

20.2   MATERIALS AND METHODS

The dynamic stress analysis and optimization for three-point linkage system of Mahindra B275DI tractor were used. The study was concentrated mainly on tractor lower links since the force experienced by lower links are comparatively higher than the top link.

20.2.1   MODELLING OF EXISTING LOWER LINK

After gathering all the data, the two-dimensional (2D) diagram was constructed as given in Figure 20.2. Using this 2D diagram, a

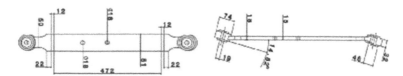

FIGURE 20.2 The front and side views of three-point linkage system of tractor (All the dimensions are in mm).

three-dimensional (3D) solid model of the existing lower links of the three-point linkage system was prepared with the help of CATIA.

20.2.2 MATERIAL PROPERTIES OF LOWER LINK

The designed model was transferred to generative structural analysis in CATIA for stress analysis of the model. The property of the material used for analysis of alloy steel is given in Table 20.1. The element and material properties are applied through CATIA control panel, located in a material library. The loads and boundary conditions were applied during analysis.

TABLE 20.1 Properties of Alloy Steel Used for Analysis.

Parameter	Value
Density, kg/m^3	7850
Poisson ratio	0.27
Thermal expansion, /°K	2×10^6
Yield strength, N/m^2	3.66×10^8
Young modulus, N/m^2	2×10^{11}

20.2.2.1 LOADS AND BOUNDARY CONDITIONS

Lower link of three-point linkage (Fig. 20.3) was constructed for stress analysis. The loads and boundary conditions applied on the lower link of three-point linkage are shown in Figure 20.4. Force F_1 of 8000 N is applied on one end of the lower link and force F_2 of 2500 N is applied on the point where lift rod is attached to the lower link. The other end of lower links is restrained. After application of the loads, the generative structural analysis was carried out with the help of CATIA, which performs static analysis of linear elastic systems with small displacements.

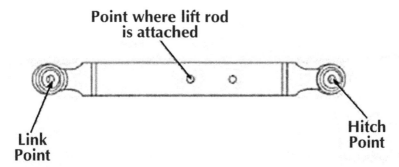

Point where lift rod is attached

Link Point

Hitch Point

FIGURE 20.3   Lower link of three-point linkage.

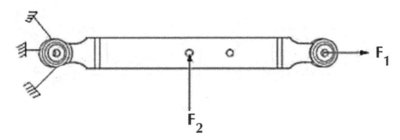

F_1

F_2

FIGURE 20.4   Loads and boundary conditions.

The 3D solid model and stress distribution after application of prede-termined load conditions in existing lower link of a three-point linkage system are shown in Figure 20.5. The critical point having maximum von Mises stress in the lower link was traced in the von Mises stress contours and the value of maximum von Mises stress was recorded.

20.2.3   ANALYSIS OF EXISTING LOWER LINKS

The critical sections on the existing lower links were located with the help of von Mises stress contour. The critical section was then strengthened by varying two parameters viz. width and thickness of the lower link. The analysis was carried out with consideration of dependent and independent parameters of the lower links.

   Independent parameters

- Width
- Thickness

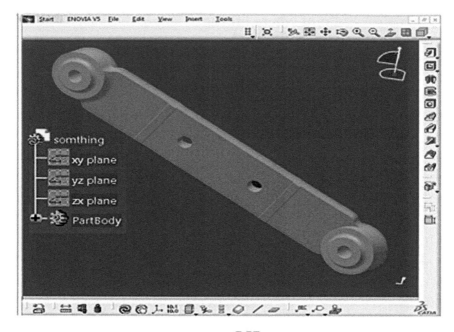

TOP

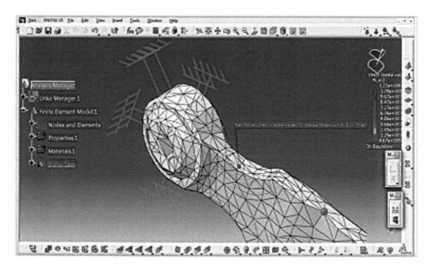

Bottom

FIGURE 20.5   Tractor lower link: top—solid model; bottom—mesh model.

Dependent parameters

* Area
* Volume
* Maximum stress

The values of the independent parameters for width were taken for the different thicknesses as shown in Table 20.2. With different combinations of width and thickness, 3D solid model was developed and dependent parameters viz. area, volume, and maximum stress of the lower link were determined with the help of CATIA generative structural analysis, which performs static analysis of the linear elastic systems with small displacements. The critical point having maximum von Mises stress in the lower link was located in the von Mises stress contours and the value of maximum von Mises stress was recorded for each combination of thickness and width.

TABLE 20.2　Lower Link Thickness and Width Considered for 3D Modeling and Analysis.

Thickness (mm)	Width (mm)
13	79, 81, 83
14	79, 81, 83
15	79, 81, 83
16	77, 79, 81
17	75, 77, 79, 81

The critical sections on the links were located with the help of von Mises stress contour. To optimize the critical section of the existing lower link, a new modified lower link was designed. Using this 2D diagram (Fig. 20.6), a 3D solid model of the modified lower links was constructed with the help of CATIA. The 3D CAD model of the lower links of the three-point linkage system is also shown in Figure 20.7.

Element and material properties are applied through CATIA control panel, located in the material library. Three-point linkage system of existing lower was analyzed with the help of CATIA software. The maximum stress value was located near to the ball and socket joint. With consideration to minimize the maximum stress value of the existing lower

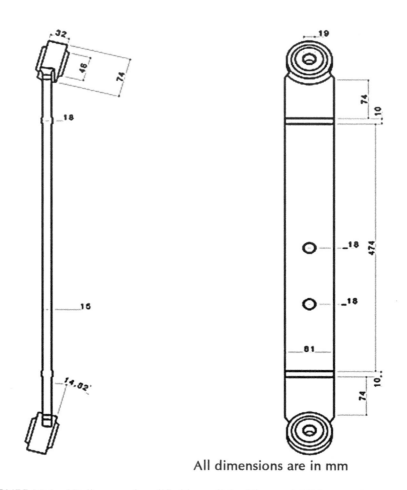

All dimensions are in mm

FIGURE 20.6   2D diagram of modified lower link of three-point linkage system.

links, a new modified design was constructed on a trial and error basis without increasing the total mass of the link. Keeping this in mind, the same element and material properties were used through CATIA control panel, as that for the existing lower links. Now analysis of the modified links was carried out with CATIA generative structural analysis, which performs static analysis of linear elastic systems with small displacements. Stress in the modified lower link was visualized as shown in Figure 20.8.

The critical point having maximum von Mises stress (Fig. 20.9) in the modified lower link was located in the von Mises stress contours and the value of maximum von Mises stress was recorded. For the different width

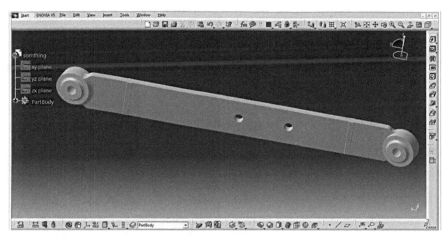

FIGURE 20.7   3D solid CAD model of modified lower link of three-point linkage system.

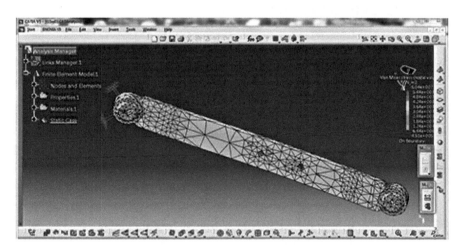

FIGURE 20.8   Stress distributions in the modified lower link.

and thickness of 3D solid model dependent parameters viz. area, volume, and maximum stress of the modified lower link were determined with the help of CATIA generative structural analysis, which performs static analysis of linear elastic systems with small displacements. The values of the independent parameters for width were 79, 81, and 83 mm and for a corresponding thickness of 13–17 mm, respectively.

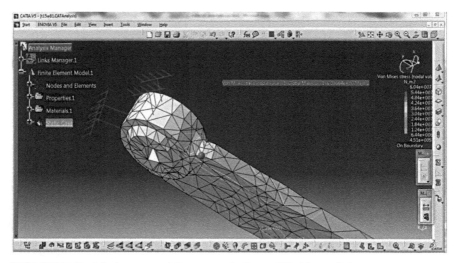

FIGURE 20.9   Maximum von Mises stress in the modified lower link.

20.3   RESULTS AND DISCUSSION

The existing lower link was analyzed by varying independent parameters and observing the dependent parameters. The independent parameters were width and thickness of the lower link. The dependent parameters were area, volume, and maximum stress of the lower link. The values of the independent parameters for width were 79, 81, and 83 mm, and thickness of 13–17 mm, respectively. The results of analysis of the existing lower links are presented in Table 20.3. The mass of the existing lower links was affected by width at different thickness. It was observed that in the present design of 15 mm thickness and 81 mm width the mass of a lower link is 8.152 kg. For a constant thickness when width was increased from 79 to 83 mm, the mass was increased. It was also perceived that when thickness was increased from 13 to 17 mm, mass was increased with increasing width. Mass of the existing lower links of the three-point linkage system for thickness 13–17 mm varied from 7.275–7.547, 7.645–7.920, 8.048–8.281, 8.297–8.542, and 8.501–8.889 kg at 79, 81, and 83 mm thickness, respectively. It was observed that the mass was 8.152 kg for the present design of 15 mm thickness and 81 mm width. It was found that for a constant thickness when width was increased from 79 to 81 mm, the mass increased. Further, it was also found that when thickness is increased from 13 to 17 mm, mass was increased with increasing width.

TABLE 20.3   Analysis of Existing Lower Link of Three-point Linkage System.

Thickness (mm)	Width (mm)	Area (m²)	Volume (m³)	Mass (kg)	Maximum stress (N/m²)
13	77	0.147	0.000,910,547	7.148	6.78×10^7
	79	0.150	0.000,926,717	7.275	9.124×10^7
	81	0.153	0.000,941,052	7.387	9.763×10^7
	83	0.156	0.000,961,361	7.547	9.198×10^7
14	79	0.152	0.000,973,837	7.645	1.077×10^8
	81	0.154	0.00,099,013	7.773	8.208×10^7
	83	0.157	0.001	7.920	8.646×10^7
15	79	0.153	0.001	8.048	1.093×10^8
	81	0.152	0.001	8.152	1.207×10^8
	83	0.154	0.001	8.281	1.138×10^8
16	77	0.154	0.001	8.297	9.698×10^7
	79	0.156	0.001	8.405	9.043×10^7
	81	0.157	0.001	8.542	8.549×10^7
17	75	0.153	0.001	8.501	9.228×10^7
	77	0.154	0.001	8.682	8.469×10^7
	79	0.156	0.001	8.712	8.620×10^7
	81	0.157	0.001	8.889	9.468×10^7

Areas of the modified lower links of the three-point linkage system at various thickness and width are presented in Table 20.4. Area of the modified lower links of the three-point linkage system for thickness 13 and 14 mm varied between 0.157–0.162 and 0.159–0.165 mm² at 79, 81, and 83 mm thickness, respectively. It was observed that in the present design of 15 mm thickness and 81 mm width after modification the area was found to be 0.163 mm². For a constant thickness when width was increased from 79 to 81 mm, the area was increased. Mass of the modified lower links of the three-point linkage system for thickness 13 and 14 mm varied between 7.276–7.532 and 8.006–8.303 kg at 79, 81, and 83 mm thickness, respectively. It was perceived that in the present modified design of 15 mm thickness and 81 mm width the mass weighed 8.568 kg. It was observed that for a constant thickness when width was increased from 79 to 81 mm, the

TABLE 20.4   Analysis of Modified Lower Link of Three-Point Linkage System.

Thickness (mm)	Width (mm)	Area (m²)	Volume (m²)	Mass (kg)	Maximum stress (N/m²)
13	79	0.157	0.000,926,843	7.276	7.55×10^7
	81	0.160	0.00,094,336	7.405	6.88×10^7
	83	0.162	0.000,959,534	7.532	6.283×10^7
14	79	0.159	0.001	8.006	7.194×10^7
	81	0.162	0.001	8.155	6.851×10^7
	83	0.165	0.001	8.303	6.259×10^7
15	79	0.161	0.001	8.409	5.728×10^7
	81	0.163	0.001	8.568	6.036×10^7
	83	0.166	0.001	8.728	6.108×10^7

mass was increased. It was also noted that when thickness was increased from 13 to 14 mm, mass also increased with increasing width.

The comparison of existing and modified lower links of the three-point linkage system showed that for the same mass, the value of maximum stress in modified link was found lower as compared to the existing link. The difference of the maximum stress in existing model and selected model was found to be 37.44%. The difference of maximum mass in existing model and selected model was found to be 3.61%.

20.5   SUMMARY

The computer-aided engineering analysis and design optimization of the three-point linkage system of Mahindra B275-DI tractor was done on the basis of finite-element method by using CATIA software. In order to calculate stress in lower link, the forces were applied to the lower link and then it was modeled, meshed, and loaded in CATIA software. The maximum stress was observed near the ball and socket joint of the lower link. It was observed that in existing lower link of 15 mm thickness and 81 mm width the maximum stress was 1.207×10^8 N/m². Maximum stress increase as mass is increased and after sometime it starts to decrease. The differences of maximum stress and maximum mass in existing and modified link were found to be less considering same with and thickness. The following conclusions can be drawn from this study:

- The existing design of the lower link for a constant thickness when width was increased from 79 to 83 mm, the mass was increased, when thickness is increased from 13 to 17 mm, mass was increased with increasing width.
- A linear decreasing pattern was observed for the effect of width on maximum stress of the existing lower link for different thickness.
- The maximum stress was located near to the ball and socket joint of tractor's lower links in the three-point linkage systems in both existing and modified design.
- It was noticed that there was a linear relationship between thickness and mass. For a constant width as thickness was increased the mass was linearly increased and vice versa.
- The maximum stress and mass for modified design of the lower link were found 6.036×10^7 N/m^2 and 8.568 kg, respectively. It was observed as mass increased, stress decreased and after sometime it started increasing.
- The maximum stress in the modified link was lower as compared to the existing link and the difference of maximum stress was found to be 37.44%.

KEYWORDS

- **Agricultural tractor hitch**
- **computer-aided design**
- **dynamic stress analysis**
- **finite element analysis**
- **maximum stress**
- **three-point linkage system**
- **von Mises stress**

REFERENCES

1. Anonymous. Different Types of Tractors Available in India/abroad and its Importance in Agriculture; 2014. http://ecoursesonline.iasri.res.in/mod/page/view.php?id=126174 (accessed Jan 2015)

2. Bentaher, H.; Hamza, E.; Kantchev, G.; Maalej, A.; Arnold, W. Three-Point Hitch-Mechanism Instrumentation for Tillage Power Optimization. *Biosyst. Eng.* **2008,** *100*(1), 24–30.

3. Ling-Feng, H. U. The Application Present Situation and Prospect of CAD Nology in Agriculture Machinery Domain. *J. Weifang Univ.* **2006,** *6,* 84–86.

4. Liuxuan, M.; Xuhong, C.; Xiaohai, L.; Junfa, W.; Congyu, Q. Computer Aided Design and Simulation Analysis for Matsune Shovel of Stubble Harvester. In: *Mechanic Automation and Control Engineering (MACE)*, International Conference IEEE, 2010; pp 381–383.

5. Dimas, E.; Briassoulis, D. 3-D Geometric Modelling Based on NURBS: A Review. *Adv. Eng. Software* **1999,** *30*(9), 741–751.

GLOSSARY OF TECHNICAL TERMS

3-D surface plot is a chart that shows a three-dimensional surface that connects set of data points.

Amphiphilic molecule refers to an amphiphilic molecule is represented by a coarse grained model, which contains a hydrophobic tail and a hydrophilic head group such as proteins.

Beta lactam antibiotics (B-lactam antibiotics) are a class of broad-spectrum antibiotics, consisting of all antibiotic agents that contain a β-lactam ring in their molecular structures.

Betadine antibiotic is used to treat minor wounds (e.g., cuts, scrapes, burns) and to help prevent or treat mild skin infections.

Bioactive lipids are fatty acids which provide health benefits, which include butyric acid, linolenic acid, linoleic acid, and medium chain fatty acids.

Biochemical oxygen demand (BOD) is the amount of dissolved oxygen needed (i.e., demanded) by aerobic biological organisms to break down organic material present in a given water sample at certain temperature over a specific time period.

Biosensor is an analytical device which converts a biological response into an electrical signal

Biosynthesis (also called biogenesis or anabolism) is a multi-step, enzyme-catalyzed process where substrates are converted into more complex products in living organisms. In biosynthesis, simple compounds are modified, converted into other compounds, or joined together to form macromolecules.

Blastospore is an asexual fungal spore produced by budding.

Carbon nanotubes are one dimensional carbon materials with aspect ratio greater than thousand. They are cylinders composed of rolled-up graphite planes with diameters in nanometer scale.

Chemical oxygen demand (COD) test is commonly used to indirectly measure the amount of organic compounds in water. Most applications of COD determine the amount of organic pollutants found in surface water (e.g. lakes and rivers) or wastewater, making COD a useful measure of water quality. It is expressed in milligrams per liter (mg/L), which indicates the mass of oxygen consumed per liter of solution.

Chlamydospore is the thick-walled big resting spore of several kinds of fungi.

Coacervates are electrostatically stabilized viscous colloidal droplets.

Coliforms are commonly used indicator of sanitary quality of foods and water. They are defined as rod-shaped Gram-negative non-spore forming and motile or non-motile bacteria which can ferment lactose with the production of acid and gas when incubated at 35-37°C. Coliforms can be found in the aquatic environment, in soil and on vegetation; they are universally present in large numbers in the faces of warm-blooded animals.

Colony-forming unit (CFU) is a unit used to estimate the number of viable bacteria in a sample. Viable is defined as the ability to multiply via binary fission under the controlled conditions.

Constraint refers to a condition of an optimization problem that the solution must satisfy. There are several types of constraints—primarily equality constraints, inequality constraints, and integer constraints. The set of candidate solutions that satisfy all constraints is called the feasible set.

Degrees of freedom is the number of levels of the factor minus 1, for the main effects of factor.

Detoxification is the physiological or medicinal removal of toxic substances from a living organism, including the human body, which is mainly carried out by the liver.

Drug delivery refers to approaches, formulations, technologies, and systems for transporting a pharmaceutical compound/any biological agent in the body as needed to safely achieve its desired therapeutic effect.

Drug targeting is a method of delivering medication to a patient in a manner that increases the concentration of the therapeutic/diagnostic agent in some parts of the body relative to others.

Durham vial or Durham tubes are used in microbiology to detect production of gas by microorganisms. They are simply smaller test tubes inserted upside down in another test tube. This small tube is initially filled with the solution in which the microorganism is to be grown. If gas is produced after inoculation and incubation, a visible gas bubble will be trapped inside the small tube. The initial air gap produced when the tube is inserted upside down is lost during sterilization, usually performed at 121°C for 15 or so minutes.

Edible films are thin layer of material which can be consumed and provides a barrier to oxygen, moisture, and solute movement for the food. The material can be a complete food coating or can be disposed as a continuous layer between food components. Edible films can be formed as food coatings and free-standing films, and have the potential to be used with food as gas aroma barrier.

Electrospray ionization (ESI) is a technique used in mass spectrometry to produce ions using an electrospray in which a high voltage is applied to a liquid to create an aerosol.

Emulsifying ointment is a mixture of paraffin oils. It is a greasy moisturizer that provides a layer of oil on the surface of the skin to prevent water evaporating from the skin surface.

Emulsion is a mixture of two immiscible liquid phases dispersed one in another. An amphiphilic surfactant is used to stabilize the interface of immiscible liquids. In food milk (oil-in-water), butter (water-in-oil) are examples for emulsions.

Encapsulation is a process in which micro or nano bioactive materials (core) are enveloped with a continuous film of polymeric material (the shell) to produce capsules in the micrometer or millimeter range. Encapsulation can be achieved using a number of chemical or physical techniques, including interfacial polymerization, extrusion spheronization, and spray drying.

Enzyme-linked immunosorbent assay (ELISHA) is a test that uses antibodies and color change to identify a substance.

Eosin Methylene Blue Agar (EMB Agar) is a both selective and differential culture medium. It is selective culture medium for gram-negative bacteria (selects against gram positive bacteria) and is commonly used for the isolation and differentiation of coliforms and fecal coliforms.

Facultative lagoons are a type of stabilization pond used for biological treatment of industrial and domestic wastewater. Sewage or organic waste from food processing may be catabolized in a system of constructed ponds where adequate space is available to provide an average waste retention time exceeding a month. A series of ponds prevents mixing of untreated waste with treated wastewater and allows better control of waste residence time for uniform treatment efficiency.

Extracellular means a process carried out of the cell.

Fermentation is a metabolic process that converts sugar to acids, gases, or alcohol.

Fouling is a process of accumulation of unwanted deposits on heat transfer surface. The foulant layer imposes an additional resistance to heat transfer and the narrowing of flow area due to the presence of deposit.

Fullerene is a molecule of carbon in the form of a hollow sphere, ellipsoid, tube, and many other shapes.

Fullerene soot is a fine powder composed of a mix of C60 and C70 fullerenes in a ratio of roughly 22% C60 to 76% C70.

Graphene is an allotrope of carbon in the form of a two-dimensional, atomic-scale, honey-comb lattice.

Herbalism (also herbology or herbal medicine) is the use of plants for medicinal purposes, and the study of botany for such use.

Horizontal gene transfer (HGT) is the movement of genetic material between unicellular and/or multicellular organisms other than via vertical transmission. HGT is synonymous with lateral gene transfer (LGT) and the terms are interchangeable.

Intracellular means located or occurring within the cell.

Jaggery is a coarse dark brown sugar made in India by evaporation of the date, cane juice, palm sap.

Kinetic parameters are estimated to study the behavior of a process.

Linear programming (LP) is a method to achieve the best outcome (such as maximum profit or lowest cost) in a linear mathematical model whose requirements are represented by linear constraints or relationships.

Liposome is a sphere-shaped vesicle with a membrane composed of phospholipid bilayer used to deliver the functional compounds into targeted sites.

Liquid chromatography–mass spectrometry (LC-MS) is an analytical chemistry technique that combines the physical separation capabilities of liquid chromatography (or HPLC) with the mass analysis capabilities of mass spectrometry (MS). LC-MS is a powerful technique that has very high sensitivity, making it useful in the separation, general detection and potential identification of chemicals of particular masses in complex mixtures.

MacConkey Broth is used for cultivating Gram-negative, lactose-fermenting bacilli in water, foods and pharmaceutical raw materials as a presumptive test for coliform organisms.

Maillard browning is a non-enzymatic browning reaction between carbonyl and amino groups in food which gives flavor and color during frying, roasting and baking.

Mass spectrometry is an analytical technique that ionizes chemical species and sorts the ions based on their mass to charge ratio.

Metabolomics is the scientific study of the set of metabolites present within an organism, cell, or tissue.

Methicillin-resistant Staphylococcus aureus (MRSA) is a bacterium responsible for several difficult-to-treat infections in humans. MRSA is any strain of *Staphylococcus aureus* that has developed, through horizontal gene transfer and natural selection, multi- resistance to beta-lactam antibiotics, which include the penicillins (methicillin, dicloxacillin, nafcillin, oxacillin, etc.) and the cephalosporins.

Microaray is a set of DNA sequences representing the entire set of genes/ transcripts of an organism, arranged in a grid pattern.

Microbial Hazard occurs when food becomes contaminated by microorganisms found in the air, food, water, soil, animals and the human body. Many microorganisms are helpful and necessary for life itself. However, given the right conditions, some microorganisms may cause a foodborne illness.

Micronutrients are nutrients required by organisms throughout life in small quantities to orchestrate a range of physiological functions.

microRNA (miRNA) refers to small non-coding RNA molecule (containing about 22 nucleotides) found in plants, animals and some viruses, that functions in RNA silencing and post-transcriptional regulation of gene expression.

Mimosa pudica is also called sensitive plant, sleepy plant, Dormilones or shy plant. It is a creeping annual or perennial herb of the pea family *Fabaceae* often grown for its curiosity value: the compound leaves fold inward and droop when touched or shaken, defending themselves from harm, and re-open a few minutes later.

Minimum inhibitory concentration (MIC) is the lowest concentration of a chemical that prevents visible growth of a bacterium (in other words, at which it has bacteriostatic activity); whereas the minimum bactericidal concentration (MBC) is the concentration that results in microbial death (In other words, the concentration at which it is bactericidal)

Most probable number (MPN) is composed of three phases: the presumptive, confirmed, and completed phases. While the MPN method does not exactly measure the number of coliforms present in a sample, it does give an estimate and can determine whether or not the water is below the safe threshold for potable water.

Muller-Hinton agar is a microbiological growth medium that is commonly used for antibiotic susceptibility testing.

Multi Drug Resistance (MDR) is antimicrobial resistance shown by a species of microorganism to multiple antimicrobial drugs.

Multi-walled carbon nanotubes (MWCNTs) consist of multiple rolled layers (concentric tubes) of graphene.

Nano-Biotechnology is the application of nanotechnology in biological fields. Nanotechnology is a multidisciplinary field that currently recruits approach, technology and facility available in conventional as well as advanced avenues of engineering, physics, chemistry and biology.

Nanoparticles refer to particles between 1 and 100 nm in size.

Nanotechnology is the manipulation of matter on an atomic, molecular, and supramolecular (nano) scale.

Next generation sequencing (NGS) is used to describe a number of different modern DNA sequencing technologies including Illumina

(Solexa) sequencing, Roche 454 sequencing, Ion torrent, PacBio, SOLiD sequencing etc., which allow us to sequence nucleic acids much more quickly and cheaply than the previously used sequencing techniques.

Non-convex function: If any two points lying in the function are connected by a line segment, and if that line segment does not entirely lies above or on the function, then the function is considered a nonconvex function.

Nutrient agar is a general purpose medium supporting growth of a wide range of non-fastidious organisms. It typically contains (mass/volume): 0.5% Peptone—this provides organic nitrogen. 0.3% beef extract/yeast extract—the water-soluble content of these contribute vitamins, carbohydrates, nitrogen, and salts.

Over-the-counter (OTC) drugs are medicines sold directly to a consumer without a prescription from a healthcare professional, as compared to prescription drugs, which may be sold only to consumers possessing a valid prescription.

Pareto chart of effects displays the t-value on the y-axis, in Design-Expert® v8 software.

Pathogen is an infectious agent such as a virus, bacterium, prion, a fungus, or even another micro-organism which can cause disease.

Pathogenicity islands are a distinct class of genomic islands acquired by microorganisms through horizontal gene transfer.

Pathway is a linked series of chemical reactions occurring within a cell, in biochemistry.

Phytotoxicity refers to a toxic effect by a compound on plant growth, caused by a wide variety of compounds, including trace metals, salinity, pesticides, phytotoxins or nanoparticles.

Polymerase chain reaction (PCR) is a technique used in molecular biology to amplify a single copy or a few copies of a piece of DNA across several orders of magnitude, generating thousands to millions of copies of a particular DNA sequence.

Process optimization is the discipline of adjusting a *process* so as to *optimize* some specified set of parameters without violating some constraint.

Proteomics is scientific study of proteome- the entire complement of proteins that is expressed by a cell, tissue, or organism at a given time.

Pullulan is a polysaccharide polymer consisting of maltotriose units, also known as α-1,4-; α-1,6-glucan.

Quadratic programming (QP) is a special type of mathematical optimization problem - specifically, the problem of optimizing (minimizing or maximizing) a quadratic function of several variables subject to linear constraints on these variables. It is a particular type of nonlinear programming.

Quantitative real time polymerase chain reaction (qPCR) is the amplification of DNA with a polymerase chain reaction (PCR) is monitored in real time, which is used for nucleic acid (DNA, RNA) quantification.

Quantum dot refers to a semiconductor crystal/particle of nanometer dimensions with distinctive conductive properties determined by its size, also sometimes referred to as artificial atom.

Relative humidity (abbreviated RH) is the ratio of the partial pressure of water vapor to the equilibrium vapor pressure of water at a given temperature.

Risk assessment is a systematic process of evaluating the potential risks that may be involved in a projected activity or undertaking.

Risk is the potential of gaining or losing something of value. Values can be gained or lost when taking risk resulting from a given action or inaction, foreseen or unforeseen. Risk can also be defined as the intentional interaction with uncertainty. Uncertainty is a potential, unpredictable, and uncontrollable outcome; risk is a consequence of action taken in spite of uncertainty.

Samadera indica is an evergreen shrub or tree growing up to 20 m tall. The bole is up to 39 cm in diameter. The tree is gathered from the wild to treat a range of medical conditions. It is also used locally as an insecticide.

Serotype or serovar is a distinct variation within a species of bacteria or virus or among immune cells of different individuals.

Simplex method (SM) is a competitive method for solving linear programming problems. This method was developed by George Dantzig. This method follows the edge of the feasible region.

Single-walled carbon nanotubes (SWCNTs) have a diameter of close to 1 nm, and can be many millions of times longer. The structure of a SWNT can be conceptualized by wrapping a one-atom-thick layer of graphite called graphene into a seamless cylinder.

Soxhlet extractor is a piece of laboratory apparatus invented in 1879 by Franz von Soxhlet. The Soxhlet extraction is used when the desired compound has a limited solubility in a solvent, and the impurity is insoluble in that solvent.

Sterilization is removal of all microorganisms and other pathogens from an object or surface by treating it with chemicals or subjecting it to high heat or radiation.

Surface plasmons are coherent delocalized electron oscillations that exist at the interface between any two materials where the real part of the dielectric function changes sign across the interface

Sustainable agriculture refers to production of food, fiber, or other plant or animal products using farming techniques that protect the environment, public health, human communities, and animal welfare.

Total coliform count (TCC) refers to a large group of Gram-negative, rod-shaped bacteria that share several characteristics. The group includes thermo tolerant coliforms and bacteria of fecal origin, as well as some bacteria that may be isolated from environmental sources.

Toxicogenomics is a field of science that deals with the collection, interpretation, and storage of information about gene and protein activity within particular cell or tissue of an organism in response to toxic substances.

Transcriptomics is the study of the transcriptome—the complete set of mRNA or transcripts that are produced by the genome, under specific circumstances or in a specific cell—using high-throughput methods.

Two-dimensional gel electrophoresis (2-DE) refers to a powerful and widely used method for the analysis of complex protein mixtures, which separate proteins according to their isoelectric points (pI) (first-dimension or isoelectric focusing), followed by second-dimension SDS-polyacrylamide gel electrophoresis (SDS-PAGE) that separates proteins according to their molecular weights.

Viable but nonculturable (VBNC) bacteria are bacteria that are in a state of very low metabolic activity and do not divide, but are alive and have the ability to become culturable once resuscitated.

Yield is ratio of mass of maximum product produced to the mass of substrate utilized.

INDEX

Milton Keynes UK
Ingram Content Group UK Ltd.
UKHW022043141024
449569UK00022B/795